Jie Chen • Arjun K. Gupta

Parametric Statistical Change Point Analysis

With Applications to Genetics, Medicine, and Finance

Second Edition

 Birkhäuser

Jie Chen
Department of Mathematics and Statistics
University of Missouri-Kansas City
Kansas City, MO 64110
USA
ChenJ@umkc.edu

Arjun K. Gupta
Department of Mathematics and Statistics
Bowling Green State University
Bowling Green, OH 43403
USA
gupta@bgsu.edu

ISBN 978-0-8176-4800-8   e-ISBN 978-0-8176-4801-5
DOI 10.1007/978-0-8176-4801-5
Springer New York Dordrecht Heidelberg London

Library of Congress Control Number: 2011941117

Mathematics Subject Classification (2010): 62-02, 62F03, 62F05, 62F10, 62P10, 62P20, 62P30, 62-07, 62P99

Printed on acid-free paper

Birkhäuser Boston is part of Springer Science+Business Media (www.birkhauser.com)

To my parents and the memory of my grandparents. JC
To the memory of my parents and my grandparents. AKG

Preface to the Second Edition

This edition adds a considerable amount of new information on recent change point research results. In light of numerous book reviews of the first edition of this monograph and favorable comments that we have received from researchers during the last ten years, we have added many new results and applications of change point analysis throughout this new edition.

The main additions consist of (i) two new sections of applications of the underlying change point models in analyzing the array Comparative Genomic Hybridization (aCGH) data for DNA copy number changes in Chapter 2, (ii) a new Chapter 7 (the original Chapter 7 becomes Chapter 8 in this edition) on change points in hazard functions, (iii) a new Chapter 9 on other practical change point models such as the epidemic change point model and a smooth-and-abrupt change point model, and (iv) a number of examples of applications throughout the other chapters.

Change point analysis has been an active research area since its inauguration in the early 1950s. The authors acknowledge that there are many other works and several approaches in change point analysis that are important but not included in this monograph due to the approaches that the authors have chosen to present here. One of the primary goals of this edition is to present readers and practitioners with a systematic way of detecting change points in the particular models demonstrated in this book.

The first author started to branch out her statistical research to the analysis of high throughput data resulting from biomedical and life science research in 2003. Since then, she has actively collaborated with many scientists on modeling gene expression data, for example, resulting from blood stem-cell study and somitogenesis study. She and her collaborators envisioned the problem of modeling array data resulting from DNA copy number studies as a change point problem in the mean and variance parameters of a sequence of normal random variables, and started modeling such DNA copy number experimental data with her medical collaborators in 2006. After her many years of experience in modeling biomedical data, she is keen to add such exciting applications of change point analysis to the DNA copy number data to this volume. Meanwhile, the second author and his collaborators have studied several other change point models, especially the epidemic change point model, and this work is also added to this volume. In the context of the above-mentioned new work, the two authors were thus motivated to write this new edition, emphasizing those change point models that have major applications in modeling biomedical research data.

Every attempt was made to correct various misprints and errors from the first edition. The authors are indebted to the many readers who communicated their findings of some of these errata. Special thanks are due to Mr. Paul Plummer, one of the doctoral students of the first author, for carefully reading the first edition and noting many of the misprints and errors; to Miss Xue Bai, one of the graduate research assistants of the first author, for obtaining the computational and graphical results using the R-package called DNAcopy on the example presented in Section 2.1 of Chapter 2 and in Chapter 9; and to Dr. Fanglong Dong for proofreading this volume. The authors would like to thank Richard Scheines and Changwon Yoo for noting some misprints in Chapter 4 of the first edition, as well as Dr. Asoka Ramanayake for her help with Chapter 9 of this edition. Finally, the authors would also like to thank Professor Jiahua Chen for many conversations on various occasions regarding change point analysis and its applications.

The first author would like to especially thank her husband, Dr. Ke Xia, for his encouragement and support during the process of writing this second edition, and she would also like to thank her two daughters, Rowena and Gracelynn, who were patient and cooperative when their mother was busy writing this book and unable to play with them during their spare time. The second author would like to thank his wife Meera and his daughters, Alka, Mita, and Nisha for their support.

The authors also wish to thank the following publishers, who have granted them permission to reproduce their own results that were first published in proprietary scholarly journals (of course, full citations of these publications are given in the text): Wiley-VCH Verlag GmbH & Co. KGaA, Taylor & Francis, Pushpa Publishing House, and Susan Rivers' Culture Institute.

Finally, the authors wish to thank the two anonymous reviewers for their comments and suggestions, as well as the editors and staff of Birkhäuser for their assistance and patience during the preparation of the second edition of this monograph.

Jie Chen
Kansas City, MO

A. K. Gupta
Bowling Green, OH

June, 2011

Preface to the First Edition

Recently there has been a keen interest in the statistical analysis of change point detection and estimation. Mainly, it is because change point problems can be encountered in many disciplines such as economics, finance, medicine, psychology, geology, literature, and so on, and even in our daily lives. From the statistical point of view, a change point is a place or time point such that the observations follow one distribution up to that point and follow another distribution after that point. Multiple change points problems can also be defined similarly. So, the change point(s) problem is twofold: one is to decide if there is any change (often viewed as an hypothesis testing problem), and another is to locate the change point when there is a change present (often viewed as an estimation problem).

The earliest change point study can be traced back to the 1950s. During the following period of forty-some years, numerous articles have been published in various journals and proceedings. Many of them cover the topic of single change point in the means of a sequence of independently normally distributed random variables. Another popularly covered topic is a change point in regression models such as linear regression and autoregression. The methods used are mainly likelihood ratio, nonparametric, and Bayesian. A few authors also considered the change point problem in other model settings such as gamma and exponential.

It is impossible to discuss and include all relevant change point(s) problems in a one-volume work. In this monograph, therefore, only the most important results in change point(s) analysis have been collected and new interesting results have been added. In other words, this volume is an in-depth study of the change point problem in general, and is also a detailed exposure of change point analysis of the most commonly used statistical models. There are seven chapters devoted to different aspects/models. Multivariate normal models and univariate normal models are discussed in much detail about change point(s) in mean vectors (means), covariance matrices (variances), or in mean vectors and covariance matrices (means and variances). Regression, gamma, exponential, and discrete models are also discussed for change point(s) analysis. The methodologies involved are mainly (classical) likelihood ratio, Bayesian, and information criterion approaches. Some other methods are also discussed. An annotated bibliography is given at the end of this volume.

This research monograph should appeal to theoretical as well as applied statisticians. It should also appeal to economists, quality control managers, or graduate students who are interested in change point(s), or any other investigators who might encounter the change point problem. It can be used as a reference book concerning different aspects of change point problems.

The authors are thankful to many authors who provided their manuscripts. They are also very thankful to Professor Larry Q. Eifler for the technical help in the typesetting of this manuscript. The first author would like to thank her husband, Ke Xia, for his typing of this monograph, his support, and encouragement in the preparation of this book. Finally thanks are due to the publisher for providing all their help.

Jie Chen
Kansas City, MO

A. K. Gupta
Bowling Green, OH

February, 2000

Contents

Chapter 1
Preliminaries

1.1 Introduction

The world is filled with changes. An awareness of those changes can help people avoid unnecessary losses and to harness beneficial transitions. In many practical situations, a statistician is faced with the problem of detecting the number of change points or jumps and their locations. This is known as the change point problem. Enormous practical problems can be found in many disciplines.

Example 1.1 Stock Market Analysis: The daily U.S. stock market records show that the stock price for any company fluctuates daily. Although the fluctuation of any stock price is normal according to economic theory, there are some shiftings that are abnormal and worth the investor's special attention. Here, we may raise the question of whether the Iraqi invasion of Kuwait in 1989 caused a statistically significant change in the U.S. stock market.

Example 1.2 Quality Control: In a continuous production process, the quality of the products is expected to remain stable. However, for several reasons, the process might fail to produce products of the same quality. One therefore wants to find out if there is a point where the quality of the products starts to deteriorate.

Example 1.3 Traffic Mortality Rate: In 1987, the speed limit on many expressways in the United States had been increased from 55 miles per hour to 65 miles per hour. Would this increased speed limit cause any problem in highway traveling? One may, therefore, search for the change in the traffic accident death rate after the relaxation of the 55 mile per hour speed limit.

Example 1.4 Geology Data Analysis: The measurements of core samples obtained from different geological sites are very important to the geologist. If there is a significant change in the measurements of the same type of core obtained from different sites, then the geologist might want to know further on what site the change took place, and whether there are oil, gold, or underground mines.

Example 1.5 Genetics Data Analysis: The measurements of gene expression for a cell line may contain changes that indicate the dynamics of gene regulation, and these changes are very valuable to scientists for biomedical research. The DNA copy number variations are very important in cancer research and identifying these variations often calls for the establishment of an appropriate change point model.

Among the several examples mentioned above, Example 1.2 is a so-called online change point problem or an online surveillance change problem, and all other examples fall into the category of offline change point detection or retrospective change point analysis. The online change point problems are popularly presented in statistical quality control, public health surveillance, and signal processing (Mei, 2006). Fearnhead and Liu (2007) studied an online multiple change point problem, and Wu (2007) proposed a sequential method for detecting and identifying sparse change segments in the mean process. We concentrate on the offline or retrospective change point analysis in this monograph, and simply use the phrase "change point analysis" throughout.

Usually statistical inference about (offline) change points has two aspects. The first is to detect if there is any change in the sequence of observed random variables. The second is to estimate the number of changes and their corresponding locations. The problem of testing and estimating change points has attracted much attention in the literature since it was originally addressed in the field of quality control.

1.2 Problems

Let $\mathbf{x}_1, \mathbf{x}_2, \ldots, \mathbf{x}_n$ be a sequence of independent random vectors (variables) with probability distribution functions $F_1, F_2, \ldots, F_n$, respectively. Then in general, the change point problem is to test the following null hypothesis,

$$H_0 : F_1 = F_2 = \cdots = F_n$$

versus the alternative:

$$H_1 : F_1 = \cdots = F_{k_1} \neq F_{k_1+1} = \cdots = F_{k_2} \neq F_{k_2+1}$$
$$= \cdots F_{k_q} \neq F_{k_q+1} \cdots = F_n,$$

where $1 < k_1 < k_2 < \cdots < k_q < n, q$ is the unknown number of change points and $k_1, k_2, \ldots, k_q$ are the respective unknown positions that have to be estimated. If the distributions $F_1, F_2, \ldots, F_n$ belong to a common parametric family $F(\theta)$, where $\theta \in \mathbf{R}^p$, then the change point problem is to test the null hypothesis about the population parameters $\theta_i, i = 1, \ldots, n$:

$$H_0 : \theta_1 = \theta_2 = \cdots = \theta_n = \theta \quad \text{(unknown)} \tag{1.1}$$

versus the alternative hypothesis:

$$H_1 : \theta_1 = \cdots = \theta_{k_1} \neq \theta_{k_1+1} = \cdots = \theta_{k_2}$$
$$\neq \cdots \neq \theta_{k_{q-1}} = \cdots = \theta_{k_q} \neq \theta_{k_q+1} \cdots = \theta_n, \tag{1.2}$$

where q and $k_1, k_2, \ldots, k_q$ have to be estimated. These hypotheses together reveal the aspects of change point inference: determining if any change point exists in the process and estimating the number and position(s) of change point(s).

A special multiple change points problem is the epidemic change point problem, which is defined by testing the following null hypothesis,

$$H_0 : \theta_1 = \theta_2 = \cdots = \theta_n = \theta \quad \text{(unknown)}$$

versus the alternative:

$$H_1 : \theta_1 = \cdots = \theta_k = \alpha \neq \theta_{k+1} = \cdots = \theta_t = \beta \neq \theta_{t+1} = \cdots = \theta_n = \alpha,$$

where $1 \leq k < t \leq n$, and α and β are unknown.

The epidemic change point problem is of great practical interest, especially in quality control and medical studies. The reader is referred to Levin and Kline (1985), and Ramanayake (1998).

The change point problem can be considered in many model settings such as parametric, nonparametric, regression, times series, sequential, and Bayesian among others.

1.3 Underlying Models and Methodology

The frequently used methods for change point inference in the literature are the maximum likelihood ratio test, Bayesian test, nonparametric test, stochastic process, information-theoretic approach, and so on.

Chernoff and Zacks (1964) derived a Bayesian estimator of the current mean for a priori uniform distribution on the whole real line using a quadratic loss function. Sen and Srivastava (1975a,b) derived the exact and asymptotic distribution of their test statistic for testing a single change in the mean of a sequence of normal random variables. Later (1973, 1980) they generalized their results to the multivariate case. Hawkins (1977) and Worsley (1979) derived the null distribution for the case of known and unknown variances of a single change in the mean. Srivastava and Worsley (1986) studied the multiple changes in the multivariate normal mean and approximated the null distribution of the likelihood ratio test statistic based on Bonferroni inequality. Recently, Guan (2004) considered a change point problem from a semiparametric point of view. Gurevich and Vexler (2005) investigated the change point problem in logistic regression. Horuath et al. (2004) studied the change point problem for linear models. Ramanayake (2004) provided tests for change point detection in the shape parameter of gamma distributions. Goldenshluyer, Tsbakov, and Zeev (2006) provided optimal nonparametric change point estimation for indirect noisy observations. Osorio and Galea (2006) investigated the change point problem for Student-t linear regression models. Kirch and Steinebach (2006) gave an approximation of the critical values for the change point test obtained through a permutation method. Juruskova (2007) considered the change point problem for a Weibull distribution of three parameters. Wu (2008) provided a simultaneous change point analysis and variable solution in a regression problem. Vexler et al. (2009) provided a study of a classification problem in the context of change point analysis. Ramanayake and Gupta (2010) considered the problem of detecting a change point in an exponential distribution with repeated values.

For more works related to the change point(s) problem, the reader is referred to Hawkins (1992), Joseph and Wolfson (1993), Parzen (1992), Yao (1993), Zacks (1991), Vlachonikolis and Vasdekis (1994), Chen (1995), Gupta and Chen (1996), Chen and Gupta (1997), Ramanayake (1998), Gupta and Ramanayake (2001), Chen and Gupta (1998, 2003), Hall et al. (2003), Kelly, Lindsey, and Thin (2004), Chen et al. (2006), and Ning and Gupta (2009).

A survey of the change point studies indicates that most of the previous works were concentrated on the case of a single change point in the random sequence. The problem of multiple change points, however, has not been considered by many authors. Inclán and Tiao (1994) used the CUSUM method to test and estimate the multiple change points problem. Chen and Gupta (1995) derived the asymptotic null distribution of the likelihood procedure (see Lehmann, 1986) statistic for the simultaneous change in the mean vector and covariance of a sequence of normal random vectors. Later (1997) they studied the multiple change points problem for the changes in the variance of a sequence of normal random variables.

In order to detect the number of change points and their locations in a multidimensional random process, Vostrikova (1981) proposed a method, known as the binary segmentation procedure, and proved its consistency. This binary segmentation procedure has been widely used in detecting multiple change points, and it has the merits of detecting the number of change points and their positions simultaneously and saving a lot of computation time. It is summarized as the following.

Suppose $\mathbf{x}_1, \mathbf{x}_2, \ldots, \mathbf{x}_n$ is a sequence of independent random vectors (variables) with probability distribution functions $F_1(\theta_1), F_2(\theta_2), \ldots, F_n(\theta_n)$, respectively. If we tested (1.1) versus (1.2), we would use a detection method along with the binary segmentation technique to uncover all possible change points. A general description of the binary segmentation technique in the detection of the changes can be summarized in the following steps.

Step 1. Test for no change point versus one change point; that is, test the null hypothesis given by (1.1) versus the following alternative.

$$H_1 : \theta_1 = \cdots = \theta_k \neq \theta_{k+1} = \cdots = \theta_n, \tag{1.3}$$

where k is the location of the single change point at this stage. If H_0 is not rejected, then stop. There is no change point. If H_0 is rejected, then there is a change point and we go to Step 2.

Step 2. Test the two subsequences before and after the change point found in Step 1 separately for a change.

Step 3. Repeat the process until no further subsequences have change points.

Step 4. The collection of change point locations found by steps 1–3 is denoted by $\{\widehat{k}_1, \widehat{k}_2, \ldots, \widehat{k}_q\}$, and the estimated total number of change points is then q.

As the reader may notice, the majority of the models proposed for change point problems in the current literature are the normal models (univariate or multivariate). This is because the normal model is the most common model in practice. The change point problems under many other models, however, are also very important and are also studied in this monograph. In Chapters 2 and 3, univariate normal and multivariate normal models are considered. In Chapter 4, a regression model is discussed. The gamma and exponential models are discussed in Chapters 5 and 6, respectively. Chapter 7 is devoted to the topic of change points in the hazard function. Some discrete models are discussed in Chapter 8, and the epidemic change point model and the smooth-and-abrupt change point model are given in Chapter 9 with applications.

Chapter 2
Univariate Normal Model

Let $x_1, x_2, \ldots, x_n$ be a sequence of independent normal random variables with parameters $(\mu_1, \sigma_1^2), (\mu_2, \sigma_2^2), \ldots, (\mu_n, \sigma_n^2)$, respectively. In this chapter, different types of change point problems with regard to the mean, variance, and mean and variance are discussed. For simplicity and illustration purposes, we familiarize readers with the three types of changes in the normal sequence by presenting the following three figures, where Figure 2.1 represents a sequence of normal observations with a mean change, Figure 2.2 shows a variance change in the normal observations, and Figure 2.3 indicates a mean and variance change in the sequence of normal observations.

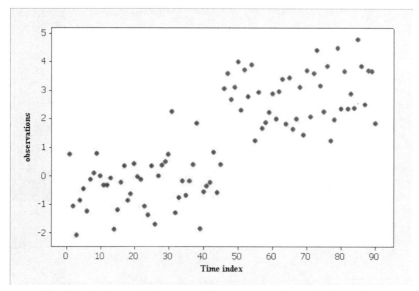

Fig. 2.1 A change in the mean of the sequence of normal observations

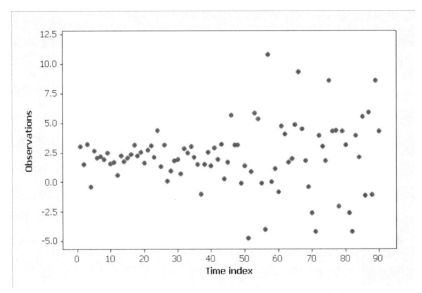

Fig. 2.2 A change in the variance of the sequence of normal observations

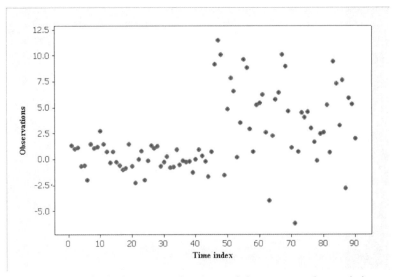

Fig. 2.3 A change in both the mean and variance of the sequence of normal observations

2.1 Mean Change

Suppose that each x_i is normally distributed with mean μ_i and common variance $\sigma^2, i = 1, 2, \ldots, n$. The interest here is about the mean change. This problem was first examined by Page (1954, 1955, 1957). Later, Chernoff and

Zacks (1964), Bhattacharya and Johnson (1968), Gardner (1969), Sen and Srivastava (1975a,b), Gupta and Chen (1996), Chen and Gupta (1997), and Chen and Gupta (1998, 2003) also contributed to the study of this problem. Throughout this section, the hypothesis of stability (the null hypothesis) is defined as

$$H_0 : \mu_1 = \mu_2 = \cdots = \mu_n = \mu.$$

The mean change problem in this one-dimensional case can be one-sided or two-sided. Only the two-sided test is addressed here. That is, the interest is to test H_0 versus

$$H_1 : \mu_1 = \cdots = \mu_k \neq \mu_{k+1} \cdots = \mu_n,$$

where k is the unknown location of the single change point. The testing procedure depends on whether the nuisance parameter σ^2 is known or unknown.

2.1.1 Variance Known

Without loss of generality, assume that $\sigma^2 = 1$. Under H_0, the likelihood function is

$$L_0(\mu) = \frac{1}{(\sqrt{2\pi})^n} e^{-\sum_{i=1}^{n}(x_i-\mu)^2/2}$$

and the maximum likelihood estimator (MLE) of μ is

$$\widehat{\mu} = \bar{x} = \frac{1}{n} \sum_{i=1}^{n} x_i.$$

Under H_1, the likelihood function is

$$L_1(\mu_1, \mu_n) = \frac{1}{(\sqrt{2\pi})^n} e^{-(\sum_{i=1}^{k}(x_i-\mu_1)^2 + \sum_{i=k+1}^{n}(x_i-\mu_n)^2)/2},$$

and the MLEs of μ_1, and μ_n are, respectively,

$$\widehat{\mu}_1 = \bar{x}_k = \frac{1}{k} \sum_{i=1}^{k} x_i, \quad \text{and} \quad \widehat{\mu}_n = \bar{x}_{n-k} = \frac{1}{n-k} \sum_{i=k+1}^{n} x_i.$$

Let

$$S_k = \sum_{i=1}^{k}(x_i - \bar{x}_k)^2 + \sum_{i=k+1}^{n}(x_i - \bar{x}_{n-k})^2,$$

$$V_k = k(\bar{x}_k - \bar{x})^2 + (n - k)(\bar{x}_{n-k} - \bar{x})^2,$$

and $S = \sum_{i=1}^{n}(x_i - \bar{x})^2$; then $V_k = S - S_k$. Simple algebra leads to $U^2 = V_{k*} = \max_{1 \le k \le n-1} V_k$ is the likelihood procedure (see Lehmann, 1986) test statistic for testing H_0 against H_1.

Hawkins (1977) derived the exact and asymptotic null distributions of the test statistic U. The following is based on his work. First, simple algebraic computation gives an alternative expression for V_k as

$$V_k = \frac{n}{k(n - k)} \left[\sum_{i=1}^{k}(x_i - \bar{x}) \right]^2.$$

Let

$$T_k = \sqrt{\frac{n}{k(n - k)}} \left[\sum_{i=1}^{k}(x_i - \bar{x}) \right];$$

then $V_k = T_k^2$ or $|T_k| = \sqrt{V_k}$. Therefore,

$$U = \sqrt{V_{k*}} = \max_{1 \le k \le n-1} \sqrt{V_k} = \max_{1 \le k \le n-1} |T_k| \tag{2.1}$$

is the equivalent likelihood-based test statistic for testing H_0 against H_1. After this computational preparation, the main theorem can be stated as follows.

Theorem 2.1 *The null probability density function of U is given by*

$$f_U(x) = 2\phi(x, 0, 1) \sum_{k=1}^{n-1} g_k(x, x)g_{n-k}(x, x),$$

where $\phi(x, 0, 1)$ is the pdf of $N(0, 1)$, $g_1(x, s) = 1$ for $x, s \ge 0$, and

$$g_k(x, s) = P[|T_i| < s, i = 1, \ldots, k - 1||T_k| = x], \tag{2.2}$$

for $x, s \ge 0$.

To prove this theorem, the following lemma is essential.

Lemma 2.2 $\{T_1, T_2, \ldots, T_{n-1}\}$ *is a Markov process.*

Proof. From the Markov process property, it suffices to show that the partial covariance $\sigma_{jk.m}$ between T_j and T_k when T_m is fixed equals zero for $j < m < k$, or equivalently the partial correlation coefficient $\rho_{jk.m}$ is zero.

For $m = 1, 2, \ldots, n-1$, and $m < k$, the correlation coefficient ρ_{mk} between T_k and T_m is

$$\rho_{mk} = \frac{n}{\sqrt{km(n-k)(n-m)}} E\left[\sum_{i=1}^{k}(x_i - \bar{x})\sum_{j=1}^{m}(x_j - \bar{x})\right]$$

$$= \frac{n}{\sqrt{km(n-k)(n-m)}} \frac{m(n-k)}{n}$$

$$= \sqrt{\frac{m(n-k)}{k(n-m)}}.$$

Then, for $j < m < k$,

$$\rho_{jk.m} = \frac{\rho_{jk} - \rho_{jm}\rho_{mk}}{\sqrt{(1 - \rho_{jm}^2)(1 - \rho_{mk}^2)}}$$

$$= \frac{\sqrt{\frac{j(n-k)}{k(n-j)}} - \sqrt{\frac{j(n-m)}{m(n-j)}\frac{m(n-k)}{k(n-m)}}}{\sqrt{\left[1 - \frac{j(n-m)}{m(n-j)}\right]\left[1 - \frac{m(n-k)}{k(n-m)}\right]}}$$

$$= 0.$$

This completes the proof of the above lemma.   □

Now, it is time to prove the theorem.

Proof of Theorem 2.1 Let A, B, C, be the following events.

$$A = \{|T_k| \in (x, x + dx)\},$$
$$B = \{|T_j| < |T_k|, j = 1, \ldots, k-1\},$$
$$C = \{|T_j| < |T_k|, j = k+1, \ldots, n-1\}.$$

$U = \max_{1 \leq k \leq n-1} |T_k|$; then

$$F_U(x + dx) - F_U(x) = P[U \in (x, x + dx)]$$

$$= P\left\{\bigcup_{k=1}^{n-1}[|T_k| \in (x, x + dx)]\bigcap[|T_k| > |T_j|, j \neq k]\right\}$$

$$= \sum_{k=1}^{n-1} P[ABC]$$

$$= \sum_{k=1}^{n-1} P[A]P[B|A]P[C|AB].$$

Next, $T_k \sim N(0,1)$, therefore $P[A] = 2\phi(x,0,1) + o(dx)$. Moreover,

$$P[B|A] = P[|T_j| < x, i = 1, \ldots, k-1 \,||T_k| = x] + o(dx)$$
$$= g_k(x,x) + o(dx).$$

Finally, from the fact that $\{T_1, T_2, \ldots, T_{n-1}\}$ is a Markovian, $\{T_1, T_2, \ldots, T_{k-1}\}$ and $\{T_{k+1}, T_2, \ldots, T_{n-1}\}$ are independent; that is, B and C are independent given $T_k = x$. Therefore, $P[C|AB] = P[C|A]$. According to the probability symmetry between B and C, similar to $P[B|A]$, we have

$$P[C|A] = g_{n-k}(x,x) + o(dx).$$

Thus, we obtain

$$P[U \in (x, x + dx)] = \sum_{k=1}^{n-1} 2\phi(x,0,1) g_k(x,x) g_{n-k}(x,x) + o(dx),$$

or

$$f_U(x) = 2\phi(x,0,1) \sum_{k=1}^{n-1} g_k(x,x) g_{n-k}(x,x).$$

This completes the proof of the theorem.   □

To be able to use the null distribution of U, one needs to know how to evaluate $g_k(x,s)$ or $g_{n-k}(x,s)$. The following theorem is given just for this purpose.

Theorem 2.3 *The function $g_k(x,s)$ is determined by the recursion:*

$$g_k(x,s) = \int_0^s g_{k-1}(y,s)[\phi(y, \rho x, \tau^2) + \phi(y, -\rho x, \tau^2)]dy,$$

where $\rho = \rho_{k-1,k}$ is the correlation coefficient between T_{k-1} and T_k, and $\tau^2 = \sqrt{1 - \rho^2}$.

Proof. From (2.2) and the facts that $\{T_1, T_2, \ldots, T_{n-1}\}$ is a Markovian, $T_k \sim N(0,1)$, the symmetry of T_k about 0, and $T_{k-1}|T_k = x \sim N(\rho x, \tau^2)$, we have:

$$g_k(x,s) = P[|T_j| < s, j = 1, \ldots, k-1 \,||T_k| = x]$$

$$= \int_{-s}^s P[|T_j| < s, j = 1, \ldots, k-2 | T_{k-1} = y \quad \text{and} \quad |T_k| = x]$$

$$d[T_{k-1} < y | T_k = x]$$

$$= \int_{-s}^s P[|T_j| < s, j = 1, \ldots, k-2 | T_{k-1} = y] d[T_{k-1} < y | T_k = x]$$

$$= \int_{-s}^{s} P[|T_j| < s, j = 1, \ldots, k-2||T_{k-1}| = y]d[T_{k-1} < y|T_k = x]$$

$$= \int_{-s}^{s} g_{k-1}(|y|, s)d[T_{k-1} < y|T_k = x]$$

$$= \int_{-s}^{s} g_{k-1}(|y|, s)\phi(y, \rho x, \tau)dy$$

$$= \int_{0}^{s} g_{k-1}(y, s)\phi(y, \rho x, \tau)dy + \int_{-s}^{0} g_{k-1}(-y, s)\phi(y, \rho x, \tau)dy$$

$$= \int_{0}^{s} g_{k-1}(y, s)[\phi(y, \rho x, \tau^2) + \phi(y, -\rho x, \tau^2)]dy. \qquad \square$$

In addition to the null distribution of U, the distribution of the location k^* of the change point has also been derived, which is given in the following theorem.

Theorem 2.4 *If k^* is the position of the change point estimated by (2.1), then for $k = 1, 2, \ldots, n$,*

$$P[k = k^*] = \int_{0}^{\infty} g_k(x, x)g_{n-k}(x, x)\phi(x, 0, 1)dx.$$

Proof. In view of the facts that $\{T_1, T_2, \ldots, T_{n-1}\}$ is a Markovian, $T_k \sim N(0, 1)$, and the symmetry of T_k about 0, we obtain:

$$P[k = k^*] = P\left\{ \sqrt{V_{k*}} = \sqrt{V_k} = \max_{1 \leq k \leq n-1} |T_k| \right\}$$

$$= P[|T_j| < |T_k|, j \neq k]$$

$$= \int_{0}^{\infty} P[|T_j| < x, j \neq k|T_k = x]dP[T_k < x]$$

$$= \int_{0}^{\infty} P[|T_j| < x, j = 1, \ldots, k-1|T_k = x]dP[T_k < x]$$

$$P[|T_j| < x, j = k+1, \ldots, n|T_k = x]$$

$$= \int_{0}^{\infty} P[|T_j| < x, j = 1, \ldots, k-1||T_k| = x]dP[T_k < x]$$

$$P[|T_j| < x, j = k+1, \ldots, n||T_k| = x]$$

$$= \int_{0}^{\infty} g_k(x, x)g_{n-k}(x, x)dP[T_k < x]$$

$$= \int_{0}^{\infty} g_k(x, x)g_{n-k}(x, x)\phi(x, 0, 1)dx. \qquad \square$$

Although the null distribution of the test statistic U has been obtained, the recursion formula requires moderate computations. Yao and Davis (1986) derived the asymptotic null distribution of U, which provides an alternative way to do formal statistical analysis when n is sufficiently large. The following is based on their work.

Let $W_k = x_1 + x_2 + \cdots + x_k, 1 \leq k \leq n$; then simple algebra leads to

$$U = \max_{1 \leq k \leq n-1} \left| \frac{W_k}{\sqrt{n}} - \frac{k}{n} \frac{W_n}{\sqrt{n}} \right| \bigg/ \left[\frac{k}{n} \left(1 - \frac{k}{n}\right) \right]^{1/2}.$$

Suppose $\{B(t); 0 \leq t < \infty\}$ is a standard Brownian motion; then under H_0, from properties of the normal random variable,

$$\{(W_k - k\mu)/\sqrt{n}; 1 \leq k \leq n\} \overset{D}{=} \left\{ B\left(\frac{k}{n}\right); 1 \leq k \leq n \right\},$$

where "$\overset{D}{=}$" means "distributed as". Furthermore,

$$U = \max_{1 \leq k \leq n-1} \left| \frac{W_k}{\sqrt{n}} - \frac{k}{n} \frac{W_n}{\sqrt{n}} \right| \bigg/ \left[\frac{k}{n} \left(1 - \frac{k}{n}\right) \right]^{1/2}.$$

$$= \max_{nt=1,\ldots,n-1} \left| \frac{W_k}{\sqrt{n}} - t \frac{W_n}{\sqrt{n}} \right| \bigg/ [t(1-t)]^{1/2}$$

$$= \max_{nt=1,\ldots,n-1} \left| \frac{W_k}{\sqrt{n}} - \frac{k\mu}{\sqrt{n}} - t \left(\frac{W_n}{\sqrt{n}} - \frac{n\mu}{\sqrt{n}} \right) \right| \bigg/ [t(1-t)]^{1/2}$$

$$\overset{D}{=} \max_{nt=1,\ldots,n-1} |B(t) - tB(1)|/[t(1-t)]^{1/2}$$

$$= \max_{nt=1,\ldots,n-1} |B_0(t)|/[t(1-t)]^{1/2},$$

where $t = k/n$, and $B_0(t) = B(t) - tB(1)$ is the Brownian bridge. The following theorem shows that the asymptotic null distribution of U is a Gumbel distribution.

Theorem 2.5 *Under H_0, for $-\infty < x < \infty$,*

$$\lim_{n \to \infty} P[a_n^{-1}(U - b_n) \leq x] = \exp\{-2\pi^{1/2}e^{-x}\},$$

where $a_n = (2 \log \log n)^{-1/2}, b_n = a_n^{-1} + \frac{1}{2}a_n \log \log \log n$.

The proof of this theorem is mainly based on the properties of Brownian motion and convergence rules from the theory of probability. The following lemmas are needed before the proof of the theorem is given.

Lemma 2.6

$$\max_{1\le nt\le [n/\log n]} \frac{|B_0(t)|}{\sqrt{t(1-t)}} - \max_{1\le nt\le [n/\log n]} \frac{|B(t)|}{\sqrt{t}} = o_p(a_n).$$

Proof. For large n and $0 < t \le 1/\log n$, clearly, $\sqrt{t/(1-t)} \le 2\sqrt{t}$, and $(1/\sqrt{1-t} - 1)/\sqrt{t} \le \sqrt{t}$. We obtain:

$$\left| \frac{|B_0(t)|}{\sqrt{t(1-t)}} - \frac{|B(t)|}{\sqrt{t}} \right| \le \left| \frac{B_0(t)}{\sqrt{t(1-t)}} - \frac{|B(t)|}{\sqrt{t}} \right|$$

$$= \left| \frac{B(t) - tB(1)}{\sqrt{t(1-t)}} - \frac{|B(t)|}{\sqrt{t}} \right|$$

$$= \left| \frac{|B(t)|}{\sqrt{t}} \left(\frac{1}{\sqrt{1-t}} - 1 \right) - \sqrt{\frac{t}{1-t}} B(1) \right|$$

$$\le \left| \frac{|B(t)|}{\sqrt{t}} \left(\frac{1}{\sqrt{1-t}} - 1 \right) \right| + \left| \sqrt{\frac{t}{1-t}} B(1) \right|$$

$$\le \sqrt{t}|B(t)| + 2\sqrt{t}|B(1)|$$

$$\le (\log n)^{-1/2}(|B(t)| + 2|B(1)|).$$

Then

$$\max_{1\le nt\le [n/\log n]} \frac{|B_0(t)|}{\sqrt{t(1-t)}} - \max_{1\le nt\le [n/\log n]} \frac{|B(t)|}{\sqrt{t}}$$

$$\le \max_{1\le nt\le [n/\log n]} \left| \frac{|B_0(t)|}{\sqrt{t(1-t)}} - \frac{|B(t)|}{\sqrt{t}} \right|$$

$$\le (\log n)^{-1/2} \max_{1\le nt\le [n/\log n]} (|B(t)| + 2|B(1)|)$$

$$= O_p((\log n)^{-1/2})$$

$$= o_p(a_n). \qquad \square$$

Lemma 2.7

$$\max_{[n/\log n]\le nt\le [n/2]} \frac{|B_0(t)|}{\sqrt{t(1-t)}} = O_p((\log\log\log n)^{1/2}).$$

Proof. Inasmuch as $B(t)$ is distributed as $N(0,t)$, from the law of iterated logarithm, for large n,

$$\frac{nB(t)}{2[nt\log\log(nt)]^{1/2}} = O_p(1).$$

Then

$$\frac{B(t)}{2[t \log\log(nt)]^{1/2}} = O_p(1),$$

because for large n, $n > \sqrt{n}$. Let t be small, say $t = 1/\sqrt{n}$; then, for large n,

$$\frac{B(t)}{2\sqrt{t \log\log(t^{-1})}} = O_p(1),$$

or with probability 1,

$$|B(t)| < 2\sqrt{t \log\log(t^{-1})}.$$

Let $t \longrightarrow 0^+$; then for $t \in \left[s, \frac{1}{2}\right]$, $\log\log(t^{-1}) < \log\log(s^{-1})$, and

$$\max_{t \in [s,1/2]} \frac{|B(t)|}{\sqrt{t}} = O_p((\log\log(s^{-1}))^{1/2}).$$

Finally,

$$\max_{[n/\log n] \leq nt \leq [n/2]} \frac{|B_0(t)|}{\sqrt{t(1-t)}}$$

$$= \max_{[n/\log n] \leq nt \leq [n/2]} \frac{|B(t) - tB(1)|}{\sqrt{t(1-t)}}$$

$$\leq \max_{[n/\log n] \leq nt \leq [n/2]} \frac{|B(t)|}{\sqrt{t(1-t)}} + \max_{[n/\log n] \leq nt \leq [n/2]} \sqrt{\frac{t}{1-t}}|B(1)|$$

$$\leq \max_{[n/\log n] \leq nt \leq [n/2]} \frac{2|B(t)|}{\sqrt{t}} + \max_{[n/\log n] \leq nt \leq [n/2]} \sqrt{\frac{t}{1-t}}|B(1)|$$

$$\leq 2 \max_{[1/\log n] \leq t \leq [1/2]} \frac{|B(t)|}{\sqrt{t}} + O_p(1)$$

$$= O_p\left(\log\log\left(\frac{1}{\log n}\right)^{-1}\right)^{1/2} + O_p(1)$$

$$= O_p((\log\log\log n)^{1/2}). \qquad \square$$

Lemma 2.8 *For* $-\infty < x < \infty$,

$$\lim_{n \to \infty} P\left[a_n^{-1}\left(\max_{1 \leq nt \leq [n/\log n]} \frac{|B(t)|}{\sqrt{t}} - b_n\right) \leq x\right] = \exp\{-\pi^{1/2}e^{-x}\}.$$

Proof. Because $\{(|B(t)|/\sqrt{t}); t = 1/n, \ldots, [n/\log n]/n\} \stackrel{D}{=} \{|B(t)|/\sqrt{t}; t = 1, \ldots, [n/\log n]\}$, and from Theorem 2 of Darling and Erdös (1956),

$$\lim_{n\to\infty} P\left[a_{[n/\log n]}^{-1}\left(\max_{t=1,\dots,[n/\log n]} \frac{|B(t)|}{\sqrt{t}} - b_{[n/\log n]}\right) \le x\right] = \exp\{-\pi^{1/2}e^{-x}\} \tag{2.3}$$

or

$$\lim_{n\to\infty} P\left[\max_{t=1,\dots,[n/\log n]} \frac{|B(t)|}{\sqrt{t}} \le a_{[n/\log n]}x + b_{[n/\log n]}\right] = \exp\{-\pi^{1/2}e^{-x}\}.$$

Now, from L'Hospital's rule, one can show that

$$a_{[n/\log n]} = a_n + o(a_n) \quad \text{and} \quad b_{[n/\log n]} = b_n + o(b_n).$$

Hence, for $-\infty < x < \infty$,

$$(2.3) = \lim_{n\to\infty} P\left[a_n^{-1}\left(\max_{1\le nt\le [n/\log n]} \frac{|B(t)|}{\sqrt{t}} - b_n\right) \le x\right] = \exp\{-\pi^{1/2}e^{-x}\}.$$

$\square$

Lemma 2.9 *The following hold as $n \to \infty$,*

(i)

$$\max_{1\le nt\le [n/2]} \frac{|B_0(t)|}{\sqrt{t(1-t)}} - \max_{1\le nt\le [n/\log n]} \frac{|B(t)|}{\sqrt{t}} = o_p(a_n).$$

(ii)

$$\max_{1\le n(1-t)\le [n/2]} \frac{|B_0(t)|}{\sqrt{t(1-t)}} - \max_{1\le n(1-t)\le [n/\log n]} \frac{|B(t)-B(1)|}{\sqrt{t}} = o_p(a_n).$$

Proof. (i) From Lemma 2.6,

$$\max_{1\le nt\le [n/\log n]} \frac{|B_0(t)|}{\sqrt{t(1-t)}} - \max_{1\le nt\le [n/\log n]} \frac{|B(t)|}{\sqrt{t}} = o_p(a_n),$$

or

$$a_n^{-1} \max_{1\le nt\le [n/\log n]} \frac{|B_0(t)|}{\sqrt{t(1-t)}} - a_n^{-1} \max_{1\le nt\le [n/\log n]} \frac{|B(t)|}{\sqrt{t}} = o_p(1). \tag{2.4}$$

Apply it to Lemma 2.8; then we have:

$$\lim_{n\to\infty} P\left[a_n^{-1}\left(\max_{1\le nt\le [n/\log n]} \frac{|B_0(t)|}{\sqrt{t(1-t)}} - b_n\right) \le x\right] = \exp\{-\pi^{1/2}e^{-x}\},$$

or

$$\lim_{n \to \infty} P\left[a_n \left(\max_{1 \le nt \le [n/\log n]} \frac{|B_0(t)|}{\sqrt{t(1-t)}}\right) \le a_n b_n + a_n^2 x\right] = \exp\{-\pi^{1/2} e^{-x}\}.$$

Letting $n \longrightarrow \infty$, and then $x \longrightarrow +\infty$, we have

$$\lim_{n \to \infty} P\left[a_n \left(\max_{1 \le nt \le [n/\log n]} \frac{|B_0(t)|}{\sqrt{t(1-t)}}\right) \longrightarrow 1\right] \longrightarrow 1;$$

that is,

$$(2\log\log n)^{-1/2} \max_{1 \le nt \le [n/\log n]} \frac{|B_0(t)|}{\sqrt{t(1-t)}}) \overset{n \to \infty}{\longrightarrow} 1, \tag{2.5}$$

in probability. From Lemma 2.7,

$$\max_{[n/\log n] \le nt \le [n/2]} \frac{|B_0(t)|}{\sqrt{t(1-t)}} = O_p((\log\log\log n)^{1/2}).$$

Then

$$a_n \max_{[n/\log n] \le nt \le [n/2]} \frac{|B_0(t)|}{\sqrt{t(1-t)}} = o_p(1).$$

Combining the above with (2.5), as $n \longrightarrow \infty$, we obtain

$$P\left[\max_{[n/\log n] \le nt \le [n/2]} \frac{|B_0(t)|}{\sqrt{t(1-t)}} \ge \max_{1 \le nt \le [n/\log n]} \frac{|B_0(t)|}{\sqrt{t(1-t)}}\right] \longrightarrow 0;$$

that is,

$$\max_{[n/\log n] \le nt \le [n/2]} \frac{|B_0(t)|}{\sqrt{t(1-t)}} < \max_{1 \le nt \le [n/\log n]} \frac{|B_0(t)|}{\sqrt{t(1-t)}}. \tag{2.6}$$

Then, according to (2.6),

$$\max_{1 \le nt \le [n/2]} \frac{|B_0(t)|}{\sqrt{t(1-t)}}$$

$$= \max\left[\max_{1 \le nt \le [n/\log n]} \frac{|B_0(t)|}{\sqrt{t(1-t)}}, \max_{[n/\log n] \le nt \le [n/2]} \frac{|B_0(t)|}{\sqrt{t(1-t)}}\right]$$

$$\overset{P}{=} \max_{1 \le nt \le [n/\log n]} \frac{|B_0(t)|}{\sqrt{t(1-t)}}.$$

Therefore, (2.4) becomes

$$a_n^{-1} \max_{1 \le nt \le [n/2]} \frac{|B_0(t)|}{\sqrt{t(1-t)}} - a_n^{-1} \max_{1 \le nt \le [n/\log n]} \frac{|B_0(t)|}{\sqrt{t(1-t)}} = o_p(1),$$

which leads to the result (i).

(ii) From the symmetry property of Brownian motion with respect to $t = \frac{1}{2}$, $B_0(t) \overset{D}{=} B_0(1-t)$ for $0 < t < 1$. Because $B(1-t) \overset{D}{=} B(1) - B(t) \sim N(0, 1-t)$, then replacing t by $1 - t$ in (i), we obtain

$$\max_{1 \le n(1-t) \le [n/2]} \frac{|B_0(t)|}{\sqrt{t(1-t)}} - \max_{1 \le n(1-t) \le [n/\log n]} \frac{|B(1-t)|}{\sqrt{t}} = o_p(a_n),$$

or

$$\max_{1 \le n(1-t) \le [n/2]} \frac{|B_0(t)|}{\sqrt{t(1-t)}} - \max_{1 \le n(1-t) \le [n/\log n]} \frac{|B(t) - B(1)|}{\sqrt{t}} = o_p(a_n).$$

$\square$

After the above preparation, we are ready to prove the theorem.

Proof of Theorem 2.5

$$P[a_n^{-1}(U - b_n) \le x | H_0]$$

$$= P[U \le a_n x + b_n | H_0]$$

$$= P\left[\max_{1 \le nt \le n-1} \frac{|B_0(t)|}{\sqrt{t(1-t)}} \le a_n x + b_n \right]$$

$$= P\left[\max_{1 \le nt \le [n/2]} \frac{|B_0(t)|}{\sqrt{t(1-t)}} \le a_n x + b_n, \right.$$

$$\left. \max_{1 \le n(1-t) \le [n/2]} \frac{|B_0(t)|}{\sqrt{t(1-t)}} \le a_n x + b_n \right]$$

$$= P\left[\max_{1 \le nt \le [n/\log n]} \frac{|B(t)|}{\sqrt{t}} \le a_n x + b_n, \right.$$

$$\left. \max_{1 \le n(1-t) \le [n/\log n]} \frac{|B(t) - B(1)|}{\sqrt{1-t}} \le a_n x + b_n \right] + o_p(1)$$

$$= P\left[\max_{1 \le nt \le [n/\log n]} \frac{|B(t)|}{\sqrt{t}} \le a_n x + b_n \right]$$

$$\cdot P\left[\max_{1 \le n(1-t) \le [n/\log n]} \frac{|B(t) - B(1)|}{\sqrt{1-t}} \le a_n x + b_n \right] + o_p(1)$$

$$\overset{n \longrightarrow \infty}{\longrightarrow} \exp(-\pi^{-1/2} e^{-x}) \cdot \exp(-\pi^{-1/2} e^{-x})$$

$$= \exp(-2\pi^{-1/2} e^{-x}).$$

2.1.2 Variance Unknown

Under H_0, the likelihood function now is

$$L_0(\mu, \sigma^2) = \frac{1}{(\sqrt{2\pi}\sigma)^n} e^{-\sum_{i=1}^{n}(x_i-\mu)^2/2\sigma^2}$$

and the MLEs of μ and σ^2 are

$$\widehat{\mu} = \overline{x} = \frac{1}{n}\sum_{i=1}^{n} x_i, \quad \text{and} \quad \widehat{\sigma}^2 = \frac{1}{n}\sum_{i=1}^{n}(x_i - \overline{x})^2,$$

respectively. Under H_1, the likelihood function is

$$L_1(\mu_1, \mu_n, \sigma_1^2) = \frac{1}{(\sqrt{2\pi})^n} e^{-\sum_{i=1}^{k}(x_i-\mu_1)^2/2\sigma_1^2 - \sum_{i=k+1}^{n}(x_i-\mu_n)^2/2\sigma_1^2},$$

and the MLEs of μ_1, μ_n, and σ_1^2 are,

$$\widehat{\mu}_1 = \overline{x}_k = \frac{1}{k}\sum_{i=1}^{k} x_i, \qquad \widehat{\mu}_n = \overline{x}_{n-k} = \frac{1}{n-k}\sum_{i=k+1}^{n} x_i,$$

and

$$\widehat{\sigma}_1^2 = \frac{1}{n}\left[\sum_{i=1}^{k}(x_i - \overline{x}_k)^2 + \sum_{i=k+1}^{n}(x_i - \overline{x}_{n-k})^2\right],$$

respectively.

Let

$$S = \sum_{i=1}^{n}(x_i - \overline{x})^2 \quad \text{and} \quad T_k^2 = \frac{k(n-k)}{n}(\overline{x}_k - \overline{x}_{n-k})^2.$$

The likelihood procedure-based test statistic is then given by

$$V = \max_{1 \le k \le n-1} \frac{|T_k|}{S}. \tag{2.7}$$

Worsley (1979) obtained the null distribution of V. His result is presented in the following.

Let $\mathbf{x} = (x_1, x_2, \ldots, x_n)$, and $\mathbf{y} = (y_1, y_2, \ldots, y_n)$, where $y_i = (x_i - \overline{x})/\sqrt{S}$, $i = 1, 2, \ldots, n$. Also, define

$$c_k = \sqrt{k/(n-k)} \quad \text{and} \quad b_{ki} = \begin{cases} n^{-1/2}c_k^{-1}, & i = 1, \ldots, k \\ -n^{-1/2}c_k, & i = k+1, \ldots, n \end{cases},$$

for $k = 1, 2, \ldots, n-1$. Let $\mathbf{b}_k$ be the vector such that $\mathbf{b}_k = (b_{k1}, \ldots, b_{kn})$. Then, $T_k = \mathbf{b}_k'\mathbf{x}$, $k = 1, \ldots, n-1$. It is easy to see that $\mathbf{b}_k'\mathbf{1} = \mathbf{0}$, where

$\mathbf{1} = (1, 1, \ldots, 1)'$ is the $n \times 1$ unit vector, hence $\mathbf{b}'_k \mathbf{y} = \mathbf{b}_k \mathbf{x}/\sqrt{S}$ and $V = \max_{1 \leq k \leq n-1} |\mathbf{b}'_k \mathbf{y}|$.

Next, let $\mathbf{y} \in R^n$. Under H_0, $\mathbf{y}'\mathbf{y} = 1$, thus $\mathbf{y}$ is uniformly distributed on the surface of the unit $(n-1)$-ball $C = \{\mathbf{y} : \mathbf{y}'\mathbf{y} = 1 \text{ and } \mathbf{1}'\mathbf{y} = 0\}$. Let $D = \{\mathbf{y} : \mathbf{1}'\mathbf{y} = 0, |\mathbf{b}'_k \mathbf{y}| \leq v, k = 1, \ldots, n-1\}$, then the event $\{V \leq v\} = \{\mathbf{y} \in C \cap D\}$. Therefore,

$$P\{V \leq v\} = \text{surface area of } C \text{ inside } D$$

$$= P\left\{\mathbf{y} \in \bigcap_{k=1}^{n-1}(A_k^+ \bigcup A_k^-)^c\right\}, \qquad (2.8)$$

where for $k = 1, \ldots, n-1$,

$$A_k^+ = \{\mathbf{y} : \mathbf{y} \in C, \mathbf{b}'_k \mathbf{y} > v\},$$

$$A_k^- = \{\mathbf{y} : \mathbf{y} \in C, -\mathbf{b}'_k \mathbf{y} > v\},$$

$$A_k = A_k^+ \bigcup A_k^-, A_k^+ \bigcap A_k^- = \phi.$$

From DeMorgan's law, (2.8) is reduced to

$$P\{V \leq v\} = 1 - P\left\{\mathbf{y} \in \bigcup_{k=1}^{n}(A_k^+ \bigcup A_k^-)\right\}$$

$$= 1 - \sum_{k=1}^{n} P\{\mathbf{y} \in A_k\} + \sum\sum_{1 \leq k_1 \leq k_2 \leq n-1} P\left\{\mathbf{y} \in A_{k_1} \bigcap A_{k_2}\right\}$$

$$+ \cdots + (-1)^p \sum\sum\cdots\sum_{1 \leq k_1 \leq k_2 \leq \cdots \leq k_p \leq n-1} P\left\{\mathbf{y} \in \bigcap_{j=1}^{p} A_{kj}\right\}$$

$$+ \cdots + (-1)^{n-1} P\left\{\mathbf{y} \in \bigcap_{j=1}^{n-1} A_j\right\}. \qquad (2.9)$$

Because $P\{\mathbf{y} \in A_k\} = P\{\mathbf{y} \in C, \text{ and } \mathbf{b}'_k \mathbf{y} > v\} + P\{\mathbf{y} \in C, \text{ and } \mathbf{b}'_k \mathbf{y} < -v\}$, and $\mathbf{b}'_k \mathbf{y} = \mathbf{b}_k \mathbf{x}/\sqrt{S} = T_k/\sqrt{S}$, then $P\{\mathbf{y} \in A_k\}$ can be calculated via the distribution of the statistic $T_k/\sqrt{S}$.

Now, under H_0, $T_k \sim N(0, \sigma^2)$, $S_k^2 = \sum_{i=1}^{k}(x_i - \bar{x}_k)^2 + \sum_{i=k+1}^{n}(x_i - \bar{x}_{n-k})^2$ is distributed as $\sigma^2 \chi_{n-2}^2$, and T_k is independent of S_k; then $T_k/[S_k/\sqrt{n-2}] \sim \chi_{n-2}^2$. But simple algebra shows that $S = S_k^2 + T_k^2$, then $T_k/S = (S_k^2/T_k^2 + 1)^{-1/2}$, hence $P\{\mathbf{y} \in A_k\}$ can be calculated via a t_{n-2} distribution.

For other terms in (2.9), we need only to consider a general one such as $P\{\mathbf{y} \in \bigcap_{j=1}^{p} \tilde{A}_{kj}\}$ for $1 < p < n-1$, with $k_1 < k_2 < \cdots < k_p$, and $\tilde{A}_{kj}$ is either A_{kj}^+ or A_{kj}^-.

Let B be a $p \times n$ matrix such that $B = (\mathbf{b}_{k1}^{*\prime}, \ldots, \mathbf{b}_{kp}^{*\prime})'$, and $\mathbf{I}_p$ be a $p \times 1$ unit vector, where

$$
\mathbf{b}_{ki}^{*\prime} = \begin{cases} b_{kj} & \text{if } \widetilde{A}_{kj} = A_{kj}^+ \\ -b_{kj} & \text{if } \widetilde{A}_{kj} = A_{kj}^- \end{cases}
$$

Then, $P\{\mathbf{y} \in \bigcap_{j=1}^p \widetilde{A}_{kj}\} = P\{B\mathbf{y} > v\mathbf{I}_p\}$. Now the following theorem gives the null probability density function of $V = B\mathbf{y}$ at v.

Theorem 2.10 *Under H_0, the pdf of $V = B\mathbf{y}$ at v is given by*

$$
f_p(v) = \begin{cases} \dfrac{\Gamma\left[\frac{n-1}{2}\right]}{\pi^{p/2}\Gamma\left[\frac{n-1-p}{2}\right]} |\Sigma|^{-1/2}[1 - v'\Sigma^{-1}v]^{(n-3-p)/2}, & \text{if } v'\Sigma^{-1}v < 1 \\ 0, & \text{otherwise} \end{cases}
$$

where $\Sigma = BB'$.

To prove this theorem, we need the following results.

Lemma 2.11 *B can be written as $B = G\Gamma$, where G is a $p \times p$ positive definite matrix, and Γ is $p \times n$ with $\Gamma\Gamma' = I_p$.*

Proof. It follows directly from Theorem 1.39 on page 11 of Gupta and Varga (1993). □

For the purpose of deriving the null distribution, WLOG, we write:

$$
H_0 : \mu_1 = \mu_2 = \cdots = \mu_n = 0.
$$

Let $J = I_n - (1/n)\mathbf{1}\mathbf{1}'$, and augment the matrix Γ in Lemma 2.11 to:

$$
Q = \begin{pmatrix} \Gamma \\ \Gamma_0 \end{pmatrix},
$$

where Γ_0 is $(n - p) \times n$ such that Q is an $n \times n$ orthogonal matrix. Also, let $M = QJQ'$, and let the first $p \times p$ principal minor of M be M_p; then we have the following Lemma 2.12.

Lemma 2.12 *$V = B\mathbf{y}$ has the pdf:*

$$
f_p(v) = \frac{\Gamma\left[\frac{n-1}{2}\right]}{\pi^{p/2}\Gamma\left[\frac{n-1-p}{2}\right]} |M_p^{-1}|^{1/2}|G^{-1}|[1 - v'G'^{-1}M_p^{-1}G^{-1}v]^{(n-3-p)/2},
$$

over the region $v'G'^{-1}M_p^{-1}G^{-1}v < 1$, and zero otherwise, where G is as in Lemma 2.11.

Proof. Clearly, $\mathbf{y} = J\mathbf{x}/\sqrt{\mathbf{x}'J\mathbf{x}}$ and $J\mathbf{1} = \mathbf{0}$. From Lemma 2.11,

$$V = B\mathbf{y} = G\Gamma\mathbf{y}$$

$$= G(I_p \vdots 0)Q\mathbf{y} \quad (0 \text{ is a } p \times (n-p) \text{ zero matrix})$$

$$= G(I_p \vdots 0)QTQ'(Q\mathbf{x}')/\sqrt{(Q\mathbf{x})'QJQ'(Q\mathbf{x})}.$$

Let $Z = Q\mathbf{x}$; then under H_0, $Z \sim N(\mathbf{0}, \sigma^2 I_n)$. Let $\mathbf{t} = \Gamma\mathbf{y} = (I_p \vdots 0)MZ$ $\sqrt{Z'MZ}$, then from Worsely (1979, 1983), the pdf of $\mathbf{t}$ is:

$$f(t_1, \ldots, t_p) = \frac{\Gamma\left[\frac{n-1}{2}\right]}{\pi^{p/2}\Gamma\left[\frac{n-1-p}{2}\right]}|M_p^{-1}|^{1/2}[1 - \mathbf{t}'M_p^{-1}\mathbf{t}]^{(n-3-p)/2}$$

for $\mathbf{t}'M_p^{-1}\mathbf{t} < 1$, and zero otherwise.

Therefore, $V = B\mathbf{y} = G\Gamma\mathbf{y} = G\mathbf{t}$ has the pdf

$$f_p(v) = \frac{\Gamma\left[\frac{n-1}{2}\right]}{\pi^{p/2}\Gamma\left[\frac{n-1-p}{2}\right]}|M_p^{-1}|^{1/2}|G^{-1}|[1 - v'G'^{-1}M_p^{-1}G^{-1}v]^{(n-3-p)/2},$$

over the region $v'G'^{-1}M_p^{-1}G^{-1}v < 1$, and zero otherwise.   □

Now it is time to prove the theorem.

Proof of Theorem 2.10 Because

$$M = QJQ'$$

$$= \begin{pmatrix} \Gamma \\ \Gamma_0 \end{pmatrix}\left(I_n - \frac{1}{n}\mathbf{11}'\right)(\Gamma' \; \Gamma_0')$$

$$= \begin{pmatrix} I_p - \frac{1}{n}\Gamma\mathbf{11}'\Gamma' & * \\ * & * \end{pmatrix},$$

we have $M_p = I_p - (1/n)\Gamma\mathbf{11}'\Gamma'$. But $\Gamma\mathbf{1} = \mathbf{0}_p$ as $B\mathbf{1} = \mathbf{0}_p$, therefore $M_p = I_p$. From Lemma 2.12, we thus obtain that the pdf of $V = B\mathbf{y}$ at v is given by

$$f_p(v) = \frac{\Gamma\left[\frac{n-1}{2}\right]}{\pi^{p/2}\Gamma\left[\frac{n-1-p}{2}\right]}|\Sigma|^{-1/2}[1 - v'\Sigma^{-1}v]^{(n-3-p)/2},$$

for $v'\Sigma^{-1}v < 1$, and zero otherwise, where $\Sigma = GM_p^{-1}G' = GG' = G\Gamma\Gamma'G' = BB'$.   □

Consequently, based on the t-distribution with $n - 2$ degrees of freedom and Theorem 2.10, the null probability function of V can be calculated through (2.9).

Remark 2.13 Ignoring the higher-order terms in (2.9), one can obtain Bonferroni approximations for the distribution of V. Worsley (1979) obtained the percentage points for the Bonferroni approximations.

2.1.3 Application to Biomedical Data

In the last decade or so, life science research has been advanced by fast developing biotechnologies. Microarray technology, among the many well-developed biotechnologies, is a breakthrough that makes it possible to quantify the expression patterns of thousands or tens of thousands of genes in various tissues, cell lines, and conditions simultaneously. Biologists, geneticists, and medical researchers now routinely use microarray technology in their specified research projects thus resulting in voluminous numerical data related to expression of each gene encoded in the genome, the content of proteins and other classes of molecules in cells and tissues, and cellular responses to stimuli, treatments, and environmental factors.

Biological and medical research (e.g., see Lucito et al. 2000) reveals that some forms of cancer are caused by somatic or inherited mutations in oncogenes and tumor suppressor genes; cancer development and genetic disorders often result in chromosomal DNA copy number changes or copy number variations (CNVs). Consequently, identification of these loci where the DNA copy number changes or CNVs have taken place will (at least partially) facilitate the development of medical diagnostic tools and treatment regimes for cancer and other genetic diseases. Due to the advancement in array technology, the array Comparative Genomic Hybridization (aCGH) technique (see Kallioniemi et al. 1992 and Pinkel et al. 1998) or single nucleotide polymorphism (SNP) arrays (see Nannya et al. 2005) are often used in experiments that are deemed to study DNA copy numbers. The resulting data are typically called aCGH and SNP array data, respectively. However, because of the random noise inherited in the imaging and hybridization process in the DNA copy number experiments, identifying statistically significant CNVs or DNA copy number changes in aCGH data and in SNP array data is challenging.

In DNA copy number experiments such as aCGH copy number experiments, differentially labeled sample and reference DNA are hybridized to DNA microarrays, and the sample intensities of the test and reference samples are obtained (Pinkel et al. 1998, Pollack et al. 1999, and Myers et al. 2004). As the reference sample is assumed or chosen to have no copy number changes, markers whose test sample intensities are significantly higher (or lower) than the reference sample intensities are corresponding to DNA copy number gains (or losses) in the test sample at those locations (Olshen et al. 2004).

Concretely, the test sample intensity at locus i on the genome is usually denoted by T_i and the corresponding reference sample intensity by R_i, and the

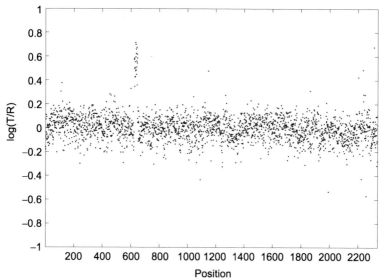

Fig. 2.4 The genome of the fibroblast cell line GM01524 of Snijders et al. (2001)

normalized log base 2 ratio of the sample and reference intensities, $\log_2 T_i/R_i$, at the ith biomarker, is one of the default outputs after the DNA copy number experiment is conducted using the aCGH technique. Here, $\log_2 T_i/R_i = 0$ indicates no DNA copy number change at locus i, $\log_2 T_i/R_i < 0$ reveals a deletion at locus i, and $\log_2 T_i/R_i > 0$ signifies duplication in the test sample at that locus. Due to various random noise, which occurs largely during the experimental and image processing stages, the $\log_2 T_i/R_i$ becomes a random variable. Ideally, this random variable is assumed to follow a Gaussian distribution of mean 0 and constant variance σ^2. Then, deviations from the constant parameters (mean and variance) presented in $\log_2 T_i/R_i$ data may indicate a copy number change. Hence, the key to identifying true DNA copy number changes becomes the problem of how to identify changes in the parameters of a normal distribution based on the observed sequence of $\log_2 T_i/R_i$. Figure 2.4 is the scatterplot of the log base 2 ratio intensities of the genome of the fibroblast cell line GM01524 obtained by Snijders et al. (2001).

Since the publication of aCGH data by many research laboratories on copy number studies for different cell lines or diseases, analyzing aCGH data has become an active research topic for scientists, data analysts, and biostatisticians among others. A recent survey on the methods developed for analyzing aCGH data can be found in Chen (2010).

Among the many methods used for aCGH data, some methods are rooted in statistical change point analysis. Olshen et al. (2004) proposed a circular binary segmentation (CBS) method to identify DNA copy number changes in an aCGH database on the mean change point model proposed in Sen and Srivastava (1975a). This CBS method is mainly the combination of the

likelihood-ratio based test for testing no change in the mean against exactly
one change in the mean with the BSP (Vostrikova, 1981) for searching multiple change points in the mean, assuming that the variance is unchanged.
The idea of the CBS method (Olshen et al. 2004) can be summarized as
follows.

Let X_i denote the normalized $\log_2 R_i/G_i$ at the ith locus along the chromosome; then $\{X_i\}$ is considered as a sequence of normal random variables
taken from $N(\mu_i, \sigma_i^2)$, respectively, for $i = 1, \ldots, n$. Consider any segment
of the sequence of the log ratio intensities $\{X_i\}$ (assumed to follow normal
distributions) to be spliced at the two ends to form a circle; the test statistic
Z_c of the CBS is based on the modified likelihood-ratio test and is specifically,

$$Z_c = \max_{1 \leq i < j \leq n} |Z_{ij}|, \qquad (2.10)$$

where Z_{ij} is the likelihood-ratio test statistic given in Sen and Srivastava
(1975a) for testing the hypothesis that the arc from $i + 1$ to j and its com-
plement have different means (i.e., there is a change point in the mean of the
assumed normal distribution for the X_is) and is given by:

$$Z_{ij} = \frac{1}{\{1/(j-i) + 1/(n-j+i)\}^{1/2}} \left\{ \frac{S_j - S_i}{j - i} - \frac{S_n - S_j + S_i}{n - j + i} \right\}, \qquad (2.11)$$

with

$$S_i = X_1 + X_2 + \cdots + X_i, \qquad 1 \leq i < j \leq n.$$

Note that Z_c allows for both a single change ($j = n$) and the epidemic
alternative ($j < n$). A change is claimed if the statistic exceeds an appropriate critical value at a given significant level based on the null distribution.
However, the null distribution of the test statistic Z_c is not attainable so far
in the literature of change point analysis. Then, as suggested in Olshen et al.
(2004), the critical value when the X_is are normal needs to be computed using
Monte Carlo simulations or the approximation given by Siegmund (1986) for
the tail probability. Once the null hypothesis of no change is rejected the
changepoint(s) is (are) estimated to be i (and j) such that $Z_c = |Z_{ij}|$ and
the procedure is applied recursively to identify all the changes in the whole
sequence of the log ratio intensities of a chromosome (usually of hundreds to
thousands of observations). The CBS algorithm is written as an R package
and is available from the R Project website.

The influence of the CBS to the analyses of aCGH data is tremendous as
the CBS method provided a statistical framework to the analysis of DNA
copy number analysis. The p-value given by the CBS for a specific locus
being a change point, however, is only obtained by a permutation method
and the calculation of such a p-value takes a long computation time when the
sequence is long, which is the case for high-density array data. Hence the CBS
method has the slowest computational speed as pointed out in Picard et al.
(2005). A recent result in Venkatraman and Olshen (2007) has improved the

Table 2.1 Computational Output for Cell Line GM07408 Using the R Package DNAcopy (Venkatraman and Olshen 2007)

Chr. ID	Start Locus	End Locus	Number Mark	Seg Mean	Statistic	p-Value
1	468.3075	72188.15	46	0.0005	6.798636	3.974580×10^{-10}
1	77228.4545	91439.90	9	−0.1613	5.206164	6.896233×10^{-6}
1	93058.0760	209867.40	63	0.0002	4.205344	7.806819×10^{-4}
1	211009.1200	240000.00	15	−0.1129	NA	NA
2	0.0000	245000.00	65	−0.0470	NA	NA
3	0.0000	218000.00	83	−0.0831	NA	NA
4	0.0000	15439.24	11	0.0549	4.091054	1.604376×10^{-3}
4	22245.2500	169000.00	126	−0.0728	2.753006	1.323652×10^{-1}
4	170170.0000	177387.29	8	0.0242	4.995123	4.854205×10^{-6}
4	178400.0000	179200.0	4	0.1962	7.467238	1.040446×10^{-12}
4	179490.1710	184000.00	12	0.0193	NA	NA
5	0.0000	198500.00	110	−0.0259	NA	NA
6	0.0000	65990.01	43	−0.0087	5.582021	7.693862×10^{-7}
6	65990.0145	103277.13	12	−0.1887	3.581999	6.122494×10^{-3}
6	104186.6075	188000.00	29	−0.0630	NA	NA
7	0.0000	161500.00	173	0.0166	NA	NA
8	0.0000	13266.95	23	0.0091	5.427554	1.043974×10^{-6}
8	16299.9040	16815.62	3	−0.2023	5.065677	7.349159×10^{-6}
8	17992.9110	41655.39	24	0.0483	6.296318	1.294907×10^{-8}
8	41932.6780	86296.08	48	−0.0553	5.116016	1.016467×10^{-5}
8	87774.9320	100534.00	15	−0.1495	7.786791	2.741869×10^{-13}
8	107253.3650	147000.00	38	0.0241	NA	NA
9	0.0000	33134.86	29	−0.1002	5.493857	1.828478×10^{-6}
9	33856.8540	115000.00	77	0.0187	NA	NA
10	0.0000	14349.90	16	0.0105	2.317172	2.252879×10^{-1}
10	15126.2185	47803.26	23	−0.0401	5.230861	3.958791×10^{-6}
10	49954.2190	69209.44	14	−0.1412	8.040106	1.713658×10^{-14}
10	69549.0280	73067.45	9	0.0718	2.372705	1.860309×10^{-1}
10	74000.0000	108902.62	25	0.0277	4.739140	4.486729×10^{-5}
10	110000.0000	117000.00	10	−0.0562	5.814313	1.408531×10^{-7}
10	118499.8495	142000.00	23	0.1067	NA	NA

computational speed of CBS. If there is an analytic formula for calculating the p-value of the change point hypothesis, the computational speed will undoubtedly be faster and more convenient.

For the application of the CBS method to the analysis of 15 fibroblast cell lines obtained in Snijders et al. (2001), readers are referred to Olshen et al. (2004) and Venkatraman and Olshen (2007). As a complete example, we present here the use of the R-package, DNAcopy (Venkatraman and Olshen 2007), on the fibroblast cell line GM07408 (Snijders et al. 2001). After the data are read into the R-package, we obtain the p-values (based on the test statistics Z_{ij}, given in (2.11)) for each segment being a change along the chromosome. These results directly output from DNAcopy are listed in Table 2.1 of this chapter.

Table 2.2 Computational Output for Cell Line GM07408 Using DNAcopy: Table 2.1
Continued

Chr. ID	Start Locus	End Locus	Number Mark	Seg Mean	Statistic	p-Value
11	0.0000	20607.03	34	−0.0076	4.551992	1.319627×10^{-4}
11	20719.3180	34420.00	14	−0.1402	5.978688	4.620736×10^{-8}
11	34420.0000	39388.78	14	0.0011	5.925536	6.168566×10^{-8}
11	39623.3960	48010.94	13	−0.1385	6.550567	2.330282×10^{-9}
11	48923.4020	87856.91	51	0.0187	7.290779	1.511135×10^{-11}
11	88570.8380	117616.94	24	−0.1376	7.615232	1.051503×10^{-12}
11	117817.0610	145000.00	30	0.0338	NA	NA
12	315.6600	142000.00	91	0.0091	NA	NA
13	5653.9480	28325.81	14	0.0122	6.231862	1.191981×10^{-8}
13	28469.3085	78470.20	22	−0.1277	5.223176	4.356758×10^{-6}
13	80645.9990	100500.00	18	−0.0093	NA	NA
14	769.5125	40288.14	30	−0.0847	6.140269	3.336396×10^{-8}
14	42901.6730	97000.00	39	0.0301	NA	NA
15	0.0000	79000.00	67	−0.0065	NA	NA
16	0.0000	52092.53	31	0.0312	5.063312	9.064410×10^{-6}
16	53000.0000	55905.26	4	−0.1804	4.543496	1.095742×10^{-4}
16	57000.0000	84000.00	30	0.0549	NA	NA
17	0.0000	52738.37	58	0.0686	5.153190	8.319006×10^{-6}
17	52804.6060	56190.77	4	−0.1529	4.598118	8.167274×10^{-5}
17	56313.4240	86000.00	28	0.0868	NA	NA
18	0.0000	86000.00	44	0.5821	NA	NA
19	0.0000	70000.00	35	0.0828	NA	NA
20	0.0000	73000.00	82	0.0459	NA	NA
21	3130.9400	19079.91	20	−0.1420	6.288501	7.767927×10^{-9}
21	19247.3920	30000.00	13	0.0274	NA	NA
22	1100.0000	33000.00	13	0.1213	NA	NA
23	0.0000	155000.00	42	0.7744	NA	NA

The plot that indicated the whole genome of GM07408 is also obtained and
is given as Figure 2.5. In Figure 2.5, the adjacent chromosomes are indicated
by green and black colors alternately, and the red line segment represents the
sample mean for each segment.

Figure 2.6 gives the collection of all 23 plots for the 23 chromosomes of
the cell line GM07408 with changes identified. It is customary to also obtain
a plot of the log ratio intensities chromosome by chromosome with mean
changes identified by the CBS method. More plots, say Figures 2.7–2.11, are
also obtained by using the same R-package for chromosomes 4, 5, 8–11, 22,
and 23 of the cell line GM07408 with changes identified.

There are changes identified by using the CBS method on each chromosome.

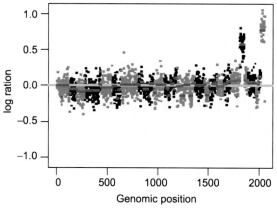

Fig. 2.5 The genome plot of the fibroblast cell line GM07408 of Snijders et al. (2001) using the *R*-package DNAcopy

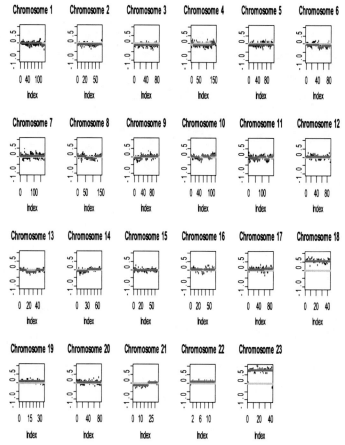

Fig. 2.6 The plots of all chromosomes on the fibroblast cell line GM07408 of Snijders et al. (2001) with mean changes indicated

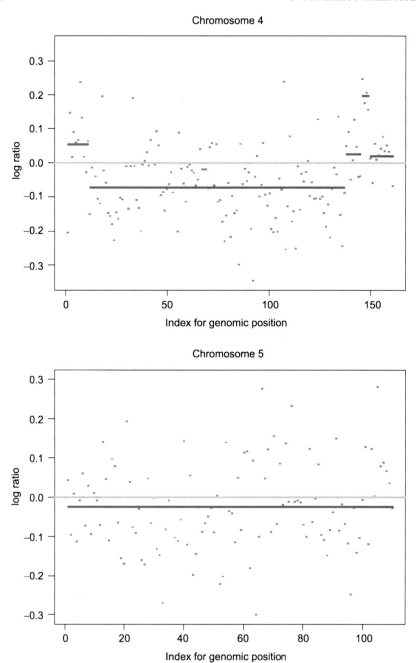

Fig. 2.7 The plot of chromosomes 4 and 5 on the fibroblast cell line GM07408 of Snijders et al. (2001) with mean changes indicated

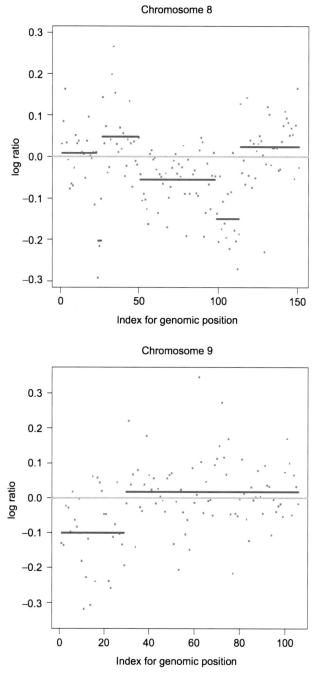

Fig. 2.8 The plot of chromosomes 8 and 9 on the fibroblast cell line GM07408 of Snijders et al. (2001) with mean changes indicated

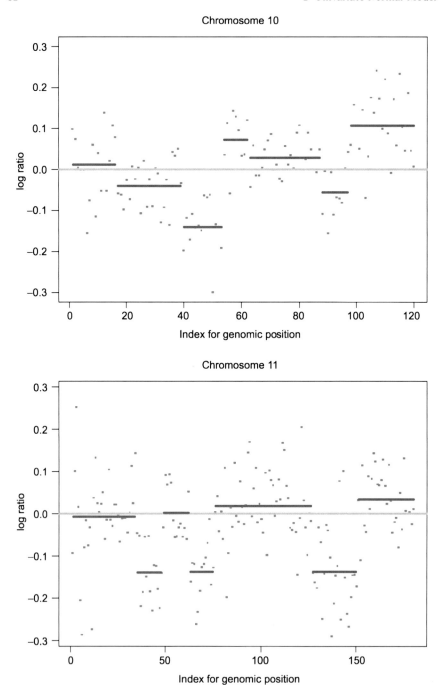

Fig. 2.9 The plot of chromosomes 10 and 11 on the fibroblast cell line GM07408 of Snijders et al. (2001) with mean changes indicated

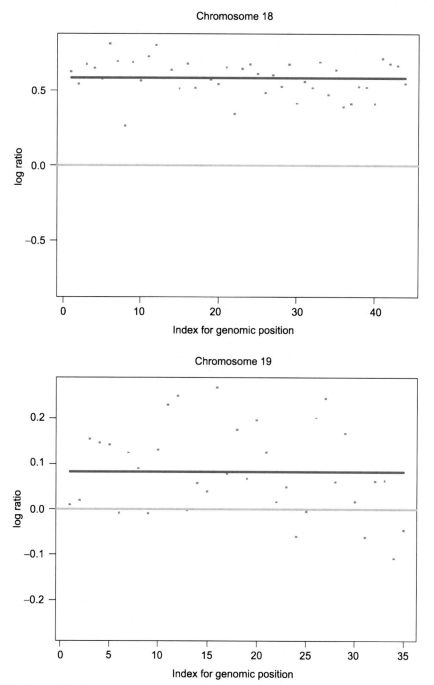

Fig. 2.10 The plot of chromosomes 18 and 19 on the fibroblast cell line GM07408 of Snijders et al. (2001) with mean changes indicated

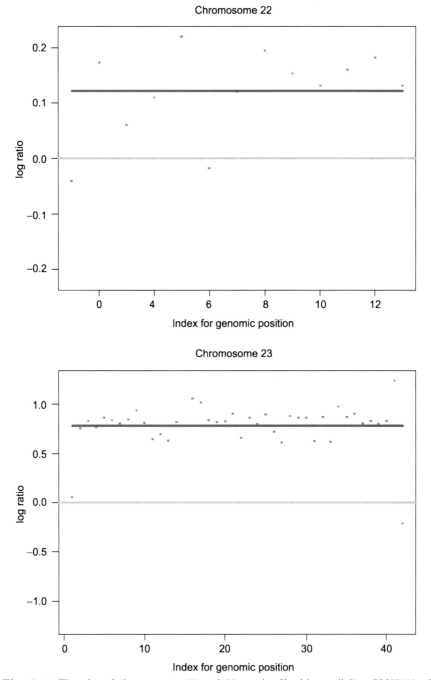

Fig. 2.11 The plot of chromosomes 22 and 23 on the fibroblast cell line GM07408 of Snijders et al. (2001) with mean changes indicated

2.2 Variance Change

Testing and estimation about mean change in a Gaussian model has been studied in Section 2.1. The corresponding problem of changes in the regression model is studied in Chapter 4. Inference about variance changes while the mean remains common has been studied by Wichern, Miller, and Hsu (1976), Hsu (1977), Davis (1979), Abraham and Wei (1984), Inclán (1993), and Chen and Gupta (1997). In this section, the variance change problem for the univariate Gaussian model using different methods is considered

Let $x_1, x_2, \ldots, x_n$ be a sequence of independent normal random variables with parameters $(\mu, \sigma_1^2), (\mu, \sigma_2^2), \ldots, (\mu, \sigma_n^2)$, respectively. Assume that μ is known. The interest here is to test the hypothesis (see Gupta and Tang, 1987):

$$H_0 : \sigma_1^2 = \sigma_2^2 = \cdots = \sigma_n^2 = \sigma^2 \text{(unknown)}, \tag{2.12}$$

versus the alternative:

$$H_A : \sigma_1^2 = \cdots = \sigma_{k_1}^2 \neq \sigma_{k_1+1}^2 = \cdots = \sigma_{k_2}^2 \neq \cdots \neq \sigma_{k_q+1}^2 = \cdots = \sigma_n^2,$$

where q is the unknown number of change points, and $1 \leq k_1 < k_2 < \cdots < k_q < n$, are the unknown positions of the change points, respectively. Using the binary segmentation procedure, as described in Chapter 1, it suffices to test and estimate the position of a single change point at each stage, that is, to test H_0 defined by (2.10) against the following alternative:

$$H_1 : \sigma_1^2 = \cdots = \sigma_k^2 \neq \sigma_{k+1}^2 = \cdots = \sigma_n^2, \tag{2.13}$$

where $1 < k < n$, is the unknown position of the single change point.

2.2.1 Likelihood-Ratio Procedure

Under H_0, the log likelihood function is:

$$\log L_0(\sigma^2) = -\frac{n}{2} \log 2\pi - \frac{n}{2} \log \sigma^2 - \frac{\sum_{i=1}^{n} (x_i - \mu)^2}{2\sigma^2}.$$

Let $\widehat{\sigma}^2$ be the MLE of σ^2 under H_0, Then

$$\widehat{\sigma}^2 = \frac{\sum_{i=1}^{n} (x_i - \mu)^2}{n},$$

and the maximum likelihood is

$$\log L_0(\widehat{\sigma}^2) = -\frac{n}{2} \log 2\pi - \frac{n}{2} \log \widehat{\sigma}^2 - \frac{n}{2}.$$

Under H_1, the log likelihood function is:

$$\log L_1(\sigma_1^2, \sigma_n^2) = -\frac{n}{2} \log 2\pi - \frac{k}{2} \log \sigma_1^2 - \frac{n-k}{2} \log \sigma_n^2$$

$$- \frac{\sum_{i=1}^{k}(x_i - \mu)^2}{2\sigma_1^2} - \frac{\sum_{i=k+1}^{n}(x_i - \mu)^2}{2\sigma_n^2}.$$

Let $\hat{\sigma}_1^2$, $\hat{\sigma}_n^2$ be the MLEs of σ_1^2, σ_n^2, respectively; then

$$\hat{\sigma}_1^2 = \frac{\sum_{i=1}^{k}(x_i - \mu)^2}{k}, \qquad \hat{\sigma}_n^2 = \frac{\sum_{i=k+1}^{n}(x_i - \mu)^2}{n-k},$$

and the maximum log likelihood is:

$$\log L_1(\hat{\sigma}_1^2, \hat{\sigma}_n^2) = -\frac{n}{2} \log 2\pi - \frac{k}{2} \log \hat{\sigma}_1^2 - \frac{n-k}{2} \log \hat{\sigma}_n^2 - \frac{n}{2}.$$

Then the likelihood-ratio procedure statistic is

$$\lambda_n = \left\{ \max_{1<k<n-1} [n \log \hat{\sigma}^2 - k \log \hat{\sigma}_1^2 - (n-k) \log \hat{\sigma}_n^2] \right\}^{1/2}. \qquad (2.14)$$

Notice that, to be able to obtain the MLEs, we can only detect changes for $2 \le k \le n-2$. According to the principle of the maximum likelihood procedure, we estimate the position k of the change point by $\hat{k}$ such that (2.12) attains its maximum at $\hat{k}$.

Next, we derive the asymptotic null distribution of λ_n. Note that, for large n,

$$\lambda_n = \left\{ \max_{1<k<n-1} [n \log \hat{\sigma}^2 - k \log \hat{\sigma}_1^2 - (n-k) \log \hat{\sigma}_n^2] \right\}^{1/2}$$

$$= \max_{1<k<n-1} \left[n \log \frac{\sum_{i=1}^{n}(x_i - \mu)^2}{n} - k \log \frac{\sum_{i=1}^{k}(x_i - \mu)^2}{k} \right.$$

$$\left. -(n-k) \log \frac{\sum_{i=k+1}^{n}(x_i - \mu)^2}{n-k} \right]^{1/2}.$$

Under H_0, $y_i = (x_i - \mu)/\sigma \sim N(0,1)$; then

$$\lambda_n \overset{D}{=} \max_{1<k<n-1} \left[n \log \frac{\sum_{i=1}^{n} y_i^2}{n} - k \log \frac{\sum_{i=1}^{k} y_i^2}{k} - (n-k) \log \frac{\sum_{i=k+1}^{n} y_i^2}{n-k} \right]^{1/2}.$$

Now, using the three-term Taylor expansion, we write

$$\xi_k = n \log \frac{\sum_{i=1}^n y_i^2}{n} - k \log \frac{\sum_{i=1}^k y_i^2}{k} - (n-k) \log \frac{\sum_{i=k+1}^n y_i^2}{n-k}$$

$$= n \left(\frac{\sum_{i=1}^n y_i^2}{n} - 1 \right) - \frac{n}{2} \left(\frac{\sum_{i=1}^n y_i^2}{n} - 1 \right)^2 + \frac{n}{3} (\theta_n^{(1)})^{-3} \left(\frac{\sum_{i=1}^n y_i^2}{n} - 1 \right)^3$$

$$- k \left(\frac{\sum_{i=1}^k y_i^2}{k} - 1 \right) + \frac{k}{2} \left(\frac{\sum_{i=1}^k y_i^2}{k} - 1 \right)^2 - \frac{k}{3} (\theta_n^{(2)})^{-3} \left(\frac{\sum_{i=1}^k y_i^2}{k} - 1 \right)^3$$

$$- (n-k) \left(\frac{\sum_{i=k+1}^n y_i^2}{n-k} - 1 \right) + \frac{n-k}{2} \left(\frac{\sum_{i=k+1}^n y_i^2}{n-k} - 1 \right)^2$$

$$- \frac{n-k}{3} (\theta_n^{(3)})^{-3} \left(\frac{\sum_{i=k+1}^n y_i^2}{n-k} - 1 \right)^3$$

$$= -\frac{1}{2n} \left[\sum_{i=1}^n (y_i^2 - 1) \right]^2 + \frac{1}{2k} \left[\sum_{i=1}^k (y_i^2 - 1) \right]^2 + \frac{1}{2(n-k)} \left[\sum_{i=k+1}^n (y_i^2 - 1) \right]^2$$

$$+ \frac{n}{3} (\theta_n^{(1)})^{-3} \left(\frac{\sum_{i=1}^n y_i^2}{n} - 1 \right)^3 - \frac{k}{3} (\theta_n^{(2)})^{-3} \left(\frac{\sum_{i=1}^k y_i^2}{k} - 1 \right)^3$$

$$- \frac{n-k}{3} (\theta_n^{(3)})^{-3} \left(\frac{\sum_{i=k+1}^n y_i^2}{n-k} - 1 \right)^3 ,$$

where $|\theta_n^{(1)} - 1| < |\sum_{i=1}^n y_i^2/n - 1|$, $|\theta_n^{(2)} - 1| < |\sum_{i=1}^k y_i^2/k - 1|$, and $|\theta_n^{(3)} - 1| < |\sum_{i=k+1}^n y_i^2/(n-k) - 1|$. Denote $\xi_k = W_k + Q_k + R_k$, where

$$W_k = -\frac{1}{2n} \left[\sum_{i=1}^n (y_i^2 - 1) \right]^2 + \frac{1}{2k} \left[\sum_{i=1}^k (y_i^2 - 1) \right]^2 + \frac{1}{2(n-k)} \left[\sum_{i=k+1}^n (y_i^2 - 1) \right]^2 ,$$

$$Q_k = \frac{n}{3} (\theta_n^{(1)})^{-3} \left(\frac{\sum_{i=1}^n y_i^2}{n} - 1 \right)^3 - \frac{k}{3} (\theta_n^{(2)})^{-3} \left(\frac{\sum_{i=1}^k y_i^2}{k} - 1 \right)^3 ,$$

$$R_k = -\frac{n-k}{3} (\theta_n^{(3)})^{-3} \left(\frac{\sum_{i=k+1}^n y_i^2}{n-k} - 1 \right)^3 .$$

Before having our main theorem, we first give the following lemmas which are needed in the sequel.

Lemma 2.14

(i) $\max_{1<k<n} k^{1/2}(\log\log k)^{-3/2}|Q_k| = O_p(1)$,

(ii) $\max_{1<k<n}(n-k)^{1/2}(\log\log(n-k))^{-3/2}|R_k| = O_p(1)$.

Proof. (i) From the law of iterated logarithm, we obtain:

$$\frac{\left|(\theta_n^{(1)})^{-1}n\left(\frac{\sum_{i=1}^n y_i^2}{n}-1\right)\right|}{(n\log\log n)^{1/2}} = O_p(1).$$

Then,

$$\frac{\left|(\theta_n^{(1)})^{-3}n^3\left(\frac{\sum_{i=1}^n y_i^2}{n}-1\right)^3\right|}{(n\log\log n)^{3/2}} = O_p(1);$$

that is,

$$n^{1/2}(\log\log n)^{-3/2}\left|(\theta_n^{(1)})^{-3}n\left(\frac{\sum_{i=1}^n y_i^2}{n}-1\right)^3\right| = O_p(1).$$

But, $k^{1/2}(\log\log k)^{-3/2} \le n^{1/2}(\log\log n)^{-3/2}$ for $1<k<n$, hence

$$\max_{1<k<n} k^{1/2}(\log\log k)^{-3/2}\left|(\theta_n^{(1)})^{-3}n\left(\frac{\sum_{i=1}^n y_i^2}{n}-1\right)^3\right| = O_p(1).$$

Similarly, we obtain

$$\max_{1<k<n} k^{1/2}(\log\log k)^{-3/2}\left|(\theta_n^{(2)})^{-3}k\left(\frac{\sum_{i=1}^k y_i^2}{k}-1\right)^3\right| = O_p(1).$$

That is, $\max_{1<k<n} k^{1/2}(\log\log k)^{-3/2}|Q_k| = O_p(1)$, which completes the proof of (i).

  (ii) Similar to the proof of (i).   □

Lemma 2.15 *Let* $a(\log n) = (2\log\log n)^{1/2}$, *and* $b(\log n) = 2\log\log n + \frac{1}{2}\log\log\log n - \log\Gamma(\frac{1}{2})$; *then for all* $x \in R$, *as* $n \longrightarrow \infty$, *the following hold.*

(i) $a^2(\log n)\max_{1<k<\log n} W_k - (x+b(\log n))^2 \xrightarrow{P} -\infty$.

(ii) $a^2(\log n)\max_{1<k<\log n} \xi_k - (x+b(\log n))^2 \xrightarrow{P} -\infty$.

(iii) $a^2(\log n)\max_{n-\log n<k<n} W_k - (x+b(\log n))^2 \xrightarrow{P} -\infty$.

(iv) $a^2(\log n)\max_{n-\log n<k<n} \xi_k - (x+b(\log n))^2 \xrightarrow{P} -\infty$.

Proof. (i) Because as $k \longrightarrow \infty$,

$$\frac{1}{k} \left[\sum_{i=1}^{k} (y_i^2 - 1) \right]^2 \xrightarrow{P} \frac{1}{2},$$

then

$$\frac{\left[\sum_{i=1}^{k} (y_i^2 - 1) \right]^2}{k \log k} \xrightarrow{P} 0,$$

as $k \longrightarrow \infty$. Thus, there exists a constant c, $0 < c < 1$, such that for large k,

$$0 < \frac{\left[\sum_{i=1}^{k} (y_i^2 - 1) \right]^2}{k \log k} < 1 - c.$$

Now,

$$\frac{a^2(\log n) \max_{1 < k < \log n} \frac{1}{k} \left[\sum_{i=1}^{k} (y_i^2 - 1) \right]^2}{[b(\log n)]^2}$$

$$\leq \frac{(2 \log \log n) \max_{1 < k < \log n} \frac{1}{k} \left[\sum_{i=1}^{k} (y_i^2 - 1) \right]^2}{[2 \log \log n]^2}$$

$$\leq \max_{1 < k < \log n} \frac{\left[\sum_{i=1}^{k} (y_i^2 - 1) \right]^2}{k \log \log n}$$

$$\leq \max_{1 < k < \log n} \frac{\left[\sum_{i=1}^{k} (y_i^2 - 1) \right]^2}{k \log k}$$

$$< 1 - c.$$

Hence, as $n \longrightarrow \infty$,

$$a^2(\log n) \max_{1 < k < \log n} \frac{1}{2k} \left[\sum_{i=1}^{k} (y_i^2 - 1) \right]^2 - (x + b(\log n))^2 \xrightarrow{P} -\infty.$$

Similarly, as $n \longrightarrow \infty$, we obtain

$$a^2(\log n) \max_{1 < k < \log n} \frac{1}{2(n-k)} \left[\sum_{i=k+1}^{n} (y_i^2 - 1) \right]^2 - (x + b(\log n))^2 \xrightarrow{P} -\infty.$$

Moreover, because as $n \longrightarrow \infty$,

$$\frac{1}{2n} \left[\sum_{i=1}^{n} (y_i^2 - 1) \right]^2 \xrightarrow{P} 1 \quad \text{and} \quad a^2(\log n) \longrightarrow \infty,$$

as $n \longrightarrow \infty$,

$$\frac{a^2(\log n)}{2n} \left[\sum_{i=1}^{n}(y_i^2 - 1) \right]^2 \xrightarrow{P} \infty.$$

Consequently, as $n \longrightarrow \infty$,

$$a^2(\log n) - \frac{1}{2n} \left[\sum_{i=1}^{n}(y_i^2 - 1) \right]^2 - (x + b(\log n))^2 \xrightarrow{P} -\infty.$$

Therefore, as $n \longrightarrow \infty$,

$$a^2(\log n) \max_{1 < k < \log n} W_k - (x + b(\log n))^2 \xrightarrow{P} -\infty.$$

Proceeding similarly as above, we can obtain (ii)–(iv). □

Lemma 2.16 *As $n \longrightarrow \infty$, the following hold.*

(i) $a^2(\log n) \max_{\log n < k < n - \log n} |\xi_k - W_k| = o_p(1)$.
(ii) $a^2(\log n) \max_{1 < k < n/\log n} \left| \frac{1}{(n-k)} \left[\sum_{i=k+1}^{n}(y_i^2-1) \right]^2 - \frac{1}{n} \left[\sum_{i=1}^{n}(y_i^2-1) \right]^2 \right| = o_p(1)$.

Proof. (i) Clearly, $\xi_k - W_k = Q_k + R_k$. Now,

$$0 \le a^2(\log n) \max_{\log n < k < n - \log n} |Q_k + R_k|$$

$$\le 2 \log \log n \max_{\log n < k < n - \log n} |Q_k| + 2 \log \log n \max_{\log n < k < n - \log n} |R_k|$$

$$= 2 \log \log n \max_{\log n < k < n - \log n} \frac{(\log \log k)^{3/2}}{k^{1/2}} (\log \log k)^{-3/2} |Q_k|$$

$$+ \; 2 \log \log n \max_{\log n < k < n - \log n} \frac{(\log \log(n - k))^{3/2}}{(n - k)^{1/2}} (\log \log(n - k))^{-3/2} |R_k|$$

$$\le \frac{2(\log \log n)^{5/2}}{(\log n)^{1/2}} \max_{\log n < k < n - \log n} k^{1/2} (\log \log k)^{-3/2} |Q_k|$$

$$+ \; \frac{2(\log \log n)^{5/2}}{(\log n)^{1/2}} \max_{\log n < k < n - \log n} (n - k)^{1/2} (\log \log(n - k))^{-3/2} |R_k|$$

$$\xrightarrow{P} 0,$$

as $n \longrightarrow \infty$, hence

$$\lim_{n \to \infty} a^2(\log n) \max_{\log n < k < n - \log n} |\xi_k - W_k| \overset{P}{=} 0;$$

that is, (i) holds. (ii) For all n and k,

$$E\left\{\frac{1}{(n-k)}\left[\sum_{i=k+1}^{n}(y_i^2-1)\right]^2-\frac{1}{n}\left[\sum_{i=1}^{n}(y_i^2-1)\right]^2\right\}=0,$$

therefore we have,

$$E\left\{a^2(\log n)\left[\left[\sum_{i=k+1}^{n}(y_i^2-1)\right]^2-\frac{1}{n}\left[\sum_{i=1}^{n}(y_i^2-1)\right]^2\right]\right\}=0$$

for all n and k. Hence, as $n \longrightarrow \infty$,

$$a^2(\log n)\left|\left[\sum_{i=k+1}^{n}(y_i^2-1)\right]^2-\frac{1}{n}\left[\sum_{i=1}^{n}(y_i^2-1)\right]^2\right|\xrightarrow{P} 0,$$

for all k and $1 < k < n/\log n$. That is,

$$a^2(\log n)\max_{1<k<n/\log n}\left|\frac{1}{(n-k)}\left[\sum_{i=k+1}^{n}(y_i^2-1)\right]^2-\frac{1}{n}\left[\sum_{i=1}^{n}(y_i^2-1)\right]^2\right|=o_p(1).$$

$\square$

Lemma 2.17 *For all $x \in R$, as $n \longrightarrow \infty$,*

$$a^2(\log n)\max_{n/\log n<k<n-n/\log n}W_k-(x+b(\log n))^2\xrightarrow{P} -\infty.$$

Proof. Recall that

$$W_k=\frac{1}{2k}\left[\sum_{i=1}^{k}(y_i^2-1)\right]^2+\frac{1}{2(n-k)}\left[\sum_{i=k+1}^{n}(y_i^2-1)\right]^2-\frac{1}{2n}\left[\sum_{i=1}^{n}(y_i^2-1)\right]^2.$$

Let's consider the first term of W_k.

From Theorem 2 of Darling and Erdös (1956), we have for $x \in R$, as $n \longrightarrow \infty$,

$$P\left[\max_{n/\log n<k<n-n/\log n}\frac{1}{2k}\left[\sum_{i=1}^{k}(y_i^2-1)\right]^2\right.$$

$$\left.< (2\log\log n)^{1/2}+\frac{\log\log\log n}{2(\log\log n)^{1/2}}+\frac{x}{2(\log\log n)^{1/2}}\right]\longrightarrow \exp\left(-\frac{1}{\sqrt{\pi}}e^{-x}\right).$$

Therefore, as $n \longrightarrow \infty$,

$$P\left[\frac{a^2(\log n)\max_{n/\log n < k < n - n/\log n}\frac{1}{2k}\left[\sum_{i=1}^{k}(y_i^2 - 1)\right]^2}{(x + b(\log n))^2}\right.$$

$$\left. < \frac{1}{2}\left(\frac{2\log\log n + \frac{1}{2}\log\log\log n + x}{x + b(\log n)}\right)^2\right]$$

$$\longrightarrow \exp\left(-\frac{1}{\sqrt{\pi}}e^{-x}\right).$$

Because $b(\log n) = 2\log\log n + \frac{1}{2}\log\log\log n - \log\Gamma\left(\frac{1}{2}\right)$, it follows that as $n \longrightarrow \infty$,

$$P\left[\frac{a^2(\log n)\max_{n/\log n < k < n - n/\log n}\frac{1}{2k}\left[\sum_{i=1}^{k}(y_i^2 - 1)\right]^2}{(x + b(\log n))^2}\right.$$

$$\left. < \left(\frac{1}{\sqrt{2}} + \frac{\log\Gamma(1/2)}{\sqrt{2}(x + b(\log n))}\right)^2\right]$$

$$\longrightarrow \exp\left(-\frac{1}{\sqrt{\pi}}e^{-x}\right).$$

Choose n sufficiently large, such that

$$\left(\frac{1}{\sqrt{2}} + \frac{\log\Gamma(1/2)}{\sqrt{2}(x + b(\log n))}\right)^2 < 1 - M, \quad \text{for } 0 < M < 1,$$

therefore, as $n \longrightarrow \infty$,

$$P\left[\frac{a^2(\log n)\max_{n/\log n < k < n - n/\log n}\frac{1}{2k}\left[\sum_{i=1}^{k}(y_i^2 - 1)\right]^2}{(x + b(\log n))^2} < 1 - M\right]$$

$$\longrightarrow \exp\left(-\frac{1}{\sqrt{\pi}}e^{-x}\right).$$

Now, letting $x \longrightarrow \infty$, as $n \longrightarrow \infty$, we obtain:

$$a^2(\log n)\max_{n/\log n < k < n - n/\log n}\frac{1}{2k}\left[\sum_{i=1}^{k}(y_i^2 - 1)\right]^2 - (x + b(\log n))^2 \xrightarrow{P} -\infty,$$

and similarly,

$$a^2(\log n) \max_{n/\log n < k < n - n/\log n} \frac{1}{2(n-k)} \left[\sum_{i=k+1}^{n} (y_i^2 - 1) \right]^2$$

$$- (x + b(\log n))^2 \xrightarrow{P} -\infty.$$

For the third term of W_k (i.e., $-(1/2n)\left[\sum_{i=1}^{n}(y_i^2-1)\right]^2$) because as $n \longrightarrow \infty$, $-(1/2n)\left[\sum_{i=1}^{n}(y_i^2-1)\right]^2 \xrightarrow{P} -1$; therefore, as $n \longrightarrow \infty$,

$$a^2(\log n) \max_{n/\log n < k < n - n/\log n} \left(-\frac{1}{2n}\left[\sum_{i=1}^{n}(y_i^2-1) \right]^2 \right) - (x + b(\log n))^2 \xrightarrow{P} -\infty.$$

This completes the proof of the lemma. $\qquad\square$

Lemma 2.18 *Let* $\{(z_i^{(1)}, \ldots, z_i^{(d)}), 1 \le i < \infty\}$ *be independently and identically distributed random vectors, and define* $S^{(j)}(k) = \sum_{i=1}^{k} z_i^{(i)}, 1 \le i \le d$. *Assume that* $E[z_i^{(1)}] = E[z_i^{(2)}] = \cdots = E[z_i^{(d)}] = 0$, *the covariance matrix of* $(z_i^{(1)}, \ldots, z_i^{(d)})$ *is the identity matrix, and* $\max_{1 \le j \le d} E|z_i^{(j)}|^r < \infty$ *for some* $r > 2$. *Then as* $n \longrightarrow \infty$,

$$a(\log n) \max_{1 \le j \le d} \left(\sum_{j=1}^{d} [k^{-1/2} S^{(j)}(k)]^2 \right)^{1/2} - b_d(\log n) \xrightarrow{D} y^*,$$

where y^* *has cdf* $F_{y^*}(x) = \exp\{-e^{-x}\}$, $a(x) = (2\log x)^{1/2}$, $b_d(x) = 2\log x + (d/2)\log\log x - \log\Gamma(d/2)$, *and* "$\xrightarrow{D}$" *means "convergence in distribution".*

Proof. See Horváth (1993). $\qquad\square$

Theorem 2.19 *Under the null hypothesis* H_0, *as* $n \longrightarrow \infty, k \longrightarrow \infty$, *such that* $(k/n) \longrightarrow \infty$; *then for all* $x \in R$,

$$\lim_{n \to \infty} P[a(\log n)\lambda_n - b(\log n) \le x] = \exp\{-2e^{-x}\},$$

where $a(\log n)$ *and* $b(\log n)$ *are defined in Lemma* 2.15.

Proof. From Lemma 2.14 (i) and (ii), we have

$$\max_{1 < k < n} \xi_k \overset{D}{=} \max_{\log n \le k \le n - \log n} \xi_k.$$

From Lemma 2.15 (i), it is seen that

$$\max_{1 < k < n} \xi_k \overset{D}{=} \max_{\log n \le k \le n - \log n} W_k.$$

From Lemma 2.15 (ii), we thus have

$$\max_{\log n \leq k \leq n/\log n} W_k \overset{D}{=} \max_{\log n \leq k \leq n/\log n} \frac{1}{2k} \left[\sum_{i=1}^{k} (y_i^2 - 1) \right]^2. \qquad (2.15)$$

In view of Lemma 2.16,

$$\max_{\log n \leq k \leq n - \log n} W_k \overset{D}{=} \max \left\{ \left[\max_{\log n \leq k \leq n/\log n} W_k \right], \left[\max_{n - n/\log n \leq k \leq n - \log n} W_k \right] \right\}. \qquad (2.16)$$

Next, applying Lemma 2.16 to (2.16), we obtain

$$\max_{n - n/\log n \leq k \leq n - \log n} W_k \overset{D}{=} \max_{n - n/\log n \leq k \leq n - \log n} \frac{1}{2(n - k)} \left[\sum_{i=k+1}^{n} (y_i^2 - 1) \right]^2. \qquad (2.17)$$

Combining (2.15) through (2.17), we obtain

$$\max_{\log n \leq k \leq n/\log n} W_k \overset{D}{=} \max \left\{ \max_{1 < k \leq n/\log n} \frac{1}{2k} \left[\sum_{i=1}^{k} (y_i^2 - 1) \right]^2, \right.$$
$$\left. \max_{n - n/\log n \leq k < n} \frac{1}{2(n - k)} \left[\sum_{i=k+1}^{n} (y_i^2 - 1) \right]^2 \right\}.$$

Therefore,

$$\lim_{n \to \infty} P[a(\log n)\lambda_n - b(\log n) \leq x]$$

$$= \lim_{n \to \infty} P \left[a(\log n) \max_{1 < k < n-1} \xi_k^{1/2} - b(\log n) \leq x \right]$$

$$= \lim_{n \to \infty} P \left[a^2(\log n) \max_{1 < k < n-1} \xi_k \leq [x + b(\log n)]^2 \right]$$

$$= \lim_{n \to \infty} P \left[a^2(\log n) \max_{\log n \leq k \leq n - \log n} W_k \leq [x + b(\log n)]^2 \right]$$

$$= \lim_{n \to \infty} P \left[a^2(\log n) \max \left\{ \max_{1 < k \leq n/\log n} \frac{1}{2k} \left[\sum_{i=1}^{k} (y_i^2 - 1) \right]^2, \right. \right.$$

$$\left. \left. \max_{n - n/\log n \leq k < n} \frac{1}{2(n - k)} \left[\sum_{i=k+1}^{n} (y_i^2 - 1) \right]^2 \right\} \leq [x + b(\log n)]^2 \right]. \qquad (2.18)$$

Inasmuch as $\{y_i, 1 \leq i \leq k, 1 \leq k \leq n/\log n\}$ and $\{y_j, k+1 \leq j \leq n, n - n/\log n \leq k \leq n\}$ are independent, (2.18) simplifies to

$$\lim_{n \to \infty} P\left[a^2(\log n) \max_{1 < k \leq n/\log n} \frac{1}{2k}\left[\sum_{i=1}^{k}(y_i^2 - 1)\right]^2 \leq [x + b(\log n)]^2\right] *$$

$$\lim_{n \to \infty} P\left[\max_{n-n/\log n \leq k < n} \frac{1}{2(n-k)}\left[\sum_{i=k+1}^{n}(y_i^2 - 1)\right]^2 \leq [x + b(\log n)]^2\right]$$

$$= \lim_{n \to \infty} P\left[a(\log n) \max_{1 < k \leq n/\log n} \frac{1}{\sqrt{2k}}\left|\sum_{i=1}^{k}(y_i^2 - 1)\right| - b(\log n) \leq x\right] *$$

$$\lim_{n \to \infty} P\left[a(\log n) \max_{1 < k \leq n/\log n} \frac{1}{\sqrt{2(n-k)}}\left|\sum_{i=k+1}^{n}(y_i^2 - 1)\right| - b(\log n) \leq x\right].$$

$$(2.19)$$

Denote the first term of (2.19) by (a) and the second by (b). Let's consider (a) first. Let $v_i = ((y_i^2 - 1)/\sqrt{2})$, $1 \leq i < \infty$; we see that $\{v_i, 1 \leq i < \infty\}$ is a sequence of iid random variables, with $E[v_i] = 0$, $\text{Var}[v_i] = 1$, and $E|v_i|^r < \infty$ for $r > 2$. Let $S(k) = \sum_{i=1}^{k} v_i$; it is easy to see that $k^{-1/2}S(k) = (1/2k)\left[\sum_{i=1}^{k}(y_i^2 - 1)\right]^2$. Then from Lemma 2.18, as $n \longrightarrow \infty$,

$$a(\log n) \max_{1 < k \leq n/\log n} \frac{1}{\sqrt{2k}}\left|\sum_{i=1}^{k}(y_i^2 - 1)\right| - b(\log n) \xrightarrow{D} y^*,$$

where y^* has cdf $F_{y^*}(x) = \exp\{-e^{-x}\}$; therefore, (a) $= \exp\{-e^{-x}\}$. Similarly, we can obtain: (b) $= \exp\{-e^{-x}\}$. Then, (2.19) $= \exp\{-2e^{-x}\}$, which completes the proof of the theorem. $\qquad\square$

Remark 2.20 In many real situations, it is more likely that μ remains common but unknown instead of being known. Under these circumstances, the likelihood procedure can still be applied. Under H_0, the maximum log likelihood is easily obtained as

$$\log L_0(\widehat{\sigma}^2, \widehat{\mu}) = -\frac{n}{2}\log 2\pi - \frac{n}{2}\log\widehat{\sigma}^2 - \frac{n}{2},$$

where $\widehat{\sigma}^2 = (\sum_{i=1}^{n}(x_i - \overline{x})^2)/n$ and $\widehat{\mu} = \overline{x}$ are the MLEs of σ^2 and μ, respectively. Under H_1, the log likelihood function is

$$\log L_1(\mu, \sigma_1^2, \sigma_n^2) = -\frac{n}{2} \log 2\pi - \frac{k}{2} \log \sigma_1^2 - \frac{n-k}{2} \log \sigma_n^2$$

$$- \frac{\sum_{i=1}^{k}(x_i - \mu)^2}{2\sigma_1^2} - \frac{\sum_{i=k+1}^{n}(x_i - \mu)^2}{2\sigma_n^2},$$

and the likelihood equations are:

$$\begin{cases} \sigma_n^2 \sum_{i=1}^{k}(x_i - \mu)^2 + \sigma_1^2 \sum_{i=k+1}^{n}(x_i - \mu)^2 = 0 \\ \sigma_1^2 = \frac{1}{k} \sum_{i=1}^{k}(x_i - \mu)^2 \\ \sigma_n^2 = \frac{1}{n-k} \sum_{i=k+1}^{n}(x_i - \mu)^2 \end{cases}$$

where the solutions of μ, σ_1^2, and σ_n^2 are the MLEs $\hat{\mu}, \hat{\sigma}_1^2$, and $\hat{\sigma}_n^2$, respectively. Unfortunately, solving this system of equations will not give us the closed forms for $\hat{\mu}, \hat{\sigma}_1^2$, and $\hat{\sigma}_n^2$. However, we can use Newton's iteration method, or some other iteration methods, to obtain an approximate solution. Under the regularity conditions (Dennis and Schnable, 1983), the solution will yield the unique MLE. Then the log maximum likelihood under H_1 can be expressed as

$$\log L_1(\hat{\mu}, \hat{\sigma}_1^2, \hat{\sigma}_n^2) = -\frac{n}{2} \log 2\pi - \frac{k}{2} \log \hat{\sigma}_1^2 - \frac{n-k}{2} \log \hat{\sigma}_n^2 - \frac{n}{2},$$

where $\hat{\mu}, \hat{\sigma}_1^2$, and $\hat{\sigma}_n^2$ are the numerical solutions of the above system of equations, and $2 \le k \le n - 2$.

2.2.2 Informational Approach

In 1973, Hirotugu Akaike introduced the Akaike Information Criterion (AIC) for model selection in statistics (Akaike, 1973). Since then, this criterion has profoundly influenced developments in statistical analysis, particularly in time series, analysis of outliers (Kitagawa, 1979), robustness, regression analysis, multivariate analysis (e.g., see Bozdogan, Sclove, and Gupta, 1994), and so on. On the basis of Akaike's work, many authors have further introduced various information criteria and used them in many other fields such as econometrics, psychometrics, control theory, and decision theory.

Suppose $x_1, x_2, \ldots, x_n$ is a sequence of independent and identically distributed random variables with probability density function $f(\cdot|)$, where f is a model with K parameters; that is,

$$Model(K) : \{f(\cdot|\theta) : \theta = (\theta_1, \theta_2, \ldots, \theta_K), \theta \in \Theta_K\}.$$

It is assumed that there are no constraints on the parameters and hence the number of free parameters in the model is K. The restricted parameter space is given by

$$\Theta_k = \{\theta \in \Theta_K | \theta_{k+1} = \theta_{k+2} = \cdots = \theta_K = 0\}$$

and the corresponding model is denoted by model (k).

To view the change point hypothesis testing of the null hypotheis given by (1.1) against the alternative hypothesis given by (1.2) in a model selection context, we target to select a "best" model from a collection of models corresponding to (1.1) and (1.2). Specifically, corresponding to the alternative hypothesis (1.2) of q change points, it is equivalent to state that: $X_1, \ldots, X_{k_1} \sim$ iid $f(\theta_1)$, $X_{k_1+1}, \ldots, X_{k_2} \sim$ iid $f(\theta_2), \ldots, X_{k_{q-1}+1}, \ldots, X_{k_q} \sim$ iid $f(\theta_{q-1})$, $X_{k_q+1}, \ldots, X_{k_n} \sim$ iid $f(\theta_q)$, where $1 < k_1 < k_2 < \cdots < k_q < n$, q is assumed to be the unknown number of change points and $k_1, k_2, \ldots, k_q$ are the respective unknown change point positions.

Akaike (1973) proposed the following information criterion,

$$\text{AIC}(k) = -2\log L(\widehat{\Theta}_k) + 2k, k = 1, 2, \ldots, K,$$

where $L(\widehat{\Theta}_k)$ is the maximum likelihood for model (k), as a measure of model evaluation. A model that minimizes the AIC (Minimum AIC estimate, MAICE) is considered to be the most appropriate model. However, the MAICE is not an asymptotically consistent estimator of model order (e.g., see Schwarz, 1978). Some authors made efforts to modify the information criterion without violating Akaike's original principles. For more details of the various kinds of modifications, the reader is referred to Bozdogan (1987), Hannan and Quinn (1979), Zhao, Krishnaiah and Bai (1986a, 1986b), and Rao and Wu (1989).

One of the modifications is the Schwarz Information Criterion, denoted as SIC, and proposed by Schwarz in 1978. It is expressed as

$$\text{SIC}(k) = -2\log L(\widehat{\Theta}_k) + k\log n, k = 1, 2, \ldots, K.$$

Apparently, the difference between AIC and SIC is in the penalty term, instead of $2k$, it is $k \log n$. However, SIC gives an asymptotically consistent estimate of the order of the true model. The SIC has been applied to change point analysis for different underlying models by many authors in the literature. Recently, Chen and Gupta (2003) and Pan and Chen (2006) proposed a new information criterion named the modified information criterion (MIC) for studying change point models, which demonstrated that the penalty term in SIC for change point problems should be defined according to the nature of change point problems. Here, for historical reasons, the SIC is employed to find the change point.

(i) SICs of the Change Point Inference

According to the information criterion principle, we are going to estimate the position of the change point k by $\widehat{k}$ such that $\text{SIC}(\widehat{k})$ is the minimal. To be specific, corresponding to the H_0 defined by (2.10), is one SIC, denoted by $SIC(n)$, which is found as

$$SIC(n) = n \log 2\pi + n \log \widehat{\sigma}^2 + n + \log n, \tag{2.20}$$

where $\widehat{\sigma}^2 = (\sum_{i=1}^{n}(x_i - \mu)^2)/n$ is the MLE of σ^2 under H_0. Corresponding to the H_1 defined by (2.11), are the $n - 3$ SICs, denoted by $SIC(k)$ for $2 \leq k \leq n - 2$, which are found as

$$SIC(k) = n \log 2\pi + k \log \widehat{\sigma}_1^2 + (n - k) \log \widehat{\sigma}_n^2 + n + 2 \log n, \tag{2.21}$$

where $\widehat{\sigma}_1^2 = (\sum_{i=1}^{k}(x_i - \mu)^2)/k$ and $\widehat{\sigma}_n^2 = ((\sum_{i=k+1}^{n}(x_i - \mu)^2)/(n - k))$ are the MLEs of σ_1^2 and σ_n^2, respectively, under H_1.

Notice that to be able to obtain the MLEs, we can only detect changes that are located between the second and $(n - 2)$ positions. According to the information criterion principle, we accept H_0 if

$$SIC(n) < \min_{2 \leq k \leq n-2} SIC(k),$$

and accept H_1 if

$$SIC(n) > SIC(k)$$

for some k, and estimate the position of the change point by $\widehat{k}$ such that

$$SIC(\widehat{k}) = \min_{2 \leq k \leq n-2} SIC(k). \tag{2.22}$$

On the one hand, we point out (see Gupta and Chen, 1996) that information criteria, such as SIC, provide a remarkable way for exploratory data analysis with no need to resort to either the distribution or the significance level α. However, when the SICs are very close, one may question that the small difference among the SICs might be caused by the fluctuation of the data, and therefore there may be no change at all. To make the conclusion about change point statistically convincing, we introduce the significance level α and its associated critical value c_α, where $c_\alpha \geq 0$. Instead of accepting H_0 when $SIC(n) < \min_{2 \leq k \leq n-2} SIC(k)$, we now accept H_0 if

$$SIC(n) < \min_{2 \leq k \leq n-2} SIC(k) + c_\alpha,$$

where c_α is determined from

$$1 - \alpha = P[SIC(n) < \min_{2 \leq k \leq n-2} SIC(k) + c_\alpha | H_0 \text{ holds}].$$

By using Theorem 2.19, the approximate c_α values can be obtained as follows.

$$1 - \alpha = P[\text{SIC}(n) < \min_{2 \leq k \leq n-2} \text{SIC}(k) + c_\alpha | H_0 \text{ holds}]$$

$$= P[\lambda_n^2 < \log n + c_\alpha | H_0 \text{ holds}]$$

$$= P[0 < \lambda_n < (\log n + c_\alpha)^{1/2} | H_0 \text{ holds}]$$

$$= P\big[-b(\log n) < a(\log n)\lambda_n - b(\log n)$$

$$< a(\log n)(\log n + c_\alpha)^{1/2} - b(\log n) | H_0 \text{ holds}\big]$$

$$\cong \exp\{-2 \exp[-a(\log n)(\log n + c_\alpha)^{1/2} + b(\log n)]\}$$

$$- \exp\{-2 \exp[b(\log n)]\},$$

and solving for c_α, we obtain:

$$c_\alpha \cong \left\{ -\frac{1}{a(\log n)} \log \log[1 - \alpha + \exp(-2e^{b(\log n)})]^{-1/2} + \frac{b(\log n)}{a(\log n)} \right\}^2$$

$$- \log n. \tag{2.23}$$

For a different significance level α ($\alpha = 0.01, 0.025, 0.05, 0.1$), and sample sizes n ($n = 13, 14, \ldots, 200$), the approximate values of c_α have been calculated according to (2.23) and tabulated in Table 2.3.

(ii) Unbiased SICs

To derive the information criterion AIC, Akaike (1973) used $\log L(\widehat{\theta})$ as an estimate of $J = E_{\widehat{\theta}}[\int f(\mathbf{y}|\theta_0) \log f(\mathbf{y}|\widehat{\theta}) dy]$, where $f(\mathbf{y}|\theta_0)$ is the probability density of the future observations $\mathbf{y} = (y_1, \ldots, y_n)$ of the same size and distribution as the xs, $\mathbf{x} = (x_1, \ldots, x_n)$, and x and y are independent. The expectation is taken under the distribution of x when H_0 is true; that is, $\theta_0 \in \Theta_{H_0}$. Unfortunately, $\log L(\widehat{\theta})$ is not an unbiased estimator of J. When the sample size n is finite, Sugiura (1978) proposed unbiased versions, the finite corrections of AIC, for different model selection problems.

In this section, we derive the unbiased versions of our SIC under our H_0 defined by (2.10) and H_1 defined by (2.11), denoted by $u - \text{SIC}(n)$, and $u - \text{SIC}(k)$, respectively.

(1) Unbiased SIC under $H_0 : u - \text{SIC}(n)$

Under H_0, let $\mathbf{y} = (y_1, y_2, \ldots, y_n)$ be a sample of the same size and distributions as $\mathbf{x}$, $\mathbf{x} = (x_1, x_2, \ldots, x_n)$, and that $\mathbf{y}$ be independent of $\mathbf{x}$.

Table 2.3 Approximate Critical Values of SIC

n/α	0.010	0.025	0.050	0.100
13	20.927	14.570	10.496	6.946
14	20.431	14.340	10.375	6.895
15	20.077	14.165	10.279	6.852
16	19.807	14.023	10.199	6.816
17	19.589	13.903	10.130	6.783
18	19.405	13.799	10.068	6.753
19	19.247	13.706	10.012	6.725
20	19.106	13.623	9.961	6.698
21	18.980	13.546	9.914	6.673
22	18.860	13.476	9.870	6.649
23	18.759	13.411	9.829	6.626
24	18.661	13.350	9.790	6.605
25	18.569	13.293	9.753	6.583
26	18.484	13.239	9.718	6.563
27	18.404	13.188	9.685	6.543
28	18.328	13.140	9.653	6.524
29	18.257	13.094	9.622	6.506
30	18.189	13.050	9.593	6.488
35	17.895	12.858	9.463	6.406
40	17.656	12.699	9.352	6.333
45	17.456	12.564	9.256	6.268
50	17.284	12.446	9.171	6.208
55	17.134	12.342	9.095	6.154
60	17.001	12.249	9.026	6.104
70	16.773	12.088	8.904	6.014
80	16.584	11.951	8.800	5.934
90	16.422	11.832	8.708	5.863
100	16.280	11.728	8.626	5.799
120	16.043	11.550	8.484	5.686
140	15.848	11.402	8.365	5.589
160	15.684	11.276	8.261	5.504
180	15.542	11.165	8.170	5.428
200	15.416	11.067	8.088	5.359

$$J = E_{\widehat{\theta}} \left[\int f(\mathbf{y}|\theta_0) \log f(\mathbf{y}|\widehat{\theta}) dy \right]$$

$$= E_{\widehat{\theta}} \left[E_y \left\{ -\frac{n}{2} \log 2\pi - \frac{n}{2} \log \widehat{\sigma}^2 - \frac{1}{2\widehat{\sigma}^2} \sum_{i=1}^{n} (y_i - \mu)^2 \right\} \right]$$

$$= E_{\widehat{\theta}} \left[-\frac{n}{2} \log 2\pi - \frac{n}{2} \log \widehat{\sigma}^2 - \frac{n}{2} + \frac{n}{2} - E_y \left\{ \frac{1}{2\widehat{\sigma}^2} \sum_{i=1}^{n} (y_i - \mu)^2 \right\} \right],$$

where $\widehat{\sigma}^2 = \sum_{i=1}^{n}(x_i - \mu)^2/n$, and μ is known. Because $\sum_{i=1}^{n}(y_i - \mu)^2 \sim \sigma^2 \chi_n^2$, that is, $n\widehat{\sigma}^2/\sigma^2 \sim \chi_n^2$, we get:

$$J = E_{\widehat{\theta}}\left[\log L_0(\widehat{\sigma}^2) + \frac{n}{2} - \frac{n\sigma^2}{2\widehat{\sigma}^2}\right]$$

$$= E_{\widehat{\theta}}[\log L_0(\widehat{\sigma}^2)] + \frac{n}{2} - \frac{n^2}{2}\frac{1}{n-2}$$

$$= E_{\widehat{\theta}}[\log L_0(\widehat{\sigma}^2)] - \frac{n}{n-2}.$$

Clearly, $-2\log L_0(\widehat{\sigma}^2) + 2n/(n-2)$ is unbiased for $-2J$. Therefore, the unbiased $u - \mathrm{SIC}(n)$ is obtained as

$$u - \mathrm{SIC}(n) = -2\log L_0(\widehat{\sigma}^2) + \frac{2n}{n-2}$$

$$= \mathrm{SIC}(n) + \frac{2n}{n-2} - \log n.$$

(2) Unbiased SIC under $H_1 : u - \mathrm{SIC}(k)$

Under H_1, let $\mathbf{y} = (y_1, y_2, \ldots, y_n)$ be a sample of the same size and distributions as $\mathbf{x}$, $\mathbf{x} = (x_1, x_2, \ldots, x_n)$, and that $\mathbf{y}$ be independent of $\mathbf{x}$. That is, $y_1, y, \ldots, y_k$ are iid $N(\mu, \sigma_1^2)$, and $y_{k+1}, y_{k+2}, \ldots, y_n$ are iid $N(\mu, \sigma_n^2)$.

$$J = E_{\widehat{\theta}}[E_y\{\log L_1(\widehat{\sigma}_1^2, \widehat{\sigma}_n^2, Y)\}]$$

$$= E_{\widehat{\theta}}\left[E_y\left\{-\frac{n}{2}\log 2\pi - \frac{k}{2}\log\widehat{\sigma}_1^2 - \frac{n-k}{2}\log\widehat{\sigma}_n^2\right.\right.$$

$$\left.\left. -\frac{1}{2\widehat{\sigma}_1^2}\sum_{i=1}^{k}(y_i - \mu)^2 - \frac{1}{2\widehat{\sigma}_n^2}\sum_{i=k+1}^{n}(y_i - \mu)^2\right\}\right],$$

where

$$\widehat{\sigma}_1^2 = \frac{\sum_{i=1}^{k}(x_i - \mu)^2}{k}, \qquad \widehat{\sigma}_n^2 = \frac{\sum_{i=k+1}^{n}(x_i - \mu)^2}{n-k},$$

and μ is known.

$\sum_{i=1}^{k}(y_i - \mu)^2 \sim \sigma_1^2\chi_k^2$, $\sum_{i=k+1}^{n}(y_i - \mu)^2 \sim \sigma_n^2\chi_{n-k}^2$, $k\widehat{\sigma}_1^2/\sigma_1^2 \sim \chi_k^2$, and $(n-k)\widehat{\sigma}_n^2/\sigma_n^2 \sim \chi_{n-k}^2$ therefore we get

$$J = E_{\widehat{\theta}}[\log L_1(\widehat{\sigma}_1^2, \widehat{\sigma}_n^2)] + \frac{n}{2} - E_{\widehat{\theta}}[\frac{k\sigma_1^2}{2\widehat{\sigma}_1^2} + \frac{(n-k)\sigma_n^2}{2\widehat{\sigma}_n^2}]$$

$$= E_{\widehat{\theta}}[\log L_1(\widehat{\sigma}_1^2, \widehat{\sigma}_n^2)] + \frac{n}{2} - \frac{k^2}{2}\frac{1}{k-2} - \frac{(n-k)^2}{2}\frac{1}{n-k-2}$$

$$= E_{\widehat{\theta}}[\log L_1(\widehat{\sigma}_1^2, \widehat{\sigma}_n^2)] - \frac{2(nk - k^2 - n)}{(k-2)(n-k-2)}.$$

Hence, the unbiased SIC under H_1 is

$$u - \mathrm{SIC}(k) = -2 \log L_1(\widehat{\sigma}_1^2, \widehat{\sigma}_n^2) + \frac{4(nk - k^2 - n)}{(k-2)(n-k-2)},$$

for $2 \leq k \leq n - 2$.

2.2.3 Other Methods

In addition to the likelihood procedure and information approach to the variance change point problem, there are several other methods available in the literature; see Hsu (1977), Davis (1979), and Abraham and Wei (1984) for more details. Here, a Bayesian approach based on the work of Inclán (1993) is presented.

Let $x_1, x_2, \ldots, x_n$ be a sequence of independent normal random variables with parameters $(0, \sigma_1^2), (0, \sigma_2^2), \ldots, (0, \sigma_n^2)$, respectively. It is desired to test the hypothesis:

$$H_0 : \sigma_1^2 = \sigma_2^2 = \cdots = \sigma_n^2 = \sigma^2 \text{ (unknown)},$$

against the alternative:

$$H_1 : \sigma_1^2 = \cdots = \sigma_{k_1}^2 = \eta_0^2 \neq \sigma_{k_1+1}^2 = \cdots = \sigma_{k_2}^2 = \eta_1^2 \neq \cdots$$
$$\neq \sigma_{k_{q-1}+1}^2 = \cdots = \sigma_{k_q}^2 = \eta_{q-1}^2 \neq \sigma_{k_q+1}^2 = \cdots = \sigma_n^2 = \eta_q^2,$$

where q is the unknown number of change points, and $1 < k_1 < k_2 < \cdots < k_q < n$, are the unknown positions of the change points, respectively.

Let $K_{r,m}$ denote the posterior odds of r changes versus m changes. A systematic way of using the posterior odds to determine q is to calculate $K_{r,r-1}$ for $r = 1, 2, \ldots, n$. Starting with $r = 1$, where $K_{1,0}$ means one change versus no change, if $K_{1,0} > 1$, then there is at least one change. Next, compute $K_{2,1}$; if $K_{2,1} > 1$; then there are at least two changes. Keep calculating $K_{r,r-1}$ as long as $K_{r,r-1} > 1$. If $K_{r+1,r} \leq 1$, stop the process and conclude that there are r changes and estimate q by $\widehat{q} = r$.

In the following, the derivation of $K_{r,m}$ is given. Let $k_0 = 0, k_{q+1} = n$; then there are $d_j = k_{j+1} - k_j$ observations with variances $\eta_j^2, j = 0, 1, dots, q$. Let $\sigma = (\sigma_1, \ldots, \sigma_q)', \eta = (\eta_0, \ldots, \eta_q)', \mathbf{k} = (k_1, \ldots, k_q)'$, and $\mathbf{x} = (x_1, \ldots, x_n)'$. The joint density of x given $\eta, \mathbf{k}, q$ can be written as

$$f(\mathbf{x}|\eta, \mathbf{k}, q) = \frac{1}{(2\pi)^{n/2}} \prod_{i=1}^{n} \sigma_i^{-1} \exp\left\{-\frac{1}{2\sigma_i^2} x_i^2\right\}$$

$$= \frac{1}{(2\pi)^{n/2}} \prod_{j=0}^{q} \eta_j^{-d_j} \exp\left\{-\frac{1}{2\eta_j^2} \sum_{i=k_j+1}^{k_{j+1}} x_i^2\right\}. \tag{2.24}$$

Now let the prior distributions be as follows,

$$q|\theta \sim \text{Binomial } (n-1, \theta),$$

where θ is the probability of observing a change; that is,

$$f(q|\theta) = \binom{n-1}{q} \theta^q (1-\theta)^{n-1-q}, \qquad q = 0, 1, \ldots, n-1.$$

Assume that η_j's are conditionally independent drawn from the inverted gamma density and independent of q and $\mathbf{k}$:

$$f(\eta_j|c, v) = \frac{2c^{v/2}}{\Gamma(v/2)} \eta_j^{-(v+1)} \exp(-c\eta_j^2), \tag{2.25}$$

where $0 < \eta_j < \infty$, $j = 0, \ldots, q$. Assume $k_1 < k_2 < \cdots < k_q$ are equally likely:

$$f(\mathbf{k}|q) = \frac{1}{\binom{n-1}{q}}. \tag{2.26}$$

Then from (2.22) through (2.24) the joint probability density function of $\mathbf{x}$, $\eta, \mathbf{k}$ given q is obtained as

$$f(\mathbf{x}, \eta, \mathbf{k}|q) = f(\mathbf{k}|q)f(\eta|\mathbf{k}, q)f(\mathbf{x}|\eta, \mathbf{k}, q)$$

$$= f(\mathbf{k}, q)f(\eta|c, v)f(\mathbf{x}|\eta, \mathbf{k}, q)$$

$$= \frac{(2\pi)^{-n/2}}{\binom{n-1}{q}} \left(\frac{2c^{v/2}}{\Gamma(v/2)}\right)^{q+1}$$

$$\cdot \prod_{j=0}^{q} \left(\eta_j^{-(d_j+v+1)} \exp\left\{-\frac{1}{2\eta_j^2}\left[\sum_{i=k_j+1}^{k_{j+1}} x_i^2 + 2c\right]\right\}\right). \tag{2.27}$$

Inasmuch as

$$\int_0^\infty \eta_j^{-(d_j+\nu+1)} \exp\left\{-\frac{1}{2\eta_j^2}\left[\sum_{i=k_j+1}^{k_{j+1}} x_i^2 + 2c\right]\right\} d\eta_j$$

$$= \frac{\Gamma\left(\frac{d_j+\nu}{2}\right)}{2\left[\frac{1}{2}\sum_{i=k_j+1}^{k_{j+1}} x_i^2 + c\right]^{(d_j+\nu)/2}},$$

and η_j's are independent,

$$f(\mathbf{x}, \mathbf{k}|q) = \int \cdots \int f(\mathbf{x}, \eta, \mathbf{k}|q)d\eta$$

$$= \frac{(2\pi)^{-n/2}}{\binom{n-1}{q}}\left(\frac{c^{\nu/2}}{\Gamma(\nu/2)}\right)^{q+1}$$

$$\cdot \prod_{j=0}^q \left\{\Gamma\left(\frac{d_j+\nu}{2}\right)\left[\frac{1}{2}\sum_{i=k_j+1}^{k_{j+1}} x_i^2 + c\right]^{-((d_j+\nu)/2)}\right\}. \qquad (2.28)$$

Now,

$$f(\mathbf{x}|q) = \sum_{k_1}\cdots\sum_{k_q} f(\mathbf{x}, \mathbf{k}|q), \qquad (2.29)$$

where the sums are over all possible values of $\mathbf{k}: k_1 = 1, 2, \ldots, n-q; k_2 = k_1 + 1, \ldots, n-q+1; \ldots; k_{j+1} = k_j+1, \ldots, n-q+j; \ldots;$ and $k_q = k_{q-1}+1, \ldots, n-1$.
Therefore,

$$f(\mathbf{k}|\mathbf{x}, q) = \frac{f(\mathbf{x}, \mathbf{k}|q)}{f(\mathbf{x}|q)}. \qquad (2.30)$$

Because

$$f(q|\mathbf{x}) = \frac{f(q)f(\mathbf{x}|q)}{f(\mathbf{x})} \propto f(q)f(\mathbf{x}|q),$$

the posterior odds $K_{r,m}$ are given by

$$K_{r,m} = \frac{P(q=r|\mathbf{x})}{P(q=m|\mathbf{x})}$$

$$= \frac{P(q=r)f(\mathbf{x}|q=r)}{P(q=m)f(\mathbf{x}|q=m)}$$

$$= \frac{\binom{n-1}{r}\theta^r(1-\theta)^{n-1-r}}{\binom{n-1}{m}\theta^m(1-\theta)^{n-1-m}} \cdot \frac{\sum_{k_1}\cdots\sum_{k_r} f(\mathbf{x}, \mathbf{k}|q=r)}{\sum_{k_1}\cdots\sum_{k_m} f(\mathbf{x}, \mathbf{k}|q=m)}$$

$$= \frac{\binom{n-1}{r}}{\binom{n-1}{m}} \left(\frac{\theta}{1-\theta}\right)^{r-m} \cdot \frac{\sum_{k_1}\cdots\sum_{kr} \frac{(2\pi)^{-n/2}}{\binom{n-1}{r}} \left(\frac{c^{\nu/2}}{\Gamma(\nu/2)}\right)^{r+1}}{\sum_{k_1}\cdots\sum_{k_m} \frac{(2\pi)^{-n/2}}{\binom{n-1}{m}} \left(\frac{c^{\nu/2}}{\Gamma(\nu/2)}\right)^{m+1}}$$

$$\frac{\prod_{j=0}^{r} \left\{ \Gamma\left(\frac{d_j+\nu}{2}\right) \left[\frac{1}{2}\sum_{i=k_j+1}^{k_{j+1}} x_i^2 + c\right]^{-((d_j+\nu)/2)} \right\}}{\prod_{j=0}^{m} \left\{ \Gamma\left(\frac{d_j+\nu}{2}\right) \left[\frac{1}{2}\sum_{i=k_j+1}^{k_{j+1}} x_i^2 + c\right]^{-((d_j+\nu)/2)} \right\}}$$

$$= \left(\frac{\theta}{1-\theta}\right)^{r-m} \left(\frac{c^{\nu/2}}{\Gamma(\nu/2)}\right)^{r-m}$$

$$\cdot \frac{\sum_{k_1}\cdots\sum_{kr} \prod_{j=0}^{r} \left\{ \Gamma\left(\frac{d_j+\nu}{2}\right) \left[\frac{1}{2}\sum_{i=k_j+1}^{k_{j+1}} x_i^2 + c\right]^{-((d_j+\nu)/2)} \right\}}{\sum_{k_1}\cdots\sum_{km} \prod_{j=0}^{m} \left\{ \Gamma(\frac{d_j+\nu}{2}) \left[\frac{1}{2}\sum_{i=k_j+1}^{k_{j+1}} x_i^2 + c\right]^{-((d_j+\nu)/2)} \right\}} . \quad (2.31)$$

It may be noted that λ, c, ν are hyperparameters that will not be modeled. However, the values of λ, c, ν can be assigned thoughtfully according to experience. A typical assignment is: $\lambda = 1/n, \nu = 1$ or $\nu = 2$, and $c = (\upsilon + 1)/2$. For a discussion of this matter, the reader is referred to Inclán (1993).

After estimating the number of change points q, the next step is to locate the change points. One way to do it is to obtain the posterior pdf of $\mathbf{k}$ given $\mathbf{x}$ and $q = \widehat{q}$. From (2.28),

$$f(\mathbf{k}|\mathbf{x}, q = \widehat{q}) = \frac{f(\mathbf{x}, \mathbf{k}|q = \widehat{q})}{f(\mathbf{x}|q = \widehat{q})}.$$

Then, obtain the marginal distributions of each k_j, for $j = 1, 2, \ldots, \widehat{q}$. Finally, the joint mode $(\text{mode}(k_1), \text{mode}(k_2), \ldots, \text{mode}(k_{\widehat{q}}))$ gives the locations of the change points; that is,

$$\widehat{k}_1 = \text{mode}(k_1), \widehat{k}_2 = \text{mode}(k_2), \ldots, \widehat{k}_q = \text{mode}(k_{\widehat{q}}).$$

2.2.4 Application to Stock Market Data

We give an application of the SIC test procedure to searching a change point in stock prices (Chen and Gupta, 1997). Hsu (1977) analyzed the U.S. stock market return series during the period 1971–1974 using T- and G-statistics, and found that there was one variance change point which is suspected to have occurred in conjunction with the Watergate events. Later, he (Hsu,

1979) reanalyzed the stock market return series data by considering a gamma sequence and came up with the same conclusion.

Here we take the same stock market price data as in Hsu (1979), and perform the change point analysis by using the SIC procedure. Let P_t be the stock price; we first transform the data into $R_t = (P_{t+1} - P_t)/P_t, t = 1, \ldots, 161$. According to Hsu (1977), $\{R_t\}$ is a sequence of independent normal random variables with mean zero. We then test the following hypothesis based on the R_t series,

$$H_0 : \sigma_1^2 = \sigma_2^2 = \cdots = \sigma_{161}^2 = \sigma^2 \text{(unknown)},$$

versus the alternative:

$$H_1 : \sigma_1^2 = \cdots = \sigma_{k_1}^2 \neq \sigma_{k_1+1}^2 = \cdots = \sigma_{k_2}^2 \neq \cdots \neq \sigma_{k_q+1}^2 = \cdots = \sigma_{161}^2,$$

where q is the unknown number of change points, and $1 \leq k_1 < k_2 < \cdots < k_q < 161$, are the unknown positions of the change points, respectively.

Using the binary segmentation procedure along with the SIC, we are able to detect all the changes in the R_t series. According to our computations, at the first stage $\min_{1<k<160}\text{SIC}(k) = \text{SIC}(89) = -787.5745 < \text{SIC}(161) = -765.6242$. If we use the c_α in Tables 2.3–2.5, we still have $\text{SIC}(89) + c_\alpha < \text{SIC}(161)$. Hence, $t = 89$ is a variance change point for the R_t series. Transferring to the price P_t, $t + 1 = 90$ is the location of the variance change point. In other words, the stock price started to change at the 91st time point, which corresponds to the calendar week of March 19–23, 1973. Our conclusion matches Hsu's (1977, 1979) conclusion at this point.

Moreover, we continue to test the two subsequences: t from 1 to 88, and t from 89 to 161. Our computational results show that there are no further changes in the subsequence of t from 89 to 161, but there are at least two more changes in the subsequence of t from 1 to 88. One of the changes occurred during the period July 19 to August 8, 1971, and the other change occurred during the period November 15 to December 12, 1971. Going back to some historical records (e.g., Leonard, Crippen and Aronson, 1988), and looking at what happened to the U.S. economy and environment during those two periods we find that: from July 19 to August 8, 1971, several union strikes influenced the changes of the U.S. stock markets. The wage increases, resulting from several union–company negotiations, caused grave concerns about market prices. Among those strikes, the one organized by the United Transportation Union on July 26 was the biggest, and the negotiations were suspended indefinitely over a dispute on work rule changes. Responding to the suspension, U.S. gold stocks fell dramatically. The rate of wage increases in steel companies, the U.S. Postal Service, and some others was as high as 30 percent. On August 4, President Nixon said he would consider establishment of wage–price review bonds to examine the situation of American markets.

From November 15 to December 12, the most eye-catching economic event was the price increases of some important industrial products. Although the

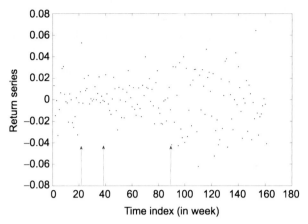

Fig. 2.12 Return series R_t of the weekly stock prices from 1971 to 1974

Nixon administration established the price commission to stabilize prices, economic conditions forced the administration to approve price increases. For example, the price commission approved steel price increases that were about triple the commission's target. Also, the commission approved price increases for the Big Three automakers, averaging nearly 3 percent.

The scatterplot of the return series R_t is given in Figure 2.12 with the identified changes indicated by arrows.

2.3 Mean and Variance Change

Let $x_1, x_2, \ldots, x_n$ be a sequence of independent normal random variables with parameters $(\mu_1, \sigma_1^2), (\mu_2, \sigma_2^2), \ldots, (\mu_n, \sigma_n^2)$, respectively. Testing and estimation about multiple mean changes in a Gaussian model have been studied in Section 2.1, and inference about multiple variance changes has been studied in Section 2.2. In this section, inference about the multiple mean and variance changes is discussed. To be specific, the interest here is to test the hypothesis (Chen and Gupta, 1999):

$$H_0 : \mu_1 = \cdots = \mu_n = \mu \quad \text{and} \quad \sigma_1^2 = \cdots = \sigma_n^2 = \sigma^2 (\mu, \sigma^2 \text{ unknown}) \quad (2.32)$$

versus the alternative:

$$H_A : \mu_1 = \cdots = \mu_{k_1} \neq \mu_{k_1+1} = \cdots = \mu_{k_2} \neq \cdots \neq \mu_{k_q+1} = \cdots = \mu_n$$

and

$$\sigma_1^2 = \cdots = \sigma_{k_1}^2 \neq \sigma_{k_1+1}^2 = \cdots = \sigma_{k_2}^2 \neq \cdots \neq \sigma_{k_q+1}^2 = \cdots = \sigma_n^2.$$

As discussed in previous sections, the binary segmentation procedure can be applied to this situation. Then it suffices to test (2.30) versus the alternative:

$$H_1 : \mu_1 = \cdots = \mu_k \neq \mu_{k+1} = \cdots = \mu_n$$

and

$$\sigma_1^2 = \cdots = \sigma_k^2 \neq \sigma_{k+1}^2 = \cdots = \sigma_n^2. \tag{2.33}$$

2.3.1 Likelihood-Ratio Procedure

Under H_0, the log likelihood function is

$$\log L_0(\mu, \sigma^2) = -\frac{n}{2} \log 2\pi - \frac{n}{2} \log \sigma^2 - \frac{1}{2\sigma^2} \sum_{i=1}^{n} (x_i - \mu)^2.$$

Denote the MLEs of μ and σ^2 by $\widehat{\mu}$ and $\widehat{\sigma}^2$; then

$$\widehat{\mu} = \overline{x} = \frac{1}{n} \sum_{i=1}^{n} x_i,$$

$$\widehat{\sigma}^2 = \frac{1}{n} \sum_{i=1}^{n} (x_i - \overline{x})^2,$$

and the maximum log likelihood is:

$$\log L_0(\widehat{\mu}, \widehat{\sigma}^2) = -\frac{n}{2} \log 2\pi - \frac{n}{2} \log \widehat{\sigma}^2 - \frac{n}{2}.$$

Under H_1, the log likelihood function is:

$$\log L_1(\mu_1, \mu_n, \sigma_1^2, \sigma_n^2) = -\frac{n}{2} \log 2\pi - \frac{k}{2} \log \sigma_1^2 - \frac{(n-k)}{2} \log \sigma_n^2$$

$$- \frac{1}{2\sigma_1^2} \sum_{i=1}^{k} (x_i - \mu_1)^2 - \frac{1}{2\sigma_n^2} \sum_{i=k+1}^{n} (x_i - \mu_n)^2.$$

Let $\widehat{\mu}_1$, $\widehat{\mu}_n$, $\widehat{\sigma}_1^2$, and $\widehat{\sigma}_n^2$ be the MLEs under H_1 of μ_1, μ_n, σ_1^2, and σ_n^2, respectively. Then

$$\widehat{\mu}_1 = \overline{x}_k = \frac{1}{k} \sum_{i=1}^{k} x_i, \qquad \widehat{\sigma}_1^2 = \frac{1}{k} \sum_{i=1}^{k} (x_i - \overline{x}_k)^2,$$

$$\widehat{\mu}_n = \overline{x}_{n-k}, \qquad \widehat{\sigma}_n^2 = \frac{1}{n-k} \sum_{i=k+1}^{n} (x_i - \overline{x}_{n-k})^2,$$

and the maximum log likelihood is

$$\log L_1(\widehat{\mu}_1, \widehat{\mu}_n, \widehat{\sigma}_1^2, \widehat{\sigma}_n^2) = -\frac{n}{2}\log 2\pi - \frac{k}{2}\log\widehat{\sigma}_1^2 - \frac{n-k}{2}\log\widehat{\sigma}_n^2 - \frac{n}{2}.$$

The likelihood-ratio procedure (Lehmann, 1986, p. 16) statistic is

$$\Lambda_n = \max_{2 \leq k \leq n-2} \frac{\widehat{\sigma}^n}{\widehat{\sigma}_1^k \widehat{\sigma}_n^{n-k}}.$$

Horváth (1993) derived the asymptotic null distribution of a function of Λ_n. The exact null distribution of Λ_n is not yet available in the literature. Therefore, in the following, Horváth's main theorem is presented and its detailed proof is given.

For large n, the asymptotic null distribution of λ_n, where

$$\lambda_n = (2\log\Lambda_n)^{1/2} = \left[\max_{2 \leq k \leq n-2}(n\log\widehat{\sigma}^2 - k\log\widehat{\sigma}_k^2 - (n-k)\log\widehat{\sigma}_{n-k}^2)\right]^{1/2}$$

is derived. However, it is convenient to simplify λ_n and to prove the following results first. Throughout, "$\overset{AD}{=}$" means "asymptotically distributed as". Under H_0;

$$\lambda_n^2 = \max_{1<k<n-1}\left[n\log\frac{\widehat{\sigma}^2}{\sigma^2} - k\log\frac{\widehat{\sigma}_k^2}{\sigma^2} - (n-k)\log\frac{\widehat{\sigma}_{n-k}^2}{\sigma^2}\right]$$

$$\overset{D}{=} \max_{1<k<n-1}\left[n\log\frac{1}{n}\chi_{n-1}^2 - k\log\frac{1}{k}\chi_{k-1}^2 - (n-k)\log\frac{1}{n-k}\chi_{n-k-1}^2\right],$$

where χ_j^2 denote the chi-square random variable with j degrees of freedom.

Let

$$\chi_{n-1}^2 = \sum_{i=1}^{n}(z_i - \bar{z})^2,$$

$$\chi_{k-1}^2 = \sum_{i=1}^{k}(z_i - \bar{z}_k)^2,$$

$$\chi_{n-k-1}^2 = \sum_{i=k+1}^{n}(z_i - \bar{z}_{n-k})^2,$$

where $z_1, \ldots, z_n$ are iid $N(0,1)$ random variables, and

$$\bar{z} = \frac{1}{n}\sum_{i=1}^{n}z_i, \qquad \bar{z}_k = \frac{1}{k}\sum_{i=1}^{k}z_i \quad \text{and} \quad \bar{z}_{n-k} = \frac{1}{n-k}\sum_{i=k+1}^{n}z_i.$$

Then

$$\lambda_n^2 \overset{D}{=} \max_{1 < k < n-1} \left[n \log \frac{1}{n} \sum_{i=1}^{n} (z_i - \bar{z})^2 - k \log \frac{1}{k} \sum_{i=1}^{k} (z_i - \bar{z}_k)^2 \right.$$

$$\left. - (n - k) \log \frac{1}{n - k} \sum_{i=k+1}^{n} (z_i - \bar{z}_{n-k})^2 \right].$$

Let

$$\xi_k = n \log \frac{1}{n} \sum_{i=1}^{n} (z_i - \bar{z})^2 - k \log \frac{1}{k} \sum_{i=1}^{k} (z_i - \bar{z}_k)^2$$

$$- (n - k) \log \frac{1}{n - k} \sum_{i=k+1}^{n} (z_i - \bar{z}_{n-k})^2.$$

Using the three-term Taylor expansion, we have

$$\xi_k = n \left[\frac{1}{n} \sum_{i=1}^{n} (z_i - \bar{z})^2 - 1 \right] - \frac{n}{2} \left[\frac{1}{n} \sum_{i=1}^{n} (z_i - \bar{z})^2 - 1 \right]^2$$

$$+ \frac{n}{3} (Q_n^{(1)})^{-3} \left[\frac{1}{n} \sum_{i=1}^{n} (z_i - \bar{z})^2 - 1 \right]^3 - k \left[\frac{1}{k} \sum_{i=1}^{k} (z_i - \bar{z}_k)^2 - 1 \right]$$

$$+ \frac{k}{2} \left[\frac{1}{k} \sum_{i=1}^{k} (z_i - \bar{z}_k)^2 - 1 \right]^2 - \frac{k}{3} (Q_k^{(2)})^{-2} \left[\frac{1}{k} \sum_{i=1}^{k} (z_i - \bar{z}_k)^2 - 1 \right]^3$$

$$- (n - k) \left[\frac{1}{n - k} \sum_{i=k+1}^{n} (z_i - \bar{z}_{n-k})^2 - 1 \right]$$

$$+ \frac{n - k}{2} \left[\frac{1}{n - k} \sum_{i=k+1}^{n} (z_i - \bar{z}_{n-k})^2 - 1 \right]^2$$

$$- \frac{n - k}{3} (Q_{n-k}^{(2)})^{-3} \left[\frac{1}{n - k} \sum_{i=k+1}^{n} (z_i - \bar{z}_{n-k})^2 + 1 \right]^2,$$

where $|Q_n^{(1)} - 1| \leq |(1/n) \sum_{i=1}^{n} (z_i - \bar{z})^2 - 1|$, $|Q_k^{(2)} - 1| \leq |(1/k) \sum_{i=1}^{k} (z_i - \bar{z}_k)^2 - 1|$, and $|Q_{n-k}^{(2)} - 1| \leq |(1/(n - k)) \sum_{i=k+1}^{n} (z_i - \bar{z}_{n-k})^2 - 1|$. After some algebraic simplification,

$$\xi_k = k\bar{z}_k^2 + (n-k)\bar{z}_{n-k}^2 - n\bar{z}^2 - \frac{1}{2n}\left[\sum_{i=1}^{n}(z_i^2-1)\right]^2 + \frac{1}{2k}\left[\sum_{i=1}^{k}(z_i^2-1)\right]^2$$

$$+ \frac{1}{2(n-k)}\left[\sum_{i=k+1}^{n}(z_i^2-1)\right]^2 + \bar{z}^2\sum_{i=1}^{n}(z_i^2-1) - \frac{n}{2}\bar{z}^4 - \bar{z}_k^2\sum_{i=1}^{k}(z_i^2-1)$$

$$+ \frac{k}{2}\bar{z}_k^4 + \frac{n}{3}(Q_n^{(1)})^{-3}(t_{n-1}-1)^3 - \frac{k}{3}(Q_k^{(2)})^{-3}(t_{k-1}-1)^3 + \frac{n-k}{2}\bar{z}_{n-k}^4$$

$$- \bar{z}_{n-k}^2\sum_{i=k+1}^{n}(z_i^2-1) + \frac{n-k}{3}(Q_{n-k}^{(2)})^{-3}(t_{n-k-1}-1)^3$$

$$= W_k^{(1)} + W_k^{(2)} + Q_k^{(1)} + Q_k^{(2)},$$

where

$$W_k^{(1)} = k\bar{z}_k^2 + (n-k)\bar{z}_{n-k}^2 - n\bar{z}^2,$$

$$W_k^{(2)} = -\frac{1}{2n}\left[\sum_{i=1}^{n}(z_i^2-1)\right]^2 + \frac{1}{2k}\left[\sum_{i=1}^{k}(z_i^2-1)\right]^2 + \frac{1}{2(n-k)}\left[\sum_{i=k+1}^{n}(z_i^2-1)\right]^2,$$

$$Q_k^{(1)} = \bar{z}\sum_{i=1}^{n}(z_i^2-1) - \frac{n}{2}\bar{z}^4 - \bar{z}_k^2\sum_{i=1}^{k}(z_i^2-1) + \frac{k}{2}\bar{z}_k^4$$

$$+ \frac{n}{3}(\theta_n^{(1)})^{-3}(t_{n-1}-1)^3 - \frac{k}{3}(\theta_k^{(2)})^{-3}(t_{k-1}-1)^3,$$

$$Q_k^{(2)} = \frac{n-k}{2}\bar{z}_{n-k}^4 - \bar{z}_{n-k}^4\sum_{i=k+1}^{n}(z_i^2-1) + \frac{n-k}{3}(\theta_{n-k}^{(3)})^{-3}(t_{n-k-1}-1)^3,$$

$$t_{n-1} = \frac{1}{n}\sum_{i=1}^{n}(z_i-\bar{z})^2,$$

$$t_{k-1} = \frac{1}{k}\sum_{i=1}^{k}(z_i-\bar{z}_k)^2, \quad \text{and}$$

$$t_{n-k-1} = \frac{1}{n-k}\sum_{i=k+1}^{n}(z_i-\bar{z}_{n-k})^2.$$

Next, we propose the following lemmas for the properties of the above-listed quantities.

Lemma 2.21

(i) $\max_{1<k<n} k^{1/2} (\log \log k)^{-(3/2)} |Q_k^{(1)}| = O_p(1).$
(ii) $\max_{1<k<n} (n-k)^{1/2} [\log \log(n-k)]^{-(3/2)} |Q_k^{(2)}| = O_p(1).$

Proof. (i) Because $kt_{k-1} = \sum_{i=1}^{k} (z_i^{(j)} - \bar{z}_k^{(j)})^2 \overset{AD}{\approx} \chi_k^2$ for large k, $E(kt_{k-1}) = k$, or $E[k(t_{k-1} - 1)] = 0$, and $\text{Var}[k(t_{k-1} - 1)] = 2k$. From the law of iterated logarithm,

$$\max_{1<k<n} \frac{|(Q_k^{(2)})^{-1} k(t_{k-1} - 1)|}{(k \log \log k)^{1/2}} = Q_p(1).$$

Hence,

$$\max_{1<k<n} \frac{|(Q_k^{(2)})^{-3} k^3 (t_{k-1} - 1)^3|}{(k \log \log k)^{3/2}} = O_p(1);$$

that is,

$$\max_{1<k<n} k^{1/2} (\log \log k)^{-(3/2)} |(Q_k^{(2)})^{-3} k(t_{k-1} - 1)^3| = O_p(1) \qquad (2.34)$$

Inasmuch as $\bar{z}_k$ is distributed as $N(0, 1/k)$, $E(k\bar{z}_k) = 0$, $\text{Var}(k\bar{z}_k) = k$. From the law of iterated logarithm,

$$\max_{1<k<n} \frac{k\bar{z}_k}{(k \log \log k)^{1/2}} = O_p(1).$$

Therefore,

$$\max_{1<k<n} \frac{k^2 \bar{z}_k^2}{(k \log \log k)} = O_p(1), \qquad (2.35)$$

and

$$\max_{1<k<n} \frac{k^4 \bar{z}_k^4}{(k \log \log k)^2} = O_p(1). \qquad (2.36)$$

From (2.35),

$$\max_{1<k<n} \left(\frac{k}{(\log \log k)} \right) \bar{z}_k^2 = O_p(1). \qquad (2.37)$$

The law of iterated logarithm also implies

$$\max_{1<k<n} \frac{\sum_{i=1}^{k} (z_i^2 - 1)}{(k \log \log k)^{1/2}} = O_p(1), \qquad (2.38)$$

therefore combining (2.37) and (2.38), we obtain

$$\max_{1<k<n} k^{1/2} (\log \log k)^{-(3/2)} \bar{z}_k^2 \sum_{i=1}^{k} (z_i^2 - 1) = O_p(1). \qquad (2.39)$$

Considering the fact that $\lim_{n\to\infty}(\log\log k/k)^{1/2} = 0$ and combining it with (2.36), we thus obtain:

$$\max_{1<k<n} k^{1/2}(\log\log k)^{-(3/2)} k\bar{z}_k^4 = O_p(1). \tag{2.40}$$

Similar to (2.35), (2.39), and (2.40), from the law of iterated logarithm, we can show that

$$n^{1/2}(\log\log n)^{-(3/2)}|(Q_n^{(1)})^{-3}n(t_{n-1}-1)^3| = O_p(1),$$

$$n^{1/2}(\log\log n)^{-(3/2)}\bar{z}_n^2 \sum_{i=1}^{n}(z_i^2 - 1) = O_p(1),$$

and

$$n^{1/2}(\log\log n)^{-(3/2)}n\bar{z}_n^4 = O_p(1).$$

Due to the inequality: $k^{1/2}(\log\log k)^{-(3/2)} \leq n^{1/2}(\log\log n)^{-(3/2)}$, for $1 < k < n$, we thus conclude:

$$\max_{1<k<n} k^{1/2}(\log\log k)^{-(3/2)}|(\theta_n^{(1)})^{-3}n(t_{n-1}-1)^3| = O_p(1), \tag{2.41}$$

$$\max_{1<k<n} k^{1/2}(\log\log k)^{-(3/2)} z_n^{(j)^2} \sum_{i=1}^{n}(z_i^2 - 1) = O_p(1), \tag{2.42}$$

and

$$\max_{1<k<n} k^{1/2}(\log\log k)^{-(3/2)}n\bar{z}_n^4 = O_p(1). \tag{2.43}$$

Also, (2.34) and (2.39) through (2.43) together give us

$$\max_{1<k<n} k^{1/2}(\log\log k)^{-(3/2)}|\theta_k^{(1)}| = O_p(1).$$

(ii) Proceeding as in (i), we obtain

$$\max_{1<k<n} (n-k)^{1/2}[\log\log(n-k)]^{-(3/2)}|(\theta_{n-k}^{(3)})^{-3}(n-k)(t_{n-k-1}^{(j)}-1)^3| = O_p(1)$$

$$\max_{1<k<n} (n-k)^{1/2}[\log\log(n-k)]^{-(3/2)} z_{n-k}^{(5)^2} \sum_{i=k+1}^{n}(z_i^{(j)^2} = O_p(1)$$

$$\max_{1<k<n} (n-k)^{1/2}[\log\log(n-k)]^{-(3/2)}(n-k)\bar{z}_{n-k}^{(j)^4} = O_p(1).$$

Hence,

$$\max_{1<k<n} (n-k)^{1/2}[\log\log(n-k)]^{-(3/2)}|\theta_k^{(2)}| = O_p(1).$$

$\square$

Lemma 2.22 *For all $x \in R$, as $n \to \infty$,*

(i) $a^2(\log n) \max_{1<k<\log n}(W_k^{(1)} + W_k^{(2)}) - [x + b(\log n)]^2 \xrightarrow{P} -\infty$,

(ii) $a^2(\log n) \max_{1<k<\log n} \xi_k - [x + b(\log n)]^2 \xrightarrow{P} -\infty$,

(iii) $a^2(\log n) \max_{n-\log n<k<n}(W_k^{(1)} + W_k^{(2)}) - [x + b(\log n)]^2 \xrightarrow{P} -\infty$,

(iv) $a^2(\log n) \max_{n-\log n<k<n} \xi_k - [x + b(\log n)]^2 \xrightarrow{P} -\infty$,

where

$$a(\log n) = (2 \log \log n)^{1/2}, \tag{2.44}$$

$$b(\log n) = 2 \log \log n + \log \log \log n. \tag{2.45}$$

Proof. (i) Recall $W_k^{(1)} = k\bar{z}_k^2 + (n-k)\bar{z}_{n-k}^2 - n\bar{z}_n^2$. Because $k\bar{z}_k^2 \sim x_1^2$, $E(k\bar{z}_k^2) = 1$, we have $k\bar{z}_k^2 \xrightarrow{P} 1$ as $k \to \infty$. But $1/\log k \to 0$ as $k \to \infty$, hence $k\bar{z}_k^2/\log k \xrightarrow{P} 0$ as $k \to \infty$. There exists a constant c, $0 < c < 1$, such that $k\bar{z}_k^2/\log k \overset{P}{<} c$ for large k. Meanwhile,

$$\frac{a^2(\log n) \max_{1<k<\log n} k\bar{z}_k^2}{[b(\log n)]^2} \leq \frac{2 \log \log n \cdot \max_{1<k<\log n} k\bar{z}_k^2}{(2 \log \log n)^2}$$

$$\leq \max_{1<k<\log n} \frac{k\bar{z}_k^2}{\log \log n}$$

$$\leq \max_{1<k<\log n} \frac{k\bar{z}_k^2}{\log k}$$

$$\overset{P}{<} c = 1 - M, \qquad 0 < M < 1.$$

Hence,

$$a^2(\log n) \max_{1<k<\log n} k\bar{z}_k^2 - [x + b(\log n)]^2 \overset{P}{<} -M[x + b(\log n)]^2;$$

that is,

$$a^2(\log n) \max_{1<k<\log n} k\bar{z}_k^2 - [x + b(\log n)]^2 \xrightarrow{P} -\infty, \quad \text{as } n \to \infty. \tag{2.46}$$

Similarly, we can show that

$$a^2(\log n) \max_{1<k<\log n} (n-k)\bar{z}_{n-k}^2 - [x + b(\log n)]^2 \xrightarrow{P} -\infty \quad \text{as } n \to \infty. \tag{2.47}$$

Because $n\bar{z}_n^2 \sim \chi_1^2$, $E(n\bar{z}_n^2) = 1$. Then $n\bar{z}_n^2 \xrightarrow{P} 1$ as $n \to \infty$, $-n\bar{z}_n^2 \xrightarrow{P} -1$ as $n \to \infty$. But $a^2(\log n) \to \infty$ and $[x + b(\log n)]^2 \to \infty$ as $n \to \infty$. Hence,

$$a^2(\log n)(-n\bar{z}_n^2) - [x + b(\log n)]^2 \xrightarrow{P} -\infty \quad \text{as } n \to \infty. \tag{2.48}$$

Combining (2.46) through (2.48), we thus obtain

$$a^2(\log n) \max_{1<k<\log n} (W_k^{(1)}) - [x + b(\log n)]^2 \xrightarrow{P} -\infty \quad \text{as } n \to \infty. \qquad (2.49)$$

Recall again

$$W_k^{(2)} = -\frac{1}{2n}\left[\sum_{i=1}^{n}(z_i^2-1)\right]^2 + \frac{1}{2k}\left[\sum_{i=1}^{k}(z_i^2-1)\right]^2 + \frac{1}{2(n-k)}\left[\sum_{i=k+1}^{n}(z_i^2-1)\right]^2.$$

Inasmuch as $z_i^2 \sim \chi_1^2$, $E\left[\sum_{i=1}^{k}(z_i^2 - 1)\right] = 0$. Then

$$E\left\{\frac{1}{k}\left[\sum_{i=1}^{k}(z_i^2-1)\right]^2\right\} = \text{Var}\left\{\frac{1}{k}\left[\sum_{i=1}^{k}(z_i^2-1)\right]^2\right\} = 2$$

as $k \to \infty$, and

$$\frac{\left[\sum_{i=1}^{k}(z_i^2-1)\right]^2}{k\log k} \xrightarrow{P} 0$$

as $n \to \infty$. Therefore, there exists a constant c, $0 < c < 1$, such that

$$0 < \frac{\left[\sum_{i=1}^{k}(z_i^2-1)\right]^2}{k\log k} < 1-c$$

for large k. Now,

$$\frac{a^2(\log n)\max_{1<k<\log n}\frac{1}{k}\left[\sum_{i=1}^{k}(z_i^2-1)\right]^2}{[b(\log n)]^2}$$

$$\leq \frac{2\log\log n \max_{1<k<\log n}\frac{1}{k}\left[\sum_{i=1}^{k}(z_i^2-1)\right]^2}{(2\log\log n)^2}$$

$$< \max_{1<k<\log n}\frac{\left[\sum_{i=1}^{k}(z_i^2-1)\right]^2}{k\log\log n}$$

$$< \max_{1<k<\log n}\frac{\left[\sum_{i=1}^{k}(z_i^2-1)\right]^2}{k\log k}$$

$$\overset{P}{<} 1-c.$$

Hence, as $n \to \infty$,

$$a^2(\log n)\max_{1<k<\log n}\frac{1}{k}\left[\sum_{i=1}^{k}(z_i^2-1)\right]^2 - [x+b(\log n)]^2 \xrightarrow{P} -\infty. \qquad (2.50)$$

Similarly, as $n \to \infty$, we have

$$a^2(\log n) \max_{1 < k < \log n} \frac{1}{n-k} \left[\sum_{i=k+1}^{n} (z_i^2 - 1) \right]^2 - [x + b(\log n)]^2 \xrightarrow{P} -\infty. \quad (2.51)$$

Moreover, $\left[\sum_{i=1}^{n} (z_i^2 - 1) \right]^2 \xrightarrow{P} 0$, as $n \to \infty$, therefore $a^2(\log n)(1/n)$ $\left[\sum_{i=1}^{n} (z_i^2 - 1) \right]^2 \xrightarrow{P} 0$ as $n \to \infty$. Hence, as $n \to \infty$,

$$a^2(\log n) \left\{ -\frac{1}{2n} \left[\sum_{i=k+1}^{n} (z_i^2 - 1) \right]^2 \right\} - [x + b(\log n)]^2 \xrightarrow{P} -\infty. \quad (2.52)$$

Then, (2.50) through (2.52) together give us, as $n \to \infty$,

$$a^2(\log n) \max_{1 < k < \log n} W_k^{(2)} - [x + b(\log n)]^2 \xrightarrow{P} -\infty. \quad (2.53)$$

From (2.49) and (2.53), as $n \to \infty$, we obtain:

$$a^2(\log n) \max_{1 < k < \log n} (W_k^{(1)} + W_k^{(2)}) - [x + b(\log n)]^2 \xrightarrow{P} -\infty.$$

This completes the proof of (i).

Now recall $\xi_k = W_k^{(1)} + W_k^{(2)} + Q_k^{(1)} + Q_k^{(2)}$ and from Lemma 2.22,

$$\max_{1 < k < \log n} k^{1/2} (\log \log k)^{-(3/2)} |Q_k^{(1)}| = O_p(1).$$

Then

$$\frac{a^2(\log n) \max_{1 < k < \log n} |Q_k^{(1)}|}{[b(\log n]^2}$$

$$\leq \frac{2 \log \log n \max_{1 < k < \log n} |Q_k^{(1)}|}{(2 \log \log n)^2}$$

$$\leq \frac{1}{\log \log n} \cdot \max_{1 < k < \log n} \frac{(\log \log k)^{3/2}}{k^{1/2}} \cdot k^{1/2} (\log \log k)^{-(3/2)} |Q_k^{(1)}|$$

$$\leq \frac{(\log \log \log n)^{(3/2)}}{\log \log n} \cdot \max_{1 < k < \log n} k^{1/2} (\log \log k)^{-(3/2)} |Q_k^{(1)}|.$$

Notice that

$$\lim_{n \to \infty} \frac{(\log \log \log n)^{3/2}}{\log \log n} = 0;$$

then, there exists a constant M, $0 < M < 1$, such that

$$0 < \frac{a^2(\log n) \max_{1<k<\log n} |Q_k^{(1)}|}{[b(\log n)]^2} < 1 - M \quad \text{for large } n.$$

Hence, as $n \to \infty$,

$$a^2(\log n) \max_{1<k<\log n} |Q_k^{(1)}| - [x + b(\log n)]^2 \xrightarrow{P} -\infty. \tag{2.54}$$

From Lemma 2.22,

$$\max_{1<k<\log n} (n-k)^{1/2} [\log\log(n-k)]^{-(3/2)} |Q_k^{(2)}| = O_p(1).$$

Hence,

$$\frac{a^2(\log n) \max_{1<k<\log n} |Q_k^{(2)}|}{[b(\log n)]^2}$$

$$\leq \frac{1}{\log\log n} \max_{1<k<\log n} |Q_k^{(2)}|$$

$$= \frac{1}{\log\log n} \max_{1<k<\log n} \frac{[\log\log(n-k)]^{3/2}}{(n-k)^{1/2}} \cdot (n-k)^{1/2} [\log\log(n-k)]^{-(3/2)} |Q_k^{(2)}|$$

$$\leq \frac{(\log\log n)^{3/2}}{(\log\log n)(n-\log n)^{1/2}} \max_{1<k<\log n} (n-k)^{1/2} [\log\log(n-k)]^{-(3/2)} |Q_k^{(2)}|$$

$$= \left(\frac{\log\log n}{n-\log n}\right)^{1/2} \max_{1<k<\log n} (n-k)^{1/2} [\log\log(n-k)]^{-(3/2)} |Q_k^{(2)}|.$$

Because $\lim_{n\to\infty} (\log\log n)/(n - \log n) = 0$, there exists a constant M, $0 < M < 1$, such that for large n,

$$0 < \frac{a^2(\log n) \max_{1<k<\log n} |Q_k^{(2)}|}{[b(\log n)]^2} < 1 - M.$$

Then, as $n \to \infty$,

$$a^2(\log n) \max_{1<k<\log n} |Q_k^{(2)}| - [x + b(\log n)]^2 \xrightarrow{P} -\infty. \tag{2.55}$$

Combining (2.54), (2.55), and (i), we thus conclude that (ii) holds.

Next, recall that

$$W_k^{(1)} = k\bar{z}_k^2 + (n-k) - \bar{z}_{n-k}^2 - n\bar{z}_n^2.$$

Because $k\bar{z}_k^2 \sim \chi_1^2$, $E(k\bar{z}_k^2)^2 = 1$, and $k\bar{z}_k^2 \xrightarrow[k\to\infty]{P} 1$, as $k \to \infty$, then

$$\max_{n-\log n<k<n} k\bar{z}_k^2 \xrightarrow{P} 1$$

as $n \to \infty$. But $\lim_{n\to\infty}(1/\log\log k) = 0$ and $k \to \infty$, as $n \to \infty$, hence there exists a constant M, $0 < M < 1$, such that for large n

$$\max_{n-\log n<k<n} \frac{k\bar{z}_k^2}{\log\log k} < 1 - M.$$

Now,

$$\frac{a^2(\log n)\max_{n-\log n<k<n} k\bar{z}_k^2}{[b(\log n)]^2} < \frac{1}{\log\log n}\max_{n-\log n<k<n} k\bar{z}_k^2$$

$$= \max_{n-\log n<k<n} \frac{k\bar{z}_k^2}{\log\log n}$$

$$< \max_{n-\log n<k<n} \frac{k\bar{z}_k^2}{\log\log k}$$

$$< 1 - M;$$

then, as $n \to \infty$,

$$a^2(\log n)\max_{n-\log n<k<n} k\bar{z}_k^2 - [x + b(\log n)]^2 \xrightarrow{P} -\infty.$$

Similarly, we have as $n \to \infty$,

$$a^2(\log n)\max_{n-\log n<k<n} (n-k)\bar{z}_{n-k}^2 - [x + b(\log n)]^2 \xrightarrow{P} -\infty,$$

and

$$a^2(\log n)\max_{n-\log n<k<n} (-n\bar{z}_n^2) - [x + b(\log n)]^2 \xrightarrow{P} -\infty \quad \text{as} \quad n \to \infty.$$

Therefore, as $n \to \infty$,

$$a^2(\log n)\max_{n-\log n<k<n} W_k^{(1)} - [x + b(\log n)]^2 \xrightarrow{P} -\infty.$$

Similarly, we can show that as $n \to \infty$,

$$a^2(\log n)\max_{n-\log n<k<n} W_k^{(2)} - [x + b(\log n)]^2 \xrightarrow{P} -\infty,$$

and (iii) is established.

To prove (iv), we start with Lemma 2.22,

$$\max_{n-\log n < k < n} k^{1/2} (\log\log k)^{-(3/2)} |Q_k^{(1)}| = O_p(1).$$

Then,

$$\frac{a^2(\log n) \max_{n-\log n < k < n} |Q_k^{(1)}|}{[b(\log n)]^2}$$

$$\leq \frac{1}{\log\log n} \max_{n-\log n < k < n} |Q_k^{(1)}|$$

$$= \frac{1}{\log\log n} \cdot \max_{n-\log n < k < n} \frac{(\log\log k)^{3/2}}{k^{1/2}} \cdot k^{1/2} (\log\log k)^{-(3/2)} |Q_k^{(1)}|$$

$$\leq \left(\frac{\log\log n}{n - \log n} \right)^{1/2} \max_{n-\log n < k < n} k^{1/2} (\log\log k)^{-(3/2)} |Q_k^{(1)}|.$$

There exists a constant M, $0 < M < 1$, such that

$$\frac{a^2(\log n) \max_{n-\log n < k < n} |Q_k^{(1)}|}{[b(\log n)]^2} < 1 - M;$$

therefore, as $n \to \infty$,

$$a^2(\log n) \max_{n-\log n < k < n} |Q_k^{(1)}| - [x + b(\log n)]^2 \xrightarrow{P} -\infty.$$

Starting with Lemma 2.22, we obtain

$$a^2(\log n) \max_{n-\log n < k < n} |Q_k^{(2)}| - [x + b(\log n)]^2 \xrightarrow{P} -\infty.$$

In view of (iii), we thus conclude that (iv) holds. $\qquad\square$

Lemma 2.23 *As $n \to \infty$, the following hold.*

(i) $a^2(\log n) \max_{\log n \leq k \leq n - \log n} |\xi_k - (W_k^{(1)} + W_k^{(2)})| = o_p(1)$.

(ii) $a^2(\log n) \max_{1 < k < n/\log n} |(n - k)\bar{z}_{n-k}^2 - n\bar{z}_n^2| = o_p(1)$, $j = 1, \ldots, m$.

(iii) $a^2(\log n) \max_{1 < k < n/\log n} \left| \frac{1}{n-k} \left[\sum_{i=k+1}^n (z_i^2 - 1) \right]^2 - \frac{1}{n} \left[\sum_{i=1}^n (z_i^2 - 1) \right]^2 \right| = o_p(1)$, $j = 1, \ldots, m$.

Proof. (i) Clearly, $\xi_k - (W_k^{(1)} + W_k^{(2)}) = Q_k^{(1)} + Q_k^{(2)}$.

$$0 \leq a^2(\log n) \max_{\log n \leq k \leq n - \log n} |Q_k^{(1)} + Q_k^{(2)}|$$

$$\leq 2\log\log n \max_{\log n \leq k \leq n - \log n} |Q_k^{(1)}| + 2\log\log n \cdot \max_{\log n \leq k \leq n - \log n} |Q_k^{(2)}|$$

$$= 2 \log \log n \cdot \max_{\log n \le k \le n - \log n} \frac{(\log \log k)^{3/2}}{k^{1/2}} \cdot k^{1/2} (\log \log k)^{-(3/2)} |Q_k^{(1)}|$$

$$+ \ 2 \log \log n \cdot \max_{\log n \le k \le n - \log n} \frac{[\log \log (n - k)]^{3/2}}{(n - k)^{1/2}} \cdot (n - k)^{1/2}$$

$$\cdot [\log \log (n - k)]^{-(3/2)} |Q_k^{(2)}|$$

$$\le \frac{2(\log \log n)^{5/2}}{(\log n)^{1/2}} \cdot \max_{\log n \le k \le n - \log n} k^{1/2} (\log \log k)^{-(3/2)} |Q_k^{(1)}|$$

$$+ \frac{2(\log \log n)^{5/2}}{(\log n)^{1/2}} \max_{\log n \le k \le n - \log n} (n - k)^{1/2} [\log \log (n - k)]^{-(3/2)} |Q_k^{(2)}|.$$

Because $\lim_{n \to \infty} (\log \log n)^{5/2} / (\ln n)^{1/2} = 0$, in view of Lemma 2.22, we then obtain

$$\lim_{n \to \infty} a^2 (\log n) \max_{\log n \le k \le n - \log n} |\xi_k - (W_k^{(1)} + W_k^{(2)})| = 0 \text{ in probability.}$$

Therefore (i) holds.

(ii) First, observe that

$$(n - k)\bar{z}_{n-k}^2 - n\bar{z}_n^2 = \frac{k}{n(n - k)} \left(\sum_{i=1}^{n} z_i \right)^2$$

$$- \frac{2}{n - k} \left(\sum_{i=1}^{n} z_i \right) \left(\sum_{i=1}^{k} z_i \right) + \frac{1}{n - k} \left(\sum_{i=1}^{k} z_i \right)^2 .$$

From the law of iterated logarithm,

$$\frac{\sum_{i=1}^{n} z_i}{(n \log \log n)^{1/2}} = O_p(1), \tag{2.56}$$

and hence,

$$\frac{\left(\sum_{i=1}^{n} z_i \right)^2}{n \log \log n} = O_p(1).$$

Furthermore,

$$0 < a^2 (\log n) \max_{1 < k < n/\log n} \frac{k}{n(n - k)} \left(\sum_{i=1}^{n} z_i \right)^2$$

$$\le 2 \log \log n \cdot \frac{\frac{n}{\log n}}{n \left(n - \frac{n}{\log n} \right)} \left(\sum_{i=1}^{n} z_i \right)$$

$$\le \frac{2(\log \log n)^2}{\log n} \cdot \frac{\sum_{i=1}^{n} z_i}{n \log \log n}.$$

Because $\lim_{n\to\infty}((\log\log n)^2/\log n) = 0$, we obtain that

$$\lim_{n\to\infty} a^2(\log n) \max_{1<k<(n/\log n)} (k/n(n-k)) \left(\sum_{i=1}^{n} z_i\right)^2 = 0 \text{ in probability.}$$

From the law of iterated logarithm again, we have

$$\max_{1<k<n/\log n} \frac{\sum_{i=1}^{k} z_i}{(k\log\log k)^{1/2}} = O_p(1). \tag{2.57}$$

Then

$$0 \le a^2(\log n) \max_{1<k<n/\log n} \frac{2}{n-k} \sum_{i=1}^{n} |z_i| \cdot \left|\sum_{i=1}^{k} z_i\right|$$

$$\le \frac{4\log\log n}{n - \frac{n}{\log n}} \max_{1<k<n/\log n} \left|\sum_{i=1}^{n} z_i\right| \cdot \max_{1<k<n/\log n} \left|\sum_{i=1}^{k} z_i\right|$$

$$= \frac{4n^{\frac{1}{2}}(\log\log n)^{3/2}}{n - \frac{n}{\log n}} \frac{|\sum_{i=1}^{n} z_i|}{(n\log\log n)^{1/2}}$$

$$\cdot \max_{1<k<(n/\log n)} (k\log\log k)^{1/2} \frac{|\sum_{i=1}^{k} z_i|}{(k\log\log k)}$$

$$\le \frac{4(\log\log n)^2(\log n)^{1/2}}{\log n - 1} \frac{|\sum_{i=1}^{n} z_i|}{(n\log\log n)^{1/2}} \cdot \max_{1<k<n/\log n} \frac{|\sum_{i=1}^{k} z_i|}{(k\log\log k)}.$$

Combining $\lim_{n\to\infty} (\log\log n)^2(\log n)^{\frac{1}{2}}/(\log n - 1) = 0$ with (2.56) and (2.57), we obtain:

$$\lim_{n\to\infty} a^2(\log n) \max_{1<k<(n/\log n)} |(n-k)\bar{z}_{n-k}^{(j)^2} - n\bar{z}_n^{(j)^2}| = 0 \text{ in probability;}$$

that is, (ii) holds.

(iii) Because

$$E\left\{\frac{1}{n-k}\left[\sum_{i=k+1}^{n}(z_i^2-1)\right]^2 - \frac{1}{n}\left[\sum_{i=1}^{n}(z_i^2-1)\right]^2\right\}$$

$$= E\left\{\frac{1}{n-k}\sum_{i=k+1}^{n}(z_i^2-1)^2 + \frac{1}{n-k}\sum_{i\ne\iota}(z_i^2-1)(z_\iota^2-1) - \frac{1}{n}\sum_{i=1}^{n}(z_i^2-1)\right]^2$$

$$- \frac{1}{n}\sum_{i\ne\iota}(z_i^2-1)(z_\iota^2-1)$$

$$= \frac{1}{n-k} \sum_{i=k+1}^{n} \mathrm{Var}(z_i^2) + \frac{1}{n-k} \sum_{i \neq \iota} E(z_i^2 - 1)E(z_\iota^2 - 1) - \frac{1}{n} \sum_{i=1}^{n} \mathrm{Var}(z_i^2)$$

$$- \frac{1}{n} \sum_{i \neq \iota} E(z_i^2 - 1)E(z_\iota^2 - 1)$$

$$= \frac{1}{n-k} \sum_{i=k+1}^{n} 2 - \frac{1}{n} \sum_{i=1}^{n} 2 = 0 \quad \text{for all } n \text{ and all } k,$$

we have, $E\langle a^2(\log n)\{(1/(n-k))[\sum_{i=k+1}^{n}(z_i^2 - 1)]^2 - (1/n)[\sum_{i=1}^{n}(z_i^2 - 1)]^2\}\rangle = 0$ for all n and all k. Hence,

$$a^2(\log n) \left| \frac{1}{n-k} \left[\sum_{i=k+1}^{n} (z_i^2 - 1) \right]^2 - \frac{1}{n} \left[\sum_{i=1}^{n} (z_i^2 - 1) \right]^2 \right| \xrightarrow{P} 0$$

as $n \to \infty$ for all k, $1 < k < n/\log n$. That is,

$$a^2(\log n) \max_{1 < k < (n/\log n)} \left| \frac{1}{n-k} \left[\sum_{i=k+1}^{n} (z_i^2 - 1) \right]^2 - \frac{1}{n} \left[\sum_{i=1}^{n} (z_i^2 - 1) \right]^2 \right| = o_p(1).$$

$\square$

Lemma 2.24 *For all* $x \in R$, *as* $n \to \infty$,

$$a^2(\log n) \max_{n/\log n < k < n - n/\log n} (W_k^{(1)} + W_k^{(2)}) - [x + b(\log n)]^2 \xrightarrow{P} -\infty.$$

Proof. Note that $W_k^{(1)} = k\bar{z}_k^2 + (n-k)\bar{z}_{n-k}^2 - n\bar{z}_n^2$. Let's consider term by term:

$$k\bar{z}_k^2 = \left(\frac{\left| \sum_{i=1}^{k} z_i \right|}{k^{1/2}} \right)^2.$$

From Theorem 2 of Darling and Erdös (1956) we have

$$P\left[\max_{n/\log n < k < n - (n/\log n)} k\bar{z}_k^2 \right.$$

$$\left. < \left[(2 \log \log n)^{1/2} + \frac{\log \log \log n}{2(2 \log \log n)^{1/2}} + \frac{x}{(2 \log \log n)^{1/2}} \right]^2 \right]$$

$$= e^{-(1/\sqrt{\pi})e^{-x}} \cdot x \in R.$$

Then,

$$P\left[\frac{a^2(\log n)\max_{(n/\log n)<k<n-(n/\log n)} k\bar{z}_k^2}{[b(\log n)+x]^2}\right.$$

$$\left.<\left[\frac{2\log\log n+\frac{1}{2}\log\log\log n+x}{b(\log n)+x}\right]^2\right]$$

$$= e^{-(1/\sqrt{\pi})e^{-x}}.$$

Because $b(\log n) = 2\log\log n + \log\log\log n$, we can choose n large enough, such that

$$\left[\frac{2\log\log n+\frac{1}{2}\log\log\log n+x}{b(\log n)+x}\right]^2 < 1-M, \qquad 0 < M < 1.$$

Therefore,

$$P\left\{\frac{a^2(\log n)\max_{(n/\log n)<k<n-(n/\log n)} k\bar{z}_k^2}{[x+b(\log n)]^2} < 1-M\right\} = e^{-(1/\sqrt{\pi})e^{-x}}.$$

Letting $x \to \infty$, we then obtain

$$P\left[a^2(\log n)\max_{n/\log n<k<n-n/\log n} k\bar{z}_k^2\right.$$

$$\left.-[x+b(\log n)]^2 < -M[x+b(\log n)]^2\right] = 1.$$

Hence, as $n \to \infty$,

$$a^2(\log n)\max_{n/\log n<k<n-n/\log n} k\bar{z}_k^2 - [x+b(\log n)]^2 \xrightarrow{P} -\infty.$$

For the next term $(n-k)\bar{z}_{n-k}^2$, observe that

$$(n-k)\bar{z}_{n-k}^2 = \left[\frac{|\sum_{i=k+1}^n z_i|}{(n-k)^{1/2}}\right]^2 \quad \text{and} \quad \frac{n}{\log n} < n-k < n-\frac{n}{\log n};$$

then, proceeding in the same manner as above, we can show that, as $n \to \infty$,

$$a^2(\log n)\max_{n/\log n<k<n-n/\log n} (n-k)\bar{z}_{n-k}^2 - [x+b(\log n)]^2 \xrightarrow{P} -\infty.$$

For the last term $-n\bar{z}_n^2$, applying the law of iterated logarithm, we have $n\bar{z}_n/(n\log\log n)^{1/2} = O_p(1)$; that is, $n\bar{z}_n^2/\log\log n = O_p(1)$. Therefore, as $n \to \infty$,

$$a^2(\log n) \max_{n/\log n < k < n - n/\log n} (-n\bar{z}_n^2) \xrightarrow{P} -\infty,$$

and

$$a^2(\log n) \max_{n/\log n < k < n - n/\log n} (-n\bar{z}_n^2) - [x + b(\log n)]^2 \xrightarrow{P} -\infty.$$

Then we conclude from all of the above that as $n \to \infty$,

$$a^2(\log n) \max_{n/\log n < k < n - n/\log n} W_k^{(1)} - [x + b(\log n)]^2 \xrightarrow{P} -\infty.$$

Similarly, we can show that as $n \to \infty$,

$$a^2(\log n) \max_{n/\log n < k < n - n/\log n} W_k^{(2)} - [x + b(\log n)]^2 \xrightarrow{P} -\infty.$$

Thus the lemma is proved.   □

Similar to Lemma 2.22(ii) and (iii), we obtain the following results.

Lemma 2.25

(i) $a^2(\log n) \max_{n-(n/\log n) < k < n} |k\bar{z}_k^2 - n\bar{z}_n^2| = O_p(1), j = 1, \ldots, m.$

(ii) $a^2(\log n) \max_{n-(n/\log n) < k < n} |(1/k) \left[\sum_{i=1}^{k} (z_i^2 - 1) \right]^2 - (1/n)$
$\left[\sum_{i=1}^{n} (z_i^2 - 1) \right]^2 |, = O_p(1), j = 1, \ldots, m.$

Proof. (i) Start with the identity:

$$k\bar{z}_k^2 - n\bar{z}_n^2 = \frac{n-k}{kn} \left(\sum_{i=1}^{n} z_i \right)^2 - \frac{2}{k} \left(\sum_{i=1}^{n} z_i \right) \left(\sum_{i=k+1}^{n} z_i \right) + \frac{1}{k} \left(\sum_{i=k+1}^{n} z_i \right)^2.$$

The law of iterated logarithm yields $\sum_{i=1}^{n} z_i/(n \log \log n)^{1/2} = 0_p(1)$. Then, $(\sum_{i=1}^{n} z_i)^2/(n \log \log n) = O_p(1)$. Moreover,

$$0 < a^2(\log n) \max_{n - \frac{n}{\log n} < k < n} \frac{n-k}{kn} \left(\sum_{i=1}^{n} z_i \right)^2$$

$$< 2 \log \log n \cdot \frac{\frac{n}{\log n}}{(n - \frac{n}{\log n})n} \left(\sum_{i=1}^{n} z_i \right)^2$$

$$= \frac{2(\log \log n)^2}{\log n - 1} \cdot \frac{(\sum_{i=1}^{n} z_i)^2}{n \log \log n} \xrightarrow{P} 0, \quad \text{as } n \to \infty.$$

Hence,

$$a^2(\log n) \max_{n-(n/\log n) < k < n} \frac{n-k}{kn} \left(\sum_{i=1}^{n} z_i \right)^2 = o_p(1).$$

Proceeding similarly, we can show that

$$a^2(\log n) \max_{n-(n/\log n)<k<n} \frac{2}{k} \left| \sum_{i=1}^{n} z_i \right| \left| \sum_{i=k+1}^{n} z_i \right| = o_p(1),$$

and

$$a^2(\log n) \max_{n-(n/\log n)<k<n} \frac{1}{k} \left(\sum_{i=k+1}^{n} z_i \right)^2 = o_p(1).$$

Thus, (i) holds.

(ii) Similar to the proof of Lemma 2.22(iii), one can easily obtain (ii) here.   □

Finally, we state without proof Lemma 2.2 of Horváth (1993).

Theorem 2.26 *Under the null hypothesis H_0, when $n \to \infty$,*

$$\lim_{n\to\infty} P[a(\log n)\lambda_n - b(\log n) \le x] = \exp\{-2e^{-x}\}$$

for $x \in R$, where $a(\log n)$ and $b(\log n)$ are defined in (2.44) and (2.45).

Proof. First, observe that

$$\{1 < k < n\} = \{1 < k \le \ln n\} \cup \{\log \le k \le n - \log n\} \cup \{n - \log n < k < n\}.$$

From Lemma 2.21(ii) and (iii), we obtain:

$$\max_{1<k<n} \xi_k \overset{D}{=} \max_{\log n \le k \le n - \log n} \xi_k.$$

From Lemma 2.22(i), then

$$\max_{1<k<n} \xi_k \overset{D}{=} \max_{\log n \le k \le n - \log n} (W_k^{(1)} + W_k^{(2)}).$$

But, for large n, we have

$$|\log n \le k \le n - \log n| = \left\{ \log n \le k \le \frac{n}{\log n} \right\}$$
$$\cup \left\{ \frac{n}{\log n} < k \le n - \frac{n}{\log n} \right\}$$
$$\cup \left\{ n - \frac{n}{\log n} < k \le n - \log n \right\},$$

and

$$\left\{ \log n \le k \le \frac{n}{\log n} \right\} \subseteq \left\{ 1 \le k \le \frac{n}{\log n} \right\}.$$

From Lemma 2.22(ii) and (iii), we have

$$\max_{\log n \le k \le (n/\log n)} (W_k^{(1)} + W_k^{(2)})$$

$$\stackrel{D}{=} \max_{\log n \le k \le (n/\log n)} \left\{ k z_k^2 + \frac{1}{2k} \left[\sum_{i=1}^{k} (z_i^2 - 1) \right]^2 \right\}. \tag{2.58}$$

In view of Lemma 2.23, we have

$$\max_{\log n \le k \le n - \log n} (W_k^{(1)} + W_k^{(2)}) \stackrel{D}{=} \left[\max_{\log n \le k \le (n/\log n)} (W_k^{(1)} + W_k^{(2)}) \right]$$

$$\vee \left[\max_{n - (n/\log n) \le k \le n - \log n} (W_k^{(1)} + W_k^{(2)}) \right], \tag{2.59}$$

where $a \vee b \equiv \max\{a, b\}$. Because

$$\left\{ n - \frac{n}{\log n} \le k \le n - \log n \right\} \subseteq \left\{ n - \frac{n}{\log n} \le k \le n \right\},$$

applying Lemma 2.24(i) and (ii), we obtain:

$$\max_{n - (n/\log n) \le k \le n - \log n} (W_k^{(1)} + W_k^{(2)})$$

$$\stackrel{D}{=} \max_{n - (n/\log n) \le k \le n - \log n} \left\{ (n - k) \bar{z}_{n-k}^2 + \frac{1}{2(n - k)} \left[\sum_{i=k+1}^{n} (z_i^2 - 1) \right]^2 \right\}. \tag{2.60}$$

Combining (2.58) through (2.60), we thus have

$$\max_{\log n \le k \le n - \log n} (W_k^{(1)} + W_k^{(2)}) \stackrel{D}{=} \max \left\{ \max_{1 \le k < (n/\log n)} \left[k \bar{z}_k^2 + \frac{1}{2k} \sum_{i=1}^{k} (z_i^2 - 1) \right], \right.$$

$$\left. \max_{n - (n/\log n) \le k < n} \left[(n - k) \bar{z}_{n-k}^2 + \frac{1}{2(n - k)} \left[\sum_{i=k+1}^{n} (z_i^2 - 1) \right]^2 \right] \right\}.$$

Then,

$$\lim_{n \to \infty} P\{a(\log n)\lambda_n - b(\log n) \le x\}$$

$$= \lim_{n \to \infty} P \left\{ a(\log n) \max_{1 < k < n-1} \xi_k^{1/2} - b(\log n) \le x \right\}$$

$$= \lim_{n\to\infty} P\left\{a^2(\log n)\max_{1<k<n}\xi_k \le [x+b(\log n)]^2\right\}$$

$$= \lim_{n\to\infty} P\left\{a^2(\log n)\max_{\log n\le k<n-\log n}(W_k^{(1)}+W_k^{(2)}) \le [x+b(\log n)]^2\right\}$$

$$= \lim_{n\to\infty} P\left\{a^2(\log n)\max\left\{\max_{1\le k<(n/\log n)}[k\bar{z}_k^2 + \frac{1}{2k}\sum_{i=1}^{k}(z_i^2-1)],\right.\right.$$

$$\max_{n-(n/\log n)\le k<n}\left[(n-k)\bar{z}_{n-k}^2 + \frac{1}{2(n-k)}\left[\sum_{i=k+1}^{n}(z_i^2-1)\right]^2\right]\right\}$$

$$\le [x+b(\log n)]^2\Bigg\}. \tag{2.61}$$

Because $\{z_i, 1 \le i < (n/\log n)\}$ and $\{z_i, n-(n/\ln n) \le i \le n\}$ are independent, (2.61) reduces to

$$\lim_{n\to\infty} P\left\{a^2(\log n)\max_{1\le k<(n/\log n)}\left(k\bar{z}_k^2 + \frac{1}{2k}\left[\sum_{i=1}^{k}(z_i^2-1)\right]^2\right) \le [x+b(\log n)]^2\right\}$$

$$\cdot \lim_{n\to\infty} P\left\{\left[a^2(\log n)\max_{n-(n/\log n)\le k<n}\left[(n-k)\bar{z}_{n-k}^2\right.\right.\right.$$

$$+\frac{1}{2(n-k)}\left[\sum_{i=k+1}^{n}(z_i^2-1)\right]^2\right] \le [x+b(\log n)]^2\Bigg\}$$

$$= \lim_{n\to\infty} P\left\{a(\log n)\max_{1\le k<(n/\log n)}\left(k\bar{z}_k^2 + \frac{1}{2k}\left[\sum_{i=1}^{k}(z_i^2-1)\right]^2\right)^{1/2} -b(\log n)\le x\right\}$$

$$\cdot \lim_{n\to\infty} P\left\{a(\log n)\max_{n-(n/\log n)\le k<n}\{(n-k)\bar{z}_{n-k}^2\right.$$

$$+\frac{1}{2(n-k)}\left[\sum_{i=k+1}^{n}(z_i^2-1)\right]^2\}^{1/2} - b(\log n) \le x\right\}. \tag{2.62}$$

Denote the first term of (2.62) by (a) and the second by (b). Let's consider (a) first. Note that

$$k\bar{z}_k^2 + \frac{1}{2k}\left[\sum_{i=1}^{k}(z_i^2-1)\right]^2 = \left(\sum_{i=1}^{k}\frac{z_i}{\sqrt{k}}\right)^2 + \left(\sum_{i=1}^{k}\frac{z_i^2-1}{\sqrt{2k}}\right)^2.$$

Let $\mathbf{v}_i = (z_i, ((z_i^2 - 1)/\sqrt{2}))$, $1 \le i < \infty$, then $\{\mathbf{v}_i, 1 \le i < \infty\}$ is a sequence of iid d-dimensional random vectors with $d = 2$, and $\mathbf{v}_i^{(1)} = z_i$, $\mathbf{v}_i^{(2)} = (z_i^2 - 1)/\sqrt{2}$. Now, $E[\mathbf{v}_i^{(1)}] = E(z_i) = o$ for all i, $E[\mathbf{v}_i^{(2)}] = E((z_i^2 - 1/\sqrt{2})) = (1 - 1)/\sqrt{2} = o$ for all i. Hence, $E[\mathbf{v}_i^{(j)}] = o$ for $j = 1, 2$ and all i. $\mathrm{Cov}(\mathbf{v}_i^{(j)}, \mathbf{v}_i^{(\iota)}) = o$ for $1 \le j \ne \iota \le 2$. If $j = 1, \iota = 2$, $\mathrm{Cov}(\mathbf{v}_i^{(j)}, \mathbf{v}_i^{(\iota)}) = E(z_i, (z_i - 1)/\sqrt{2}) = E((z_i^3 - z_i)/\sqrt{2}) = o$. Therefore, the covariance matrix of $\mathbf{v}_i$ is the 2×2 identity matrix. And clearly, $E|\mathbf{v}_i^{(j)}|^r < \infty$ for $j = 1, 2$ and $r > 2$; Let $S_i^{(j)} = \sum_{i=1}^{k} \mathbf{v}_i^{(j)}$ for $j = 1, 2$; then

$$\sum_{j=1}^{2} \left(\frac{S_i^{(j)}}{\sqrt{k}} \right)^2 = k \bar{z}_k^2 + \frac{1}{2k} \left[\sum_{i=1}^{k} (z_i^2 - 1) \right]^2.$$

In view of Lemma 2.18, we thus obtain $(a) = \exp\{-e^{-x}\}$. Similarly, $(b) = \exp\{-e^{-x}\}$. This completes the proof of the theorem. $\qquad\square$

2.3.2 Informational Approach

(i) SICs

Under H_0, the MLEs for μ and σ^2 are

$$\widehat{\mu} = \bar{x} = \frac{1}{n} \sum_{i=1}^{n} x_i \quad \text{and} \quad \widehat{\sigma}^2 = \frac{1}{n} \sum_{i=1}^{n} (x_i - \bar{x})^2,$$

respectively. Then denoting that SIC under H_0 by SIC(n), we have:

$$\mathrm{SIC}(n) = n \log 2\pi + n \log \widehat{\sigma}^2 + n + 2 \log n. \tag{2.63}$$

Under H_1, we use SIC(k) to denote the SICs, for $2 \le k \le n - 2$. Then after some simple computations, we have:

$$\mathrm{SIC}(k) = n \log 2\pi + k \log \widehat{\sigma}_1^2 + (n - k) \log \widehat{\sigma}_n^2 + n + 4 \log n, \tag{2.64}$$

where $\widehat{\sigma}_1^2 = (1/k) \sum_{i=1}^{k} (x_i - \bar{x}_k)^2$, $\bar{x}_k = (1/k) \sum_{i=1}^{k} x_i$, $\widehat{\sigma}_n^2 = (1/(n-k)) \sum_{i=k+1}^{n} (x_i - \bar{x}_{n-k})^2$, and $\bar{x}_{n-k} = (1/(n-k)) \sum_{i=k+1}^{n} x_i$ are the MLEs for $\widehat{\sigma}_1^2$, μ_1, σ_n^2, and μ_n, respectively. Now, we estimate k by $\widehat{k}$ such that

$$\mathrm{SIC}(\widehat{k}) = \min_{2 \le k \le n-2} \{\mathrm{SIC}(k)\}. \tag{2.65}$$

It is noted that in order to obtain the MLEs, we can only detect change that is located at k for $2 \le k \le n - 2$.

(ii) Asymptotic Null Distribution

Let $\Delta_n = \min_{2 \leq k \leq n-2}[\text{SIC}(k) - \text{SIC}(n)]$. The asymptotic distribution of a function of Δ_n is given in the following theorem. Note that

$$\Delta_n = - \max_{2 \leq k \leq n-2}[\text{SIC}(k) - \text{SIC}(n)]$$

$$= \lambda_n^2 + 2 \log n,$$

where

$$\lambda_n^2 = \left[\max_{2 \leq k \leq n-2} \langle n \log \widehat{\sigma}^2 - k \log \widehat{\sigma}_1^2 - (n-k) \log \widehat{\sigma}_n^2 \rangle \right]^{1/2}.$$

$\lambda_n = (2 \log n - \Delta_n)^{1/2}$, thus we have the following.

Theorem 2.27 *Under H_0, for all $x \in R$,*

$$\lim_{n \to \infty} P[a(\log n)(2 \log n - \Delta_n)^{1/2} - b(\log n) \leq x] = \exp(-2e^{-x}), \qquad (2.66)$$

where $a(\log n) = (2 \log \log n)^{1/2}$, and $b(\log n) = 2 \log \log n + \log \log \log n$.

Proof. This is an immediate corollary of Theorem 2.26. $\qquad \square$

We point out (see Gupta and Chen, 1996) that information criteria, such as SIC, provide a remarkable way for exploratory data analysis with no need to resort to either the distribution or the significant level α. On the other hand, when the SICs are very close, one may question that the small difference among the SICs might be caused by the fluctuation of the data, and therefore there may be no change at all. To make the conclusion about change point statistically convincing, we introduce the significant level α and its associated critical value c_α. Instead of accepting H_0 when $\text{SIC}(n) < \min_{2 \leq k \leq n-2} \text{SIC}(k)$, we accept H_0, if $\text{SIC}(n) < \min_{2 \leq k \leq n-2} \text{SIC}(k) + c_\alpha$, where c_α and α have the relationship:

$$1 - \alpha = P[\text{SIC}(n) < \min_{2 \leq k \leq n-2} \text{SIC}(k) + c_\alpha | H_0 \text{ holds}]. \qquad (2.67)$$

From (2.66) and (2.67)

$$1 - \alpha = P\left\{ - \max_{2 < k < n-2}[\text{SIC}(n) - \text{SIC}(k)] > -c_\alpha | H_0 \text{ holds} \right]$$

$$= P[\Delta_n > -c_\alpha | H_0 \text{ holds}]$$

$$= P[-\lambda_n^2 + 2 \log n > -c_\alpha | H_0 \text{ holds}]$$

$$= P[0 < \lambda_n < (c_\alpha + 2 \log n)^{1/2} | H_0 \text{ holds}]$$

$$= P[-b(\log n) < a(\log n)\lambda_n - b(\log n)$$
$$< a(\log n)(c_\alpha + 2\log n)^{1/2} - b(\log n)|H_0 \text{ holds}]$$
$$\cong \exp\{-2\exp[b(\log n) - a(\log n)(c_\alpha + 2\log n)^{1/2}]\}$$
$$- \exp\{-2\exp(b\log n)\}.$$

Hence,

$$\exp\{-2\exp[b(\log n) - a(\log n)(c_\alpha + 2\log n)^{1/2}]\}$$
$$\cong 1 - \alpha + \exp\{-2\exp[b(\log n)]\}.$$

Solving for c_α, we obtain:

$$c_\alpha \cong \left\{ -\frac{1}{a(\log n)} \log\log[1 - \alpha + \exp(-2\exp(b(\log n)))]^{-(1/2)} \right.$$
$$\left. + \frac{b(\log n)}{a(\log n)} \right\}^2 - 2\log n.$$

For different significant levels $\alpha = 0.01, 0.025, 0.05,$ and 0.1, and different sample sizes $n = 7, \ldots, 200$, we computed the critical values for SICs, and listed them in Table 2.4.

(iii) Unbiased SICs

Recall from previous sections, we mentioned that to derive the information criterion AIC, Akaike (1973) used $\log L(\widehat{\theta})$ as an estimate of $J = E_{\widehat{\theta}}[\int f(\mathbf{y}|\theta_0)\log f(\mathbf{y}|\widehat{\theta})d\mathbf{y}]$, where $f(\mathbf{y}|\theta_0)$ is the probability density of the future observations $\mathbf{y} = (y_1, y_2, \ldots, y_n)$ of the same size and distribution as the xs, $\mathbf{x} = (x_1, x_2, \ldots, x_n)$, and $\mathbf{x}$ and $\mathbf{y}$ are independent. The expectation is taken under the distribution of $\mathbf{x}$ when H_0 is true; that is, $\theta_0 \in H_0$. Unfortunately, $\log L(\widehat{\theta})$ is not an unbiased estimator of J. When the sample size n is finite, Sugiura (1978) proposed unbiased versions, finite corrections of AIC, for different model selection problems.

In this section, we derive the unbiased version of SIC under our H_0 and H_1, denoted by $u - \text{SIC}(H_i), i = 0, 1$.
(1) $u - \text{SIC}(H_0)$

$$J = E_{\widehat{\theta}}[E_{\theta_0|y}(\log L(\widehat{\theta}))]$$
$$= E_{\mathbf{x}}\left[E_{\mathbf{y}} \left\{ -\frac{1}{2}n\log 2\pi - \frac{n}{2}\log\widehat{\sigma}^2 - \frac{1}{2}\sum_{i=1}^{n}\frac{(y_i - \overline{x})^2}{\widehat{\sigma}^2} \right\} \right]$$
$$= E_{\mathbf{x}} \left\{ -\frac{1}{2}n\log 2\pi - \frac{n}{2}\log\widehat{\sigma}^2 - \frac{n}{2} + \frac{n}{2} - \frac{1}{2}E_{\mathbf{y}}\sum_{i=1}^{n}\frac{(y_i - \overline{x})^2}{\widehat{\sigma}^2} \right\}.$$

Table 2.4 Approximate Critical Values of SIC

n/α	0.010	0.025	0.050	0.100
7	35.699	19.631	12.909	7.758
8	25.976	17.232	11.925	7.405
9	23.948	16.423	11.540	7.262
10	23.071	15.994	11.313	7.168
11	22.524	15.691	11.139	7.087
12	22.108	15.445	10.989	7.010
13	21.763	15.233	10.854	6.936
14	21.463	15.044	10.731	6.863
15	21.198	14.873	10.617	6.793
16	20.960	14.717	10.511	6.725
17	20.744	14.574	10.411	6.660
18	20.546	14.441	10.317	6.597
19	20.364	14.317	10.228	6.536
20	20.195	14.201	10.144	6.477
21	20.038	14.092	10.064	6.420
22	19.891	13.989	9.988	6.364
23	19.753	13.892	9.916	6.311
24	19.623	13.799	9.846	6.259
25	19.501	13.711	9.779	6.209
26	19.384	13.627	9.715	6.160
27	19.274	13.547	9.653	6.113
28	19.169	13.470	9.593	6.067
29	19.069	13.397	9.536	6.023
30	18.973	13.326	9.480	5.979
35	18.548	13.008	9.227	5.778
40	18.193	12.737	9.008	5.600
45	17.888	12.501	8.814	5.439
50	17.622	12.292	8.640	5.293
55	17.386	12.104	8.482	5.160
60	17.173	11.934	8.338	5.036
70	16.804	11.635	8.082	4.815
80	16.490	11.377	7.859	4.620
90	16.218	11.151	7.662	4.446
100	15.977	10.950	7.486	4.289
120	15.567	10.604	7.179	4.015
140	15.225	10.313	6.919	3.780
160	14.933	10.061	6.693	3.574
180	14.678	9.840	6.493	3.391
200	14.451	9.643	6.313	3.227

Notice that, because the x_is and y_is are independent, and are all distributed as $N(\mu, \sigma^2)$, we have $y_i - \overline{x} \sim N(0, ((n+1)/n)\sigma^2)$. Therefore,

$$\frac{n}{n+1} \sum_{i=1}^{n} \frac{(y_i - \overline{x})^2}{\widehat{\sigma}^2} \sim \chi_n^2,$$

and hence

$$J = E_{\widehat{\theta}}\left[\log L(\widehat{\theta}) + \frac{n}{2} - \frac{1}{2}\frac{(n+1)\sigma^2}{\widehat{\sigma}^2}\right]$$

$$= E_{\widehat{\theta}}[\log L(\widehat{\theta}) + \frac{n}{2} - \frac{n+1}{2}E_{\widehat{\theta}}\left(\frac{\sigma^2}{\widehat{\sigma}^2}\right)$$

$$= E_{\widehat{\theta}}[\log L(\widehat{\theta})] + \frac{n}{2} - \frac{n+1}{2}\frac{n}{n-3}$$

$$= E_{\widehat{\theta}}[\log L(\widehat{\theta})] - \frac{2n}{n-3}.$$

Therefore, $\log L(\widehat{\theta}) - 2n/(n-3)$ is unbiased for J, or $-2\log L(\widehat{\theta}) + 4n/(n-3)$ is unbiased for $-2J$. We have

$$u - \text{SIC}(H_0) = -2\log L(\widehat{\theta}) + \frac{4n}{n-3}$$

$$= \text{SIC}(n) + \frac{4n}{n-3} - 2\log n.$$

(2) $u - \text{SIC}(H_1)$

$$J = E_{\widehat{\theta}}\left[E_{\mathbf{y}}\left\{-\frac{1}{2}n\log 2\pi - \frac{k}{2}\log\widehat{\sigma}_1^2 - \frac{n-k}{2}\log\widehat{\sigma}_n^2 - \frac{1}{2}\sum_{i=1}^{k}\frac{(y_i - \overline{x}_k)^2}{\widehat{\sigma}_1^2}\right.\right.$$

$$\left.\left. -\frac{1}{2}\sum_{i=k+1}^{n}\frac{(y_i - \overline{x}_{n-k})^2}{\widehat{\sigma}_1^2}\right\}\right]$$

$$= E_{\widehat{\theta}}\left\{-\frac{1}{2}n\log 2\pi - \frac{k}{2}\log\widehat{\sigma}_1^2 - \frac{n-k}{2}\log\widehat{\sigma}_n^2 - \frac{n}{2} + \frac{n}{2}\right.$$

$$\left. -\frac{1}{2}E_{\mathbf{y}}\left[\sum_{i=1}^{k}\frac{(y_i - \overline{x}_k)^2}{\widehat{\sigma}_1^2}\right] - \frac{1}{2}E_{\mathbf{y}}\left[\sum_{i=k+1}^{n}\frac{(y_i - \overline{x}_{n-k})^2}{\widehat{\sigma}_1^2}\right]\right\}.$$

Now,

$$y_i - \overline{x}_k \sim N(0, \frac{k+1}{k}\sigma_1^2) \quad \text{and} \quad \frac{k+1}{k}\sum_{i=1}^{k}\frac{(y_i - \overline{x}_k)^2}{\widehat{\sigma}_1^2} \sim \chi_k^2,$$

and therefore,

$$E_{\mathbf{y}}\left[\sum_{i=1}^{k}\frac{(y_i - \overline{x}_k)^2}{\widehat{\sigma}_1^2}\right] = (k+1)\frac{\sigma_1^2}{\widehat{\sigma}_1^2}.$$

Similarly,

$$E_y \left[\sum_{i=k+1}^{n} \frac{(y_i - \overline{x}_{n-k})^2}{\widehat{\sigma}_1^2} \right] = (n - k + 1) \frac{\sigma_1^2}{\widehat{\sigma}_1^2}.$$

Thus,

$$J = E_{\widehat{\theta}}[\log L(\widehat{\theta})] + \frac{n}{2} - \frac{k+1}{2} E_{\widehat{\theta}} \left[\frac{\sigma_1^2}{\widehat{\sigma}_1^2} \right] - \frac{n-k+1}{2} E_{\widehat{\theta}} \left[\frac{\sigma_1^2}{\widehat{\sigma}_1^2} \right]$$

$$= E_{\widehat{\theta}}[\log L(\widehat{\theta})] + \frac{n}{2} - \frac{k+1}{2} \frac{k}{k-3} - \frac{n-k+1}{2} \frac{n-k}{n-k-3}$$

$$= E_{\widehat{\theta}}[\log L(\widehat{\theta})] - \frac{k(k+1)(n-k-3) + (k-3)(n-k)(n-k+1)}{(k-3)(n-k-3)}$$

$$- \frac{n(k-3)(n-k-3)}{(k-3)(n-k-3)}.$$

Hence,

$$u - \mathrm{SIC}(H_1) = -2 \log L(\widehat{\theta})$$

$$+ 2 \frac{k(k+1)(n-k-3) + (k-3)(n-k)(n-k+1)}{(k-3)(n-k-3)}$$

$$- 2 \frac{n(k-3)(n-k-3)}{(k-3)(n-k-3)}.$$

(iv) Data Analysis

Example 2.1 As an application of SIC for change point analysis, we analyze the tensile strength data given in Shewhart (1931). There are 60 observations. Assume that the data are normally distributed with means $\mu_1, \mu_2, \ldots, \mu_{60}$ and variances $\sigma_1^2, \sigma_2^2, \ldots, \sigma_{60}^2$, respectively. Then we test the following hypothesis,

$$H_0 : \mu_1 = \mu_2 = \cdots = \mu_{60} = \mu \quad \text{and}$$

$$\sigma_1^2 = \sigma_2^2 = \cdots = \sigma_{60}^2 = \sigma^2,$$

versus the alternative hypothesis

$$H_1 : \mu_1 = \cdots = \mu_k \neq \mu_{k+1} = \cdots = \mu_{60} \quad \text{and}$$

$$\sigma_1^2 = \cdots = \sigma_k^2 \neq \sigma_{k+1}^2 = \cdots = \sigma_{60}^2.$$

Using the method developed here, we obtain the SIC(n), and SIC(k), for $2 \leq k \leq n - 1$, and list them in Table 2.5 along with the original data values, where the starred value is the minimum SIC value. Clearly,

Table 2.5 SIC Values for the Tensile Strength Data

x_k	k	SIC(k)	x_k	k	SIC(k)	x_k	k	SIC(k)
29314	1	–	25770	21	1180.2	29668	41	1177.7
34860	2	1181.7	23690	22	1177.3	32622	42	1179.1
36818	3	1181.4	28650	23	1177.4	32822	43	1179.6
30120	4	1181.2	32380	24	1178.6	30380	44	1178.5
34020	5	1180.7	28210	25	1177.3	38580	45	1183.2
30824	6	1180.2	34002	26	1178.9	28202	46	1178.7
35396	7	1179.8	34470	27	1179.9	29190	47	1177.9
31260	8	1179.1	29248	28	1178.8	35636	48	1182.0
32184	9	1178.2	28710	29	1178.0	34332	49	1182.6
33424	10	1177.1	29830	30	1177.9	34750	50	1183.9
37694	11	1177.8	29250	31	1177.1	40578	51	1189.3
34876	12	1176.3	27992	32	1175.6	28900	52	1184.4
24660	13	1180.2	31852	33	1176.5	34648	53	1188.1
34760	14	1180.3	27646	34	1174.1	31244	54	1186.9
38020	15	1179.9	31698	35	1174.9	33802	55	1188.1
25680	16	1180.6	30844	36	1174.3	34850	56	1188.7
25810	17	1181.5	31988	37	1174.6	36690	57	1189.6
26460	18	1181.8	36640	38	1177.6	32344	58	1187.5
28070	19	1182.0	41578	39	1181.6	34440	59	–
24640	20	1181.0	30496	40	1178.5	34650	60	1172.6*

*Indicates the minimum SIC value

$\min_{2 \le k \le 58} \text{SIC}(k) = \text{SIC}(34) = 1174.1$. Then $\text{SIC}(n) = \text{SIC}(60) = 1172.6 < \min_{2 \le k \le 58} \text{SIC}(k)$, and these two values are very close. What decision should we make? Use our Table 2.4, for any α, because $c_\alpha > 0$, then $\text{SIC}(n) < \min_{2 \le k \le 58} \text{SIC}(k) + c_\alpha$. Therefore, we fail to reject H_0, and conclude that there is no change in both mean and variance of the tensile strength. This conclusion matches the one drawn in Shewhart (1931).

2.3.3 Application to Biomedical Data

We show an application of the mean and variance change point model to the analysis of aCGH data introduced in Section 2.1.3. In Linn et al. (2003) and Olshen et al. (2004), DNA copy number changes were viewed as a mean change point model (MCM) with a fixed variance in the distributions of the sequence $\{X_i\}$. As pointed out by Hodgson et al. (2001), the aCGH technology may not guarantee the aCGH data to have a constant variance; it is more reasonable to analyze the DNA copy number changes using the mean and variance change model (MVCM) proposed in Chen and Wang (2009) for the distributions of the sequence $\{X_i\}$. Observing the following normalized log-ratio intensities obtained through aCGH experiments of Lucito et al. (2003) on breast cancer cell line SK-BR-3 (see Figure 2.13), it is evident that both mean and variance of the sequence have changed.

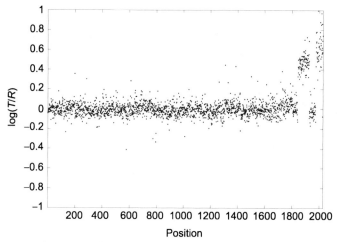

Fig. 2.13 Genome of the fibroblast cell line GM07408 Snijders et al. (2001)

The multiple DNA copy number changes in the sequence of log ratio intensities can be defined as the hypothesis testing problem stated earlier in this section (see (2.32)). Specifically, using BSP, we just need to focus on how to detect the single change (the most significant one), specified by testing (2.32) versus (2.33), each time and repeat the searching scheme of BSP to get all the significant changes. Here, μ and σ^2 are the unknown common mean and variance under the null hypothesis, and k, $1 < k < n$, is the unknown position of the single change at each single stage. For a given significance level α, when H_0 is not rejected, there is no change in the DNA copy number sequence and the search scheme stops at this stage. If H_0 is rejected at a given significance level α, there is a significant change in the DNA copy number sequence and the search scheme of the BSP continues until no more significant changes are found.

As pointed out in Chen and Wang (2009), the advantage of using the MVCM model is that MVCM leads to fewer change points than that of the mean change point model (MCM) as MCM tends to divide large segments into smaller pieces so that the homogeneous variance assumption for all segments can be met (Picard et al., 2005). Therefore, the MVCM model has the potential to give fewer false positives than MCM. Adding the variance component in the change point analysis will improve the estimation of the change point location even if just the mean shifts greatly. This is because in the MVCM model, the variances under the alternative hypothesis are estimated for each subsequence without pooling all subsequences (with possible different means) together, whereas in MCM the homogeneous variance under the alternative hypothesis is estimated by pooling all subsequences with different means together. Using either MVCM or MCM also depends on the biological experiment in which the scientists may have prior knowledge on whether there are potential variance changes. In that case, the MVCM model is proposed as an alternative to MCM when possible variance changes exist in the sequence.

To carry out the hypothesis testing of the null hypothesis (2.32), which claims no DNA copy number changes, versus the alternative hypothesis (2.33), the research hypothesis that there is a change in the mean and variance and hence a change in the DNA copy number, Chen and Wang introduced the SIC-based procedure along with an approximate p-value given by

$$p - \text{value} = 1 - \exp\{-2\exp[b(\log n) - a(\log n)\lambda_n^{1/2}]\}, \qquad (2.68)$$

where $\lambda_n = 2\log n - \Delta_n$, $\Delta_n = \min_{2 \leq k \leq n-2}[\text{SIC}(k) - \text{SIC}(n)]$, and $\text{SIC}(k)$ and $\text{SIC}(n)$ are given by (2.63) and (2.64), respectively.

The applications of the SIC method to the detection of change point loci in the 15 fibroblast cell lines (Snijders et al., 2001) and other known aCGH data are given in Chen and Wang (2009). There are also comparisons of using the mean and variance change point model with the CBS method which is based on a mean change point model in Chen and Wang (2009).

There are important aCGH copy number experiments conducted by Snijders et al. (2001) on 15 fibroblast cell lines, namely GM03563, GM00143, GM05296, GM07408, GM01750, GM03134, GM13330, GM03576, GM01535, GM07081, GM02948, GM04435, GM10315, GM13031, and GM01524, and the obtained aCGH data on the genome of all such cell lines are regarded as benchmark aCGH datasets. There are many different computational and statistical methodology research articles published on how to analyze such aCGH datasets. The change point methods, CBS and MVCM, which were used for the analysis of the fibroblast aCGH data, were compared in Chen and Wang (2009) in terms of the change loci identified, the sensitivity, and specificity of the two methods. Two applications of the SIC approach are presented below and a predetermined significant level of $\alpha = .001$ is used.

The first one is a chromosomewide copy number change search using SIC in MVCM on chromosome 4 of the fibroblast cell line GM13330. There are 167 genomic positions on which log base 2 ratio of intensities were recorded. The SIC values at all of the genomic locations were calculated according to expressions (2.63) and (2.64). The minimum SIC occurred at location index 150 with minSIC $= -299.8695$ and corresponding p-value (according to (2.68)) of 6.465314×10^{-9}. The graph of SIC values for this chromosome is given in Figure 2.14. Transferring back to the log ratio intensities, a scatterplot of the log ratio intensities of chromosome 4 of the fibroblast cell line GM13330 is provided as Figure 2.15 with the red circle indicating the change point identified.

The second application is a genomewide CNV search using SIC in MVCM on the cell line GM07408. It is found that the minimum SIC value of the whole sequence of 2027 log ratio intensity values occurs at locus 1841 with the p-value of 0.00000. The BSP is applied to the searching process. For the subsequence containing the first through the 1841st observations, the search

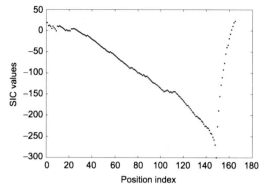

Fig. 2.14 SIC values for every locus on chromosome 4 of the fibroblast cell line GM13330

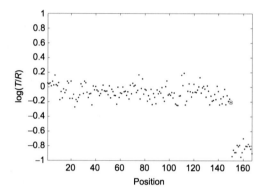

Fig. 2.15 Chromosome 4 of the fibroblast cell line GM13330 (Snijders et al., 2001)

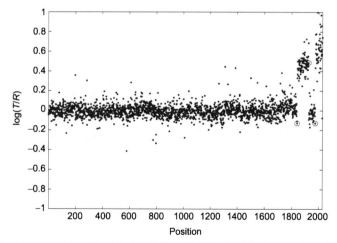

Fig. 2.16 Genome of the fibroblast cell line GM07408 with changes identified by red circles

locates no significant CNV, and for the subsequence consisting of the 1842nd through the 2027th log ratio intensity value, the minimum SIC occurs at the 1927th locus of the original sequence with the p-value of 5.246360×10^{-5}. After the identification of the 1927th change location, the subsequence is further broken into two subsubsequences and a third change is found at locus 1975 with the p-value of 2.75359×10^{-6}. These three loci are circled as red in the scatterplot, Figure 2.16, of the genome of the fibroblast cell line GM07408.

Chapter 3
Multivariate Normal Model

In Chapter 2, we have discussed the inferences about change point(s) for a univariate normal model in different situations. In this chapter, we investigate change point(s) problems when the underlying distribution is a multivariate normal distribution.

3.1 Mean Vector Change

Let $\mathbf{x}_1, \mathbf{x}_2, \ldots, \mathbf{x}_n$ be a sequence of independent m-dimensional normal random vectors with parameters $(\mu_1, \Sigma_1), (\mu_2, \Sigma_2), \ldots, (\mu_n, \Sigma_n)$, respectively. Assume $\Sigma_1 = \Sigma_2 = \cdots = \Sigma_n = \Sigma$ and Σ is unknown. We are concerned with any changes that might be presented in the sequence. In other words, it is desired to test the following hypothesis,

$$H_0 : \mu_1 = \mu_2 = \cdots = \mu_n = \mu \text{ (unknown)} \tag{3.1}$$

versus the alternative:

$$H_A : \mu_1 = \cdots = \mu_{k_1} \neq \mu_{k_1+1} = \cdots = \mu_{k_2} \neq \cdots \neq \mu_{k_q+1} = \cdots = \mu_{k_{q+1}} = \mu_n,$$

where $q, 1 \leq q \leq n-1$, is the unknown number of changes and $k_1, k_2, \ldots, k_q$ are the unknown positions of the change points. Similar to the discussion in Chapter 2, by means of the binary segmentation procedure, we just need to test the single change point hypothesis and then repeat the procedure for each subsequence. Hence, we turn to the testing of (3.1) against the alternative:

$$H_1 : \mu_1 = \cdots = \mu_k \neq \mu_{k+1} = \cdots = \mu_n \tag{3.2}$$

where k now represents the position of the single change point at each stage, $1 \leq k \leq n-1$. This single change point hypothesis about the mean vector change has been studied by some authors in the literature. Sen and

Srivastava (1973) studied the problem of a single mean vector change for a sequence of independent normal random vectors using a Bayesian test statistic, and derived the exact and asymptotic null distribution of the test statistic. Srivastava and Worsley (1986) applied likelihood ratio tests for detecting a change in the mean vector of a sequence of independent normal random vectors. Zhao, Krishnaiah, and Bai (1986a,b) studied the problem of detection of the number of signals in the presence of white noise and when the noise covariance matrix is arbitrary. Krishnaiah, Miao, and Zhao (1990) suggested a local likelihood method for estimating the change point in the multivariate normal mean. James, James, and Siegmund (1992) obtained asymptotic approximations for likelihood ratio tests and confidence regions for the change in the multivariate normal mean.

In this chapter, we discuss the detection of the change point in multivariate normal means by the likelihood-ratio procedure approach and information criterion approach.

3.1.1 Likelihood-Ratio Procedure

(i) The Test Statistic

Under H_0, the likelihood function is

$$L_0(\mu, \Sigma) = (2\pi)^{-mn/2} |\Sigma|^{-n/2} \exp\left\{ -\frac{1}{2} \sum_{i=1}^{n} (\mathbf{x}_i - \mu)' \Sigma^{-1} (\mathbf{x}_i - \mu) \right\},$$

and the MLEs of μ and Σ are

$$\widehat{\mu} \equiv \overline{\mathbf{x}} = \frac{1}{n} \sum_{i=1}^{n} \mathbf{x}_i, \qquad \widehat{\Sigma} = \frac{1}{n} \sum_{i=1}^{n} (\mathbf{x}_i - \mu)(\mathbf{x}_i - \mu)'.$$

The maximum likelihood under H_0 is then

$$L_0(\widehat{\mu}, \widehat{\Sigma}) = (2\pi)^{-mn/2} |\widehat{\Sigma}|^{-n/2} e^{-mn/2}.$$

Under H_1, the likelihood function is

$$L_1(\mu_1, \mu_n, \Sigma_1)$$
$$= (2\pi)^{-mn/2} |\Sigma_1|^{-n/2}$$
$$\cdot \exp\left\{ -\frac{1}{2} \left[\sum_{i=1}^{k} (\mathbf{x}_i - \mu_1)' \Sigma_1^{-1} (\mathbf{x}_i - \mu_1) + \sum_{i=k+1}^{n} (\mathbf{x}_i - \mu_n)' \Sigma_1^{-1} (\mathbf{x}_i - \mu_n) \right] \right\},$$

and the MLEs of μ_1, μ_n, Σ_1 for $m < k < n - m$ are

$$\widehat{\mu}_1 \equiv \overline{\mathbf{x}}_k = \frac{1}{k} \sum_{i=1}^{k} \mathbf{x}_i,$$

$$\widehat{\mu}_n \equiv \overline{\mathbf{x}}_{n-k} = \frac{1}{n-k} \sum_{i=k+1}^{n} \mathbf{x}_i,$$

$$\widehat{\Sigma}_1 = \frac{1}{n} \left[\sum_{i=1}^{k} (\mathbf{x}_i - \overline{\mathbf{x}}_k)(\mathbf{x}_i - \overline{\mathbf{x}}_k)' + \sum_{i=k+1}^{n} (\mathbf{x}_i - \overline{\mathbf{x}}_{n-k})(\mathbf{x}_i - \overline{\mathbf{x}}_{n-k})' \right].$$

Then the maximum likelihood under H_1 is

$$L_1(\widehat{\mu}_1, \widehat{\mu}_n, \widehat{\Sigma}_1) = (2\pi)^{-mn/2} |\widehat{\Sigma}_1|^{-n/2} e^{-mn/2}.$$

For fixed k, the alternative can be viewed as claiming that a sample of size k is from a normal distribution with mean vector μ_1, and an independent sample of size $n - k$ is from a normal distribution with mean vector μ_n, where $\mu_1 \neq \mu_n$. Then the Hotelling's T^2 test can be used, which is based on the likelihood-ratio test. Let the standardized difference between the two samples (before and after the change point k) be denoted by

$$\mathbf{y}_k = \sqrt{\frac{k(n-k)}{n}} (\overline{\mathbf{x}}_k - \overline{\mathbf{x}}_{n-k}),$$

and also let

$$W_k = \frac{1}{n-2} \left[\sum_{i=1}^{k} (\mathbf{x}_i - \overline{\mathbf{x}}_k)(\mathbf{x}_i - \overline{\mathbf{x}}_k)' + \sum_{i=k+1}^{n} (\mathbf{x}_i - \overline{\mathbf{x}}_{n-k})(\mathbf{x}_i - \overline{\mathbf{x}}_{n-k})' \right].$$

Then the Hotelling's T^2 test statistic for testing the hypotheses is

$$T_k^2 = \mathbf{y}_k' W_k^{-1} \mathbf{y}_k, \quad \text{for } k = 1, \ldots, n-1,$$

and H_0 is rejected for

$$\max_{1 \leq k \leq n-1} T_k^2 > c,$$

where c is a constant to be determined by the null distribution of $\max_{1 \leq k \leq n-1} T_k^2$, and the unknown position of the change point k is estimated by $\widehat{k}$ such that

$$T_{\widehat{k}}^2 = \max_{1 \leq k \leq n-1} T_k^2.$$

Many authors tried to obtain the null distribution of $\max_{1 \leq k \leq n-1} T_k^2$ in the past, and obtained different approximations. We present a relatively good approximation to the null distribution of a function of $\max_{1 \le k \le n-1} T_k^2$ due to Srivastava and Worsley (1986). We need the following basic lemmas.

Lemma 3.1 *Let* $S_k = \mathbf{y}_k' V^{-1} \mathbf{y}_k$, *with* $V = \sum_{i=1}^{n} (\mathbf{x}_i - \overline{\mathbf{x}})(\mathbf{x}_i - \overline{\mathbf{x}})'$, *for* $k = 1, \ldots, n-1$; *then*

$$S_k = \frac{T_k^2}{n - 2 + T_k^2}.$$

Proof. Inasmuch as

$$\frac{T_k^2}{n - 2 + T_k^2} = \frac{\mathbf{y}_k' W_k^{-1} \mathbf{y}_k}{n - 2 + \mathbf{y}_k' W_k^{-1} \mathbf{y}_k}$$

$$= \frac{\mathbf{y}_k' W_k^{*-1} \mathbf{y}_k}{1 + \mathbf{y}_k' W_k^{*-1} \mathbf{y}_k},$$

where $W_k^* = (n - 2) W_k$, it suffices to show that

$$\frac{\mathbf{y}_k' W_k^{*-1} \mathbf{y}_k}{1 + \mathbf{y}_k' W_k^{*-1} \mathbf{y}_k} = \mathbf{y}_k' V^{-1} \mathbf{y}_k.$$

After some algebra, it is seen that

$$\overline{\mathbf{x}} = \frac{k \overline{\mathbf{x}}_k + (n - k) \overline{\mathbf{x}}_{n-k}}{n},$$

$$\sum_{i=1}^{k} (\mathbf{x}_i - \overline{\mathbf{x}}_k)(\mathbf{x}_i - \overline{\mathbf{x}}_k)' = \sum_{i=1}^{k} (\mathbf{x}_i - \overline{\mathbf{x}})(\mathbf{x}_i - \overline{\mathbf{x}})'$$
$$- (n - k)[(\overline{\mathbf{x}} - \overline{\mathbf{x}}_{n-k})(\overline{\mathbf{x}} - \overline{\mathbf{x}}_{n-k})$$
$$+ (\overline{\mathbf{x}}_k - \overline{\mathbf{x}})(\overline{\mathbf{x}} - \overline{\mathbf{x}}_{n-k})'$$
$$- (\overline{\mathbf{x}} - \overline{\mathbf{x}}_{n-k})(\overline{\mathbf{x}} - \overline{\mathbf{x}}_{n-k})'],$$

$$\sum_{i=k+1}^{n} (\mathbf{x}_i - \overline{\mathbf{x}}_{n-k})(\mathbf{x}_i - \overline{\mathbf{x}}_{n-k})' = \sum_{i=k+1}^{n} (\mathbf{x}_i - \overline{\mathbf{x}})(\mathbf{x}_i - \overline{\mathbf{x}})'$$
$$- k[(\overline{\mathbf{x}} - \overline{\mathbf{x}}_k)(\overline{\mathbf{x}}_{n-k} - \overline{\mathbf{x}})'$$
$$+ (\overline{\mathbf{x}}_{n-k} - \overline{\mathbf{x}})(\overline{\mathbf{x}} - \overline{\mathbf{x}}_k)'$$
$$- (\overline{\mathbf{x}} - \overline{\mathbf{x}}_k)(\overline{\mathbf{x}} - \overline{\mathbf{x}}_k)'],$$

$$W_k^* = V - \mathbf{y}_k \mathbf{y}_k',$$

and hence

$$V - W_k^* - \mathbf{y}_k \mathbf{y}_k' = 0.$$

Therefore,

$$\mathbf{y}_k' W_k^{*-1} [V - W_k^* - \mathbf{y}_k \mathbf{y}_k'] V^{-1} \mathbf{y}_k = 0,$$

which leads to

$$\mathbf{y}_k' W_k^{*-1} \mathbf{y}_k - \mathbf{y}_k' V^{-1} \mathbf{y}_k - (\mathbf{y}_k' W_k^{*-1} \mathbf{y})(\mathbf{y}_k' V^{-1} \mathbf{y}_k) = 0,$$

and finally

$$\frac{\mathbf{y}_k' W_k^{*-1} \mathbf{y}_k}{1 + \mathbf{y}_k' W_k^{*-1} \mathbf{y}_k} = \mathbf{y}_k' V^{-1} \mathbf{y}_k. \qquad \square$$

S_k is increasing in T_k^2, therefore equivalently, H_0 is rejected for

$$\max_{1 \le k \le n-1} S_k > c,$$

where c is a constant to be determined by the null distribution of $\max_{1 \le k \le n-1} S_k$, and the unknown position of the change point k is estimated by $\hat{k}$ such that

$$S_{\hat{k}} = \max_{1 \le k \le n-1} S_k.$$

To be able to obtain the c values, we need to know the null distribution of $S_{\hat{k}}$.

(ii) Approximate Null Distribution of $S_{\hat{k}}$

An approximation to the null distribution of $S_{\hat{k}}$ is given in this section. We first present the following lemma.

Lemma 3.2 *Under H_0, the distribution of S_k is beta with parameters $\frac{1}{2}m$, and $\frac{1}{2}(n - m - 1)$, for $k = 1, \ldots, n - 1$.*

Proof. Let

$$Y = \frac{T_k^2}{n-2} \cdot \frac{n - m - 1}{m};$$

then from Anderson (1984), under H_0, Y is distributed as a central F-distribution with parameters m and $n - m - 1$.

Clearly,

$$S_k = \frac{mY}{n - m - 1 + mY},$$

and from the variable transformation theory, the pdf of S_k is found as

$$f_{S_k}(s) = \frac{\Gamma(\frac{n-1}{2})}{\Gamma(\frac{m}{2})\Gamma(\frac{n-m-1}{2})} s^{m/2-1}(1 - s)^{(n-m-1)/2-1} I_{(0,1)}(s).$$

Therefore the lemma is proven. $\square$

There are different approximations to the null distribution of $S_{\hat{k}}$. One popular approximation is from the Bonferroni inequality. Srivastava and

Worsley (1986) derived an approximate null distribution of $S_{\widehat{k}}$ which improves the Bonferroni inequality approximation.

Theorem 3.3 *Under H_0,*

$$P(S_{\widehat{k}} > c) \lessapprox 1 - G_{m,\nu}(c) + q_1 \sum_{k=1}^{n-2} t_k - q_2 \sum_{k=1}^{n-2} t_k^3,$$

where

$$\nu = \frac{n - m - 1}{2},$$

$$q_1 = g_{m,\nu}\{2c(1-c)/\pi\}^{1/2}\Gamma\{(m+\nu-1)/2\}/\Gamma\{(m+\nu)/2\},$$

$$q_2 = q_1\{(m^2 - 1)/c + (\nu^2 - 1)/(1 - c) - (m+\nu)(m+\nu-1)\}$$
$$/\{12(m+\nu)\},$$

$g_{m,\nu}(\cdot)$ *is the pdf of beta* $(m/2, \nu/2)$, *and* $G_{m,\nu}(\cdot)$ *is the cdf of beta* $(m/2, \nu/2)$.

Proof. The derivation of this theorem is given in Srivastava and Worsley (1986). □

3.1.2 Informational Approach

(i) Expressions of SIC

As before, the information criterion such as SIC provides a remarkable way to locate the change point when the inference of a change point is viewed as a model selection problem. We derive the SIC under H_0, denoted by SIC(n), as follows.

$$\text{SIC}(n) = mn \log 2\pi + mn + n \log |\widehat{\Sigma}| + \frac{1}{2}m(m+3) \log n,$$

where $\widehat{\Sigma}$ is the MLE of Σ obtained in previous section.

Similarly, we derive the SIC under H_1, denoted by SIC(k), for $1 \leq k \leq n$, as

$$\text{SIC}(k) = mn \log 2\pi + mn + n \log |\widehat{\Sigma}_1| + \frac{1}{2}m(m+5) \log n,$$

where $\widehat{\Sigma}_1$ is the MLE of Σ_1 obtained in the previous section.

3.1.3 *Applications to Geology and Literature Data*

In this section, we apply the results obtained in the previous section to two real-world problems of searching for mean vector change points as presented in Gupta and Chen (1996).

Example 3.1 We first apply the SIC procedure to a set of geological data and estimate the change points. The reader is referred to Chernoff (1973) for the original dataset.

Chernoff analyzed a sequence of assays of seven mineral contents of a 4500-foot core drilled from a Colorado mountain side by the "faces" method, and visually estimated the number and location of change points. Srivastava and Worsley (1986) analyzed the same data by using the likelihood-ratio test based on the maximum Hotelling's T^2 and found the change points at different significance levels. As in Srivastava and Worsley, we also choose $m = 5$ variables z_1, z_8, z_9, z_{10}, and z_{12} which are of the highest assays. So we have a sample of size $n = 53$ of five-dimensional independent normal random vectors. In our calculations, we drop constants $mn \log 2\pi$ and mn from the expressions of SIC(n) and SIC(k). The computational results are given in Table 3.1. Please note that in Table 3.1, $*$ indicates where the minimum SIC

Table 3.1 SIC Values of Example 3.1

Obs.	1–53	1–24	1–18	1–12			
1	2541.8	1051.3	697.0	453.0			
2	2538.5	1050.2	695.9	456.1			
3	2534.5	1050.3	696.7	458.0			
4	2534.4	1049.4	694.6	454.8			
5	2527.7	1043.8	690.0	450.6			
6	2515.8	1031.8	679.5	441.5*			
7	2510.3	1027.6	676.8	444.3			
8	2510.0	1029.0	681.4	449.5			
9	2506.4	1032.2	683.1	454.1			
10	2504.8	1032.9	679.5	446.4			
11	2501.1	1033.0	680.3	444.5			
12	2501.7	1032.4	675.9*	447.5			
13	2499.3	1033.7	683.8				
14	2497.6	1031.3	685.4				
15	2493.7	1035.5	694.2				
16	2494.3	1028.0	695.7				
17	2486.4	1021.2	694.6				
18	2473.4	1000.1*	686.2				
19	2464.0	1015.6					
20	2455.1	1001.7					
21	2458.9	1023.4					
22	2452.6	1042.3					
23	2446.3	1033.9					
24	2439.5*	1039.7					

Table 3.1 Continued

Obs.	1–53	25–53	25–32				
25	2454.7	1374.6	327.7				
26	2448.4	1354.4	314.0				
27	2446.5	1352.6	312.2				
28	2457.9	1352.7	304.6*				
29	2468.5	1356.6	308.4				
30	2474.4	1352.5	318.3				
31	2477.7	1339.5	316.8				
32	2475.4	1326.1*	322.3				
			33–53				
33	2466.4	1339.9	958.0				
34	2462.2	1336.4	946.0*				
				35–53	35–46	35–43	35–41
35	2469.6	1343.7	956.5	862.9	548.1	404.6	271.6
36	2474.6	1340.5	948.5	852.2	538.8	398.5	255.6
37	2481.9	1349.5	956.5	856.5	541.4	390.3	252.0
38	2495.0	1357.5	961.5	859.4	547.4	396.6	264.5
39	2507.4	1364.5	964.1	860.6	550.4	397.1	249.5*
40	2508.2	1361.6	961.3	857.4	549.3	385.2	266.6
41	2509.3	1357.8	958.2	853.6	543.8	380.7*	270.4
42	2516.5	1366.6	963.1	857.8	540.4	390.7	
43	2521.4	1368.5	957.8	852.1	531.2*	398.5	
44	2524.8	1370.3	956.8	851.2	532.6		
45	2525.3	1369.3	954.9	849.3	543.2		
46	2523.1	1366.9	954.4	847.9*	540.2		
					47–53		
47	2524.6	1367.1	957.2	850.8	272.3		
48	2521.6	1362.7	955.1	848.6	177.9*		
49	2526.5	1369.3	962.8	855.3	264.8		
50	2527.2	1368.6	960.1	852.7	251.8		
51	2531.0	1371.2	963.8	856.2	278.4		
52	2536.7	1376.6	967.7	858.9	278.7		
53	2524.2	1365.5	959.9	851.7	278.9		

occurs in each sequence, hence the corresponding observation index is the position of a change point.

From Table 3.1, it is noted that there are twelve change points that are located at the 6th, 12th, 18th, 24th, 28th, 32nd, 34th, 39th, 41st, 43rd, 46th, and 48th sites. In Table 3.2 we list the results of Chernoff, Srivastava, and Worsley, and ours for the position of change points.

From Table 3.3 it is noted that the SIC procedure is able to detect all the possible change points that Srivastava and Worsley have detected, plus some more. Chernoff's detection was visual and therefore at best approximate. Consequently, to be safe, one ought to check all the change points obtained from our procedure for any decision. Just as Chernoff pointed out, the observations from site 36 to 53 could be characterized by a different constellation of special features, and his method might have disguised some of the phenomena clearly observable. The change points we obtained (i.e., the

Table 3.2 SIC Values of Example 3.2

Obs.	1–38	1–14	2–14	
1	−2011.4	−756.2*		
2	−2011.1	−756.1	−705.8	
3	−2009.2	−742.6	−697.9	
4	−2004.0	−737.4	−695.3	
5	−2005.6	−735.7	−692.3	
6	−2009.4	−738.9	−695.8	
7	−2009.7	−738.9	−695.3	
8	−2011.8	−739.0	−695.6	
9	−2016.7	−738.2	−695.1	
10	−2019.2	−744.8	−703.3	
11	−2026.5	−750.1	−706.7	
12	−2022.1	−738.8	−696.7	
13	−2027.2	−741.2	−698.6	
14	−2030.2*	−751.3	−708.6*	
		15–38	15–23	
15	−2027.1	−1247.3	−536.1	
16	−2020.9	−1249.4	−519.9	
17	−2019.9	−1251.4	−604.2*	
18	−2024.1	−1254.8	−559.7	
19	−2017.4	−1252.7	−556.2	
20	−2016.3	−1255.0	−528.6	
21	−2012.8	−1252.6	−516.7	
22	−2012.4	−1259.5	−542.3	
23	−2018.2	−1267.7*	−521.0	
			24–38	
24	−2019.8	−1265.9	−767.3	
25	−2017.4	−1263.1	−767.3	
26	−2015.4	−1264.1	−779.9*	
				27–38
27	−2014.4	−1260.7	−769.0	−619.3
28	−2013.4	−1255.9	−763.7	−632.8
29	−2015.8	−1259.4	−777.7	−640.2
30	−2020.2	−1266.5	−771.5	−625.5
31	−2013.8	−1263.2	−678.0	−626.9
32	−2009.7	−1258.1	−762.5	−627.4
33	−2016.3	−1264.1	−765.8	−649.4*
34	−2015.1	−1256.8	−764.5	−637.5
35	−2012.0	−1256.3	−764.5	−624.8
36	−2014.5	−1259.0	−770.3	−638.6
37	−2007.8	−1250.1	−758.0	−621.3
38	−2023.2	−1261.6	−769.1	−628.6

39th, 41st, 43rd, 46th, and 48th sites) are in the group Chernoff mentioned (see Chernoff, 1979, p. 362). Furthermore, through observing the "faces" Chernoff obtained, one can visually notice that those additional sites (39, 41, 43, 46, and 48) picked up by our procedure appear to be reasonable change points.

Table 3.3 Change Points of Example 3.1

Chernoff	Srivastava & Worsley	Gupta & Chen
		6
	12 [b]	12
20	18 [a]	18
24	24 [a]	24
		28
32	32 [a]	32
35	34 [b]	34
		39
		41
		43
		46
		48

[a]Significant at level $\alpha = .01$.
[b]Significant at level $\alpha = .05$

Table 3.4 Change Point Locations of Example 3.2

Scholars (Charney)	Srivastava & Worsley	Gupta & Chen
1	2 [a]	1
14	14 [b]	14
	23 [a]	17
		23
25	26 [a]	26
	29 [a]	
35		33

[a]Significant at level $\alpha = .01$.
[b]Significant at level $\alpha = .05$

Example 3.2 In this example, we apply the results obtained in Section 3.2 to a set of data consisting of the frequencies of pronouns in 38 dramas of Shakespeare. Our purpose here is to find the changes in Shakespeare's play-writing style by examining the frequencies of different pronouns in his dramas, which are taken in the order as they appear in the First Folio edition of 1623; see Spevack (1968). In 1979, Brainerd presented a paper with a frequency table of pronouns used by Shakespeare in his dramas using discriminant analysis. Later in 1986, Srivastava and Worsley employed the statistics K_{2r} to analyze the same data and perceived change points at different significance levels. Here, we adopt Srivastava and Worsley's formulation and assume that the random variables $X_i, i = 1, \ldots, 7$, are independent. However, if we do not assume independence, our results remain unchanged. As in Example 1, here we cannot detect a change for a sequence with less than nine observations. We also drop the constants $mn \log 2\pi$ and mn from the expressions of SIC(n) and SIC(k). The SICs at each stage are displayed in Table 3.2 and our results are listed in Table 3.4 along with those of Srivastava and Worsley and other scholars for comparison. Please note that in Table 3.2, $*$ indicates where the minimum SIC occurs in each sequence, hence the corresponding observation index is the position of a change point.

From Table 3.4, we conclude that the first stage change point occurs at the 14th play (*The Winter's Tale*). Two more changes take place at the first play (*The Tempest*) and 23rd play (*The Life and Death of King Richard III*). It is noted that the method of Srivastava and Worsley missed the change at the 1st play, whereas our procedure did pick it up. Three further changes occur at the 17th, 26th, and 33rd play. In Charney (1993) *The Tempest* and *The Winter's Tale* are classified as romance, and all plays in between are classified as comedies. The plays after *The Winter's Tale* and before *Coriolanus* (the 26th play) are classified as histories except *Troilus and Cressida* (the 25th play). All plays after *Troilus and Cressida* and before *King Lear* including the next two plays (*Othello, the Moor of Venice* and *Antony and Cleopatra*) are tragedies and the last three plays (*Cymbeline, Pericles*, and *The Two Noble Kinsmen*) are romances. Therefore, our results match with the scholars' classification very well.

3.2 Covariance Change

Let $x_1, x_2, \ldots, x_n$ be a sequence of independent m-dimensional normal random vectors with parameters $(\mu_1, \Sigma_1), (\mu_2, \Sigma_2), \ldots, (\mu_n, \Sigma_n)$, respectively. Assume $\mu_1 = \mu_2 = \cdots = \mu_n = \mu$ and μ is known; then without loss of generality, take $\mu = 0$. It is desired to test the following hypothesis (see Gupta, Chattopadhyay, and Krishnaiah, 1975; Tang and Gupta, 1984):

$$H_0 : \Sigma_1 = \Sigma_2 = \cdots = \Sigma_n = \Sigma \text{ (unknown)} \qquad (3.3)$$

versus the alternative:

$$H_A : \Sigma_1 = \cdots = \Sigma_{k_1} \neq \Sigma_{k_1+1} = \cdots = \Sigma_{k_2} \neq \cdots \neq \Sigma_{k_q+1}$$
$$= \cdots = \Sigma_{k_q+1} = \Sigma_n,$$

where $q, m < q < n - m$, is the unknown number of changes and $k_1, k_2, \ldots, k_q$ are the unknown positions of the change points. Similar to the discussion in Chapter 2, by means of the binary segmentation procedure, we just need to test the single change point hypothesis and then repeat the procedure for each subsequence. Hence, we turn to the testing of (3.3) against the alternative:

$$H_1 : \Sigma_1 = \cdots = \Sigma_k \neq \Sigma_{k+1} = \cdots = \Sigma_n, \qquad (3.4)$$

where k now represents the position of the single change point at each stage, $m < k < n - m$.

3.2.1 Likelihood-Ratio Procedure

(i) The Test Statistic

Under $H_0, \mathbf{x}_1, \mathbf{x}_2, \ldots, \mathbf{x}_n$ are iid $N_m(0, \Sigma)$. The log likelihood function is:

$$\log L_0(\Sigma) = -\frac{1}{2}mn \log 2\pi - \frac{n}{2} \log |\Sigma| - \frac{1}{2}\sum_{i=1}^{n} \mathbf{x}_i' \Sigma^{-1} \mathbf{x}_i,$$

and the MLE of Σ is $\widehat{\Sigma} = (1/n)\sum_{i=1}^{n} \mathbf{x}_i\mathbf{x}_i'$. Hence the maximum log likelihood is

$$\log L_0(\widehat{\Sigma}) = -\frac{1}{2}mn \log 2\pi - \frac{n}{2} \log \left|\frac{1}{n}\sum_{i=1}^{n} x_i x_i'\right| - \frac{n}{2}.$$

Under $H_1, \mathbf{x}_1, \mathbf{x}_2, \ldots, \mathbf{x}_k$ are iid $N_m(0, \Sigma_1)$, and $\mathbf{x}_{k+1}, \mathbf{x}_{k+2}, \ldots, \mathbf{x}_n$ are iid $N_m(0, \Sigma_n)$. The log likelihood function is:

$$\log L_1(\Sigma_1, \Sigma_n) = -\frac{mn}{2} \log 2\pi - \frac{k}{2} \log |\Sigma_1| - \frac{n-k}{2} \log |\Sigma_n|$$
$$- \frac{1}{2}\left[\sum_{i=1}^{k} \mathbf{x}_i' \Sigma_1^{-1} \mathbf{x}_i + \sum_{i=k+1}^{n} \mathbf{x}_i' \Sigma_n^{-1} \mathbf{x}_i\right],$$

and the MLEs of Σ_1 and Σ_n are $\widehat{\Sigma}_1$ and $\widehat{\Sigma}_n$, respectively, where

$$\widehat{\Sigma}_1 = \frac{1}{k}\sum_{i=1}^{k} \mathbf{x}_i\mathbf{x}_i' \quad \text{and} \quad \widehat{\Sigma}_n = \frac{1}{n-k}\sum_{i=k+1}^{n} \mathbf{x}_i\mathbf{x}_i'.$$

Hence, the maximum log likelihood is:

$$\log L_1(\widehat{\Sigma}_1, \widehat{\Sigma}_n) = -\frac{mn}{2} \log 2\pi - \frac{k}{2} \log \left|\frac{1}{k}\sum_{i=1}^{k} x_i x_i'\right|$$
$$- \frac{n-k}{2} \log \left|\frac{1}{n-k}\sum_{i=k+1}^{n} x_i x_i'\right| - \frac{n}{2}.$$

Then the log likelihood procedure statistic is

$$\lambda_n = \max_{m<k<n-m} \log \left(\frac{\left|\frac{1}{n}\sum_{i=1}^{n} \mathbf{x}_i\mathbf{x}_i'\right|^n}{\left|\frac{1}{k}\sum_{i=1}^{k} x_i x_i'\right|^k \left|\frac{1}{n-k}\sum_{i=k+1}^{n} x_i x_i'\right|^{n-k}}\right)^{1/2}. \tag{3.5}$$

As before, to be able to obtain the MLEs, we can detect changes only for $m < k < n - m$. From the principle of minimum information criterion, we estimate the change point position by $\widehat{k}$ such that (3.5) attains its maximum.

(ii) Asymptotic Null Distribution of the Test Statistics

Based on Chen and Gupta (2004), we first derive the asymptotic null distribution of λ_n, where λ_n is given by (3.5). We assume that m is fixed. To prove the main theorem of this section, we need the following results.

Lemma 3.4 *Under H_0, when $n \to \infty$, $k \to \infty$ such that $(k/n) \to 0$, λ_n^2 is asymptotically distributed as*

$$\max_{m<k<n-m} \left\{ \sum_{j=1}^{m} \left[n \log \frac{\chi_n^{2(j)}}{n} - k \log \frac{\chi_k^{2(j)}}{k} - (n-k) \log \frac{\chi_{n-k}^{2(j)}}{n-k} \right] \right\},$$

where $\chi_n^{2(j)}, \chi_k^{2(j)}$, and $\chi_{n-k}^{2(j)}$ are distributed as chi-square random variables with $n, k, n - k$ degrees of freedom, respectively, and $\chi_k^{2(j)}$ and $\chi_{n-k}^{2(j)}$ are independent. Furthermore, $\{\chi_n^{2(j)}, \chi_k^{2(j)}, \chi_{n-k}^{2(j)}\}, j = 1, \dots, m$, are also independent.

Proof. From Anderson (1984, Chapter 7), we obtain:

$\left| \frac{1}{n} \sum_{i=1}^{n} x_i x_i' \right| \overset{D}{=} \frac{|\Sigma|}{n^m} \chi_n^2 \chi_{n-1}^2, \dots, \chi_{n-m+1}^2,$ where $\chi_n^2, \chi_{n-1}^2, \dots, \chi_{n-m+1}^2$ are independent,

$\left| \frac{1}{k} \sum_{i=1}^{k} x_i x_i' \right| \overset{D}{=} \frac{|\Sigma_1|}{k^m} \chi_k^2 \chi_{k-1}^2, \dots, \chi_{k-m+1}^2,$ where $\chi_k^2, \chi_{k-1}^2, \dots, \chi_{k-m+1}^2$ are independent, and

$\left| \frac{1}{n-k} \sum_{i=k+1}^{n} x_i x_i' \right| \overset{D}{=} \frac{|\Sigma_n|}{(n-k)^m} \chi_{n-k}^2 \chi_{n-k-1}^2 \cdots \chi_{n-k-m+1}^2,$ where $\chi_{n-k}^2, \chi_{n-k-1}^2, \dots, \chi_{n-k-m+1}^2$ are independent.

Under H_0, then

$$\lambda_n^2 \overset{D}{=} \max_{m<k<n-m} \left\{ \log \frac{\left| \chi_n^2 \chi_{n-1}^2 \cdots \chi_{n-m+1}^2 \right|^n}{\left| \chi_k^2 \chi_{k-1}^2 \cdots \chi_{k-m+1}^2 \right|^k \left| \chi_{n-k}^2 \chi_{n-k-1}^2 \cdots \chi_{n-k-m+1}^2 \right|^{n-k}} \right.$$

$$\left. + \log \frac{k^{mk} (n-k)^{m(n-k)}}{n^{mn}} \right\}.$$

Because $\chi_{n-\iota}^2 \overset{AD}{=} \chi_n^2, \chi_{k-\iota}^2 \overset{AD}{=} \chi_k^2, \chi_{n-k-\iota}^2 \overset{AD}{=} \chi_{n-k}^2$ as $n \to \infty, k \to \infty$ such that $(k/n) \to 0$, for $\iota = 1, 2, \dots, m-1$, where "$\overset{AD}{=}$" means "asymptotically distributed as," we have

$$\lambda_n^2 \overset{D}{=} \max_{m<k<n-1} \left\{ \sum_{j=1}^{m} [n \log \chi_n^{2(j)} - k \log \chi_k^{2(j)} - (n-k) \log \chi_{n-k}^{2(j)}] \right.$$

$$\left. + \log \frac{k^{mk}(n-k)^{m(n-k)}}{n^{mn}} \right\}$$

$$\overset{D}{=} \max_{1<k<n-1} \left\{ \sum_{j=1}^{m} \left[n \log \frac{\chi_n^{2(j)}}{n} - k \log \frac{\chi_k^{2(j)}}{k} - (n-k) \log \frac{\chi_{n-k}^{2(j)}}{n-k} \right] \right\},$$

where $\{\chi_n^{2(j)}, \chi_k^{2(j)}, \chi_{n-k}^{2(j)}\}, j = 1, \ldots, m$, are independent. $\qquad\square$

Denote $\chi_n^{2(j)} = \sum_{i=1}^{n} y_i^{(j)}, \chi_k^{2(j)} = \sum_{i=1}^{k} y_i^{(j)}, \chi_{n-k}^{2(j)} = \sum_{i=k+1}^{n} y_i^{(j)}$, where $y_i^{(j)}$s are iid chi-square random variables with 1 degree of freedom for all $i = 1, \ldots, n, j = 1, m$. Let

$$\xi_k^{(j)} = n \log \frac{\sum_{i=1}^{n} y_i^{(j)}}{n} - k \log \frac{\sum_{i=1}^{k} y_i^{(j)}}{k} - (n-k) \log \frac{\sum_{i=k+1}^{n} y_i^{(j)}}{n-k},$$

then $\lambda_n^2 = \max_{1<k<n-1} \left\{ \sum_{j=1}^{m} \xi_k^{(j)} \right\}$. Now, for each $\xi_k^{(j)}, j = 1, 2, \ldots, m$, using the three-term Taylor expansion, we obtain:

$$\xi_k^{(j)} = n \left(\frac{\sum_{i=1}^{n} y_i^{(j)}}{n} - 1 \right) - \frac{n}{2} \left(\frac{\sum_{i=1}^{n} y_i^{(j)}}{n} - 1 \right)^2 + \frac{n}{3}(\theta_n^{(1)})^{-3} \left(\frac{\sum_{i=1}^{n} y_i^{(j)}}{n} - 1 \right)^3$$

$$- k \left(\frac{\sum_{i=1}^{k} y_i^{(j)}}{k} - 1 \right) + \frac{k}{2} \left(\frac{\sum_{i=1}^{k} y_i^{(j)}}{k} - 1 \right)^2 - \frac{k}{3}(\theta_k^{(2)})^{-3} \left(\frac{\sum_{i=1}^{k} y_i^{(j)}}{k} - 1 \right)^3$$

$$- (n-k) \left(\frac{\sum_{i=k+1}^{n} y_i^{(j)}}{n-k} - 1 \right) + \frac{n-k}{2} \left(\frac{\sum_{i=k+1}^{n} y_i^{(j)}}{n-k} - 1 \right)^2$$

$$- \frac{n-k}{3}(\theta_{n-k}^{(3)})^{-3} \left(\frac{\sum_{i=k+1}^{n} y_i^{(j)}}{n-k} - 1 \right)^3,$$

where

$$|\theta_n^{(1)} - 1| < \left| \frac{\sum\limits_{i=1}^{n} y_i^{(j)}}{n} - 1 \right|, \qquad |\theta_k^{(1)} - 1| < \left| \frac{\sum\limits_{i=1}^{k} y_i^{(j)}}{k} - 1 \right|, \quad \text{and}$$

$$|\theta_{n-k}^{(3)} - 1| < \left| \frac{\sum\limits_{i=k+1}^{n} y_i^{(j)}}{n-k} - 1 \right| \quad \text{for } j = 1, 2, \ldots, m.$$

Then clearly,

$$\xi_k^{(j)} = -\frac{1}{2n}\left[\sum_{i=1}^{n}(y_i^{(j)} - 1)\right]^2 + \frac{1}{2k}\left[\sum_{i=1}^{k}(y_i^{(j)} - 1)\right]^2 + \frac{1}{2(n-k)}\left[\sum_{i=k+1}^{n}(y_i^{(j)} - 1)\right]^2$$

$$+ \frac{n}{3}(\theta_n^{(1)})^{-3}\left(\frac{\sum\limits_{i=1}^{n} y_i^{(j)}}{n} - 1\right)^3 - \frac{k}{3}(\theta_k^{(2)})^{-3}\left(\frac{\sum\limits_{i=1}^{k} y_i^{(j)}}{k} - 1\right)^3$$

$$- \frac{n-k}{3}(\theta_{n-k}^{(3)})^{-3}\left(\frac{\sum\limits_{i=k+1}^{n} y_i^{(j)}}{n-k} - 1\right)^3 .$$

Let

$$W_k^{(j)} = -\frac{1}{2n}\left[\sum_{i=1}^{n}(y_i^{(j)} - 1)\right]^2 + \frac{1}{2k}\left[\sum_{i=1}^{k}(y_i^{(j)} - 1)\right]^2$$

$$+ \frac{1}{2(n-k)}\left[\sum_{i=k+1}^{n}(y_i^{(j)} - 1)\right]^2$$

$$Q_k^{(j)} = \frac{n}{3}(\theta_n^{(1)})^{-3}\left(\frac{\sum\limits_{i=1}^{n} y_i^{(j)}}{n} - 1\right)^3 - \frac{k}{3}(\theta_k^{(2)})^{-3}\left(\frac{\sum\limits_{i=1}^{k} y_i^{(j)}}{k} - 1\right)^3$$

$$R_k^{(j)} = -\frac{n-k}{3}(\theta_{n-k}^{(3)})^{-3}\left(\frac{\sum\limits_{i=k+1}^{n} y_i^{(j)}}{n-k} - 1\right)^3 ;$$

then $\lambda_n^2 = \max_{1 < k < n-1}\left\{ \sum_{j=1}^{m} W_k^{(j)} + \sum_{j=1}^{m} Q_k^{(j)} + \sum_{j=1}^{m} R_k^{(j)} \right\}.$

For fixed m, we can establish and prove the following lemmas which are similar to the ones we proved in Section 2.2 of Chapter 2. Here, we just state those lemmas as follows.

Lemma 3.5

$(i) \max_{1<k<n} k^{1/2} (\log \log k)^{-(3/2)} \left| \sum_{j=1}^{m} Q_k^{(j)} \right| = O_p(1).$

$(ii) \max_{1<k<n} (n-k)^{1/2} [\log \log (n-k)]^{-(3/2)} \left| \sum_{j=1}^{m} R_k^{(j)} \right| = O_p(1).$

Lemma 3.6 *For all $x \in R$, as $n \to \infty$, the following hold.*

$(i) \; a^2 (\log n) \max_{1<k<\log n} \left(\sum_{j=1}^{m} W_k^{(j)} \right) - [x + b_m(\log n)]^2 \overset{P}{\to} \infty,$

$(ii) \; a^2 (\log n) \max_{1<k<\log n} \left(\sum_{j=1}^{m} \xi_k^{(j)} \right) - [x + b_m(\log n)]^2 \overset{P}{\to} \infty,$

$(iii) \; a^2 (\log n) \max_{n-\log n<k<n} \left(\sum_{j=1}^{m} W_k^{(j)} \right) - [x + b_m(\log n)]^2 \overset{P}{\to} \infty,$

$(iv) \; a^2 (\log n) \max_{n-\log n<k<n} \left(\sum_{j=1}^{m} \xi_k^{(j)} \right) - [x + b_m(\log n)]^2 \overset{P}{\to} \infty,$

where $a(\log n) = (2 \log \log n)^{1/2}$, *and*
$b_m(\log n) = 2 \log \log n + (m/2) \log \log \log n - \log \Gamma(m/2).$

Lemma 3.7 *As $n \to \infty$, the following hold.*

$(i) \; a^2 (\log n) \max_{\log n \leq k \leq n - \log n} \left| \sum_{j=1}^{m} (\xi_k^{(j)} - W_k^{(j)}) \right| = O_p(1),$

$(ii) \; a^2 (\log n) \max_{1<k<n/\log n} \left| \frac{1}{n-k} \left[\sum_{i=k+1}^{n} (y_i^{(j)} - 1) \right]^2 - \frac{1}{n} \left[\sum_{i=1}^{n} (y_i^{(j)} - 1) \right]^2 \right| =$
$O_p(1), \text{ for } j = 1, \ldots, m,$

$(iii) \; a^2 (\log n) \max_{n-n/\log n<k<n} \left| \frac{1}{k} \left[\sum_{i=1}^{k} (y_i^{(j)} - 1) \right]^2 - \frac{1}{n} \left[\sum_{i=1}^{n} (y_i^{(j)} - 1) \right]^2 \right| =$
$O_p(1), \text{ for } j = 1, \ldots, m.$

Lemma 3.8 *For all $x \in R$, as $n \to \infty$,*

$$a^2 (\log n) \max_{(n/\log n)<k<n-n/\log n} \left(\sum_{j=1}^{m} W_k^{(j)} \right) - (x + b_m(\log n))^2 \overset{P}{\to} -\infty.$$

Now, we are in a position to state and prove the main result of this section.

Theorem 3.9 *Under the null hypothesis H_0, when $n \to \infty$, $k \to \infty$ such that $(k/n) \to 0$,*

$$\lim_{n \to \infty} P\{a(\log n)\lambda_n - b_m(\log n) \leq x\} = \exp\{-2e^{-x}\},$$

for all $x \in R$, where $a(\log n)$ and $b_m(\log n)$ are defined in Lemma 3.6.

Proof. From Lemma 3.6 (ii) and (iii), we obtain

$$\max_{1<k<n} \sum_{j=1}^{m} \xi_k^{(j)} \overset{D}{=} \max_{\log n<k<n-\log n} \sum_{j=1}^{m} \xi_k^{(j)}.$$

From Lemma 3.7 (i), we have

$$\max_{1<k<n} \sum_{j=1}^{m} \xi_k^{(j)} \overset{D}{=} \max_{\log n<k<n-\log n} \sum_{j=1}^{m} W_k^{(j)}.$$

and Lemma 3.7 (ii) indicates:

$$\max_{\log n<k<n/\log n} \sum_{j=1}^{m} W_k^{(j)} \overset{D}{=} \max_{\log n<k<n-\log n} \sum_{j=1}^{m} \frac{1}{2k} \left[\sum_{i=1}^{k} (y_i^{(j)} - 1) \right]^2. \quad (3.6)$$

In view of Lemma 3.8, we obtain

$$\max_{\log n<k<n-\log n} \left(\sum_{j=1}^{m} W_k^{(j)} \right) \overset{D}{=} \left[\max_{\log n<k<(n/\log n)} \left(\sum_{j=1}^{m} W_k^{(j)} \right) \right]$$

$$\vee \left[\max_{n-(n/\log n)<k<n-\log n} \left(\sum_{j=1}^{m} W_k^{(j)} \right) \right], \quad (3.7)$$

where $a \vee b \equiv \max\{a, b\}$. Applying Lemma 3.7, we then have

$$\max_{n-n/\log n<k<n-\log n} \sum_{j=1}^{m} W_k^{(j)}$$

$$\overset{D}{=} \max_{n-n/\log n<k<n-\log n} \sum_{j=1}^{m} \frac{1}{2(n-k)} \left[\sum_{i=k+1}^{n} (y_i^{(j)} - 1) \right]^2. \quad (3.8)$$

Combining (3.6) through (3.8), we get

$$\max_{\log n < k < n - \log n} \sum_{j=1}^{m} W_k^{(j)} \overset{D}{=} \max \left\{ \max_{1 < k < n/\log n} \sum_{j=1}^{m} \frac{1}{2k} \left[\sum_{i=1}^{k} (y_i^{(j)} - 1) \right]^2 , \right.$$

$$\left. \max_{n - n/\log n < k < n} \sum_{j=1}^{m} \frac{1}{2(n-k)} \left[\sum_{i=k+1}^{n} (y_i^{(j)} - 1) \right]^2 \right\}.$$

Therefore,

$$\lim_{n \to \infty} P\{a(\log n)\lambda_n - b_m(\log n) \leq x\}$$

$$= \lim_{n \to \infty} P \left\{ a(\log n) \max_{1 < k < n} \left(\sum_{j=1}^{m} W_k^{(j)} \right)^{1/2} - b_m(\log n) \leq x \right\}$$

$$= \lim_{n \to \infty} P \left\{ a^2(\log n) \max_{1 < k < n} \sum_{j=1}^{m} \xi_k^{(1)} \leq [x + b_m(\log n)]^2 \right\}$$

$$= \lim_{n \to \infty} P \left[a^2(\log n) \max \left\{ \max_{1 < k < n/\log n} \sum_{j=1}^{m} \frac{1}{2k} \left[\sum_{i=1}^{k} (y_i^{(j)} - 1) \right]^2 , \right. \right.$$

$$\left. \left. \max_{n - n/\log n < k < n} \sum_{j=1}^{m} \frac{1}{2(n-k)} \left[\sum_{i=k+1}^{n} (y_i^{(j)} - 1) \right]^2 \right\} \leq (x + b_m(\log n))^2 \right].$$

$$(3.9)$$

$\{y_i^{(j)}, 1 \leq i \leq k, 1 \leq k \leq (n/\log n)\}$ and $\{y_i^{(j)}, k+1 \leq i \leq n, n - (n/\log n) \leq k \leq n\}$ are independent, therefore (3.9) reduces to

$$\lim_{n \to \infty} P \left\{ a^2(\log n) \max_{1 < k < n/\log n} \sum_{j=1}^{m} \frac{\left[\sum_{i=1}^{k} (y_i^{(j)} - 1) \right]^2}{2k} \leq (x + b_m(\log n))^2 \right\}$$

$$\cdot \lim_{n \to \infty} P \left\{ a^2(\log n) \max_{n - n/\log n < k < n} \sum_{j=1}^{m} \frac{\left[\sum_{i=k+1}^{n} (y_i^{(j)} - 1) \right]^2}{2(n-k)} \leq (x + b_m(\log n))^2 \right\}$$

$$= \lim_{n \to \infty} P \left\{ a(\log n) \max_{1 < k < n/\log n} \left(\sum_{j=1}^{m} \frac{\left[\sum_{i=1}^{k} (y_i^{(j)} - 1) \right]^2}{2k} \right)^{1/2} - b_m(\log n) \le x \right\}$$

$$\cdot \lim_{n \to \infty} P \left\{ a(\log n) \max_{n - n/\log n < k < n} \left(\sum_{j=1}^{m} \frac{\left[\sum_{i=k+1}^{n} (y_i^{(j)} - 1) \right]^2}{2(n-k)} \right)^{1/2} - b_m(\log n) \le x \right\}.$$

$$(3.10)$$

Denote the first term of (3.10) by (c) and the second by (d). Let's consider (c) first. Notice that

$$\frac{1}{2k} \left[\sum_{i=1}^{k} (y_i^{(j)} - 1) \right]^2 = \left[\sum_{i=1}^{k} \frac{(y_i^{(j)} - 1)}{\sqrt{2k}} \right]^2.$$

Let

$$\mathbf{v}_i = \left(\frac{y_i^{(1)} - 1}{\sqrt{2}}, \frac{y_i^{(2)} - 1}{\sqrt{2}}, \ldots, \frac{y_i^{(m)} - 1}{\sqrt{2}} \right), \qquad 1 \le i < \infty,$$

then $\{\mathbf{v}_i, 1 \le i < \infty\}$ is a sequence of iid m-dimensional random vectors with

$$\mathbf{v}_i^{(1)} = \frac{y_i^{(1)} - 1}{\sqrt{2}}, \qquad \mathbf{v}_i^{(2)} = \frac{y_i^{(2)} - 1}{\sqrt{2}}, \ldots, \mathbf{v}_i^{(m)} = \frac{y_i^{(m)} - 1}{\sqrt{2}}.$$

Now,

$$E(\mathbf{v}_i^{(j)}) = E \left[\frac{y_i^{(j)} - 1}{\sqrt{2}} \right] = 0 \quad \text{for } j = 1, \ldots, m \text{ and } i = 1, \ldots, n.$$

$$\mathrm{Var}(\mathbf{v}_i^{(j)}) = V \left[\frac{y_i^{(j)} - 1}{\sqrt{2}} \right] = \frac{1}{2} V(y_i^{(j)}) = 1, \quad \text{for } j = 1, \ldots, m \text{ and all } i.$$

$$\mathrm{Cov}(\mathbf{v}_i^{(j)}, \mathbf{v}_i^{(\iota)}) = E \left[\frac{y_i^{(j)} - 1}{\sqrt{2}} \cdot \frac{y_i^{(\iota)} - 1}{\sqrt{2}} \right] = 0, \quad \text{for } 1 \le j \ne \iota \le m.$$

Hence the covariance matrix of $\mathbf{v}_i$ is the identity matrix. And obviously, $E|\mathbf{v}_i^{(j)}|^r < \infty$ for all j and some $r > 2$. Let $S_i^{(j)} = \sum_{i=1}^{k} \mathbf{v}_i^{(j)}$ for $1 \le j \le m$, then

$$\sum_{j=1}^{m} \left(\frac{S_i^{(j)}}{\sqrt{k}} \right)^2 = \sum_{j=1}^{m} \frac{1}{2k} \left[\sum_{i=1}^{k} (y_i^{(j)} - 1) \right]^2.$$

Using Lemma 2.18 of Chapter 2, we thus obtain: $(c) = \exp\{-e^{-x}\}$. Similarly, we can show that $(d) = \exp\{-e^{-x}\}$. This completes the proof of the theorem.

$\square$

(iii) Unknown Mean Case

In practice, it is more likely that μ remains common under the two hypotheses but unknown. Under this situation, the likelihood procedure is still applicable.
     Under H_0, the maximum likelihood is easily obtained as

$$L_0(\widehat{\Sigma}) = -\frac{mn}{2} \log 2\pi - \frac{n}{2} \log |\widehat{\Sigma}| - \frac{n}{2}, \quad \text{where } \widehat{\Sigma} = \frac{1}{n} \sum_{i=1}^{n} \mathbf{x}_i \mathbf{x}_i'.$$

Under H_1, the log likelihood function is

$$\log L_1(\mu, \Sigma_1, \Sigma_n) = -\frac{mn}{2} \log 2\pi - \frac{k}{2} \log |\Sigma_1| - \frac{n-k}{2} \log |\Sigma_n|$$

$$-\frac{1}{2} \left[\sum_{i=1}^{k} (\mathbf{x}_i - \mu)' \Sigma_1^{-1} (\mathbf{x}_i - \mu) \right.$$

$$\left. + \sum_{i=k+1}^{n} (\mathbf{x}_i - \mu)' \Sigma_n^{-1} (\mathbf{x}_i - \mu) \right].$$

Differentiating $\log L_1(\mu, \Sigma_1, \Sigma_n)$ with respect to μ, Σ_1, and Σ_n, we obtain the following equations in terms of the MLEs $\widehat{\mu}, \widehat{\Sigma}_1, \widehat{\Sigma}_n$,

$$\sum_{i=1}^{k} \widehat{\Sigma}_1^{-1} (\mathbf{x}_i - \widehat{\mu}) + \sum_{i=k+1}^{k} \widehat{\Sigma}_1^{-1} (\mathbf{x}_i - \widehat{\mu}) = 0,$$

$$k I_m - \sum_{i=1}^{k} \widehat{\Sigma}_1^{-1} (\mathbf{x}_i - \widehat{\mu})(\mathbf{x}_i - \widehat{\mu})' = O_m,$$

$$(n-k) I_m - \sum_{i=1}^{k} \widehat{\Sigma}_1^{-1} (\mathbf{x}_i - \widehat{\mu})(\mathbf{x}_i - \widehat{\mu})' = O_m,$$

where I_m is the $m \times m$ identity matrix, and O_m is the $m \times m$ zero matrix. There are no closed-form solutions for $\widehat{\mu}, \widehat{\Sigma}_1, \widehat{\Sigma}_n$ from the above equations. In practice, we have to use the numerical method to obtain approximate solutions for $\widehat{\mu}, \widehat{\Sigma}_1, \widehat{\Sigma}_n$. Because $\widehat{\mu}$ is a vector, and $\widehat{\Sigma}_1, \widehat{\Sigma}_n$ are matrices, the numerical iteration may be tedious for large m. Once we obtain the unique solutions $\widehat{\mu}, \widehat{\Sigma}_1, \widehat{\Sigma}_n$, the problem of change points can be solved accordingly.

3.2.2 Informational Approach

(i) SICs Under the Two Hypotheses

Under $H_0, \mathbf{x}_1, \mathbf{x}_2, \ldots, \mathbf{x}_n$ are iid $N_m(0, \Sigma)$. The log likelihood function is

$$\log L_0(\Sigma) = -\frac{1}{2}mn \log 2\pi - \frac{n}{2} \log |\Sigma| - \frac{1}{2}\sum_{i=1}^{n} \mathbf{x}_i' \Sigma^{-1} \mathbf{x}_i,$$

and the MLE of Σ is $\widehat{\Sigma} = (1/n)\sum_{i=1}^{n} \mathbf{x}_i \mathbf{x}_i'$. Hence the maximum log likelihood is:

$$\log L_0(\widehat{\Sigma}) = -\frac{1}{2}mn \log 2\pi - \frac{n}{2} \log \left|\frac{1}{n}\sum_{i=1}^{n} \mathbf{x}_i \mathbf{x}_i'\right| - \frac{n}{2},$$

and the $\mathrm{SIC}(n)$ under H_0 is:

$$\mathrm{SIC}(n) = mn \log 2\pi + n \log \left|\frac{1}{n}\sum_{i=1}^{n} \mathbf{x}_i \mathbf{x}_i'\right| + n + \frac{m(m+1)}{2} \log n. \qquad (3.11)$$

Under $H_1, \mathbf{x}_1, \mathbf{x}_2, \ldots, \mathbf{x}_k$ are iid $N_m(0, \Sigma_1)$, and $\mathbf{x}_{k+1}, \mathbf{x}_{k+2}, \ldots, \mathbf{x}_n$ are iid $N_m(0, \Sigma_n)$. The log likelihood function is

$$\log L_1(\Sigma_1, \Sigma_2) = -\frac{mn}{2} \log 2\pi - \frac{k}{2} \log |\Sigma_1| - \frac{n-k}{2} \log |\Sigma_n|$$

$$- \frac{1}{2}\left[\sum_{i=1}^{k} x_i' \Sigma_1^{-1} x_i + \sum_{i=k+1}^{n} x_i' \Sigma_1^{-1} x_i\right],$$

and the MLEs of Σ_1 and Σ_n are $\widehat{\Sigma}_1$ and $\widehat{\Sigma}_n$, respectively, where

$$\widehat{\Sigma}_1 = \frac{1}{k}\sum_{i=1}^{k} \mathbf{x}_i \mathbf{x}_i' \quad \text{and} \quad \widehat{\Sigma}_n = \frac{1}{n-k}\sum_{i=k+1}^{n} \mathbf{x}_i \mathbf{x}_i'.$$

Hence, the maximum log likelihood is

$$
\log L_1(\widehat{\Sigma}_1, \widehat{\Sigma}_n) = -\frac{mn}{2} \log 2\pi - \frac{k}{2} \log \left| \frac{1}{k} \sum_{i=1}^{k} \mathbf{x}_i \mathbf{x}_i' \right|
$$

$$
- \frac{n-k}{2} \log \left| \frac{1}{n-k} \sum_{i=1}^{n} \mathbf{x}_i \mathbf{x}_i' \right| - \frac{n}{2},
$$

and the SIC(k) under H_1, $m < k < n - m$, is

$$
\text{SIC}(k) = mn \log 2\pi + k \log \left| \frac{1}{k} \sum_{i=1}^{k} \mathbf{x}_i \mathbf{x}_i' \right| + (n-k) \log \left| \frac{1}{n-k} \sum_{i=1}^{n} \mathbf{x}_i \mathbf{x}_i' \right|
$$

$$
+ n + m(m+1) \log n. \tag{3.12}
$$

As before, to be able to obtain the MLEs, we can detect changes only for $m < k < n - m$. From the principle of minimum information criterion, we accept H_0 if $\text{SIC}(n) < \min_{m<k<n-m} \text{SIC}(k)$, and accept H_1 if $\text{SIC}(n) > \text{SIC}(k)$ for some k and estimate the change point position by $\widehat{k}$ such that

$$
\text{SIC}(\widehat{k}) = \min_{m<k<n-m} \text{SIC}(k). \tag{3.13}
$$

The following theorem is an immediate corollary of Theorem 3.9.

Theorem 3.10 *Let* $\Delta_n = \min_{m<k<n-m}[\text{SIC}(k) - \text{SIC}(n)]$, *where* $\text{SIC}(k)$ *is defined by (3.12), and* $\text{SIC}(n)$ *by (3.11). Then under* H_0,

$$
\lim_{n \to \infty} P \left\{ a(\log n) \left[\frac{m(m+1)}{2} \log n - \Delta_n \right]^{1/2} - b_m(\log n) \leq x \right\}
$$

$$
= \exp\{-2e^{-x}\},
$$

where $a(\log n)$ *and* $b_m(\log n)$ *are defined in Lemma 3.6.*

As in the previous chapter, we modify the minimum SIC procedure by introducing the significant level α and its associated critical values c_α. Then we accept H_0 if $\text{SIC}(n) < \min_{m<k<n-m} \text{SIC}(k) + c_\alpha$, where $c_\alpha \geq 0$ and is determined by

$$
1 - \alpha = P(\text{SIC}(n) < \min_{m<k<n-m} \text{SIC}(k) + c_\alpha | H_0 \text{ holds}).
$$

Using Theorem 3.10,

$$
1 - \alpha = P(\Delta_n > -c_\alpha | H_0)
$$

$$
= P \left(\lambda_n^2 < \frac{m(m+1)}{2} \log n + c_\alpha | H_0 \right)
$$

$$= P\left(0 < \lambda_n < \left[\frac{m(m+1)}{2}\log n + c_\alpha\right]^{1/2} \Big| H_0\right)$$

$$= P\left\{-b_m(\log n) < a(\log n)\lambda_n - b_m(\log n) < a(\log n)\right.$$

$$\left. \cdot \left[\frac{m(m+1)}{2}\log n + c_\alpha\right]^{1/2} - b_m(\log n)\right\}$$

$$\cong \exp\left\{-2\exp\left\langle -a(\log n)\left[\frac{m(m+1)}{2}\log n + c_\alpha\right]^{1/2} + b_m(\log n)\right\rangle\right\}$$

$$- \exp\{-2\exp[b_m(\log n)]\}.$$

Solving for c_α c_α, we obtain:

$$c_\alpha = \left\{-\frac{1}{a(\log n)}\log\log\left[1 - \alpha + \exp(-2\exp(b_m(\log n)))\right] + \frac{b_m(\log n)}{a(\log n)}\right\}$$

$$-\frac{m(m+1)}{2}\log n.$$

For selected values of α and n, the values of c_α have been computed and are given in Table 3.5 for $m = 2$ and in Table 3.6 for $m = 3$.

(ii) Unbiased Version of SICs

As in the previous chapter, for the hypotheses (3.3) and (3.4), and finite sample size n, we derive the unbiased versions $u - \text{SIC}(n)$ and $u - \text{SIC}(k)$. $m < k < n - m$, respectively.

(1) Unbiased SIC Under $H_0 - u - \text{SIC}(n)$

Let $Y = (\mathbf{y}_1, \mathbf{y}_2, \ldots, \mathbf{y}_n)$ be a sample of the same size and distribution as the Xs, $X = (\mathbf{x}_1, \mathbf{x}_2, \ldots, \mathbf{x}_n)$, and that Y is independent of X. That is, $\mathbf{y}_1, \mathbf{y}_2, \ldots, \mathbf{y}_n$ are also iid $N_m(0, \Sigma)$.

$$J = E_{\widehat{\theta}}[E_{\theta_0|Y}(\log L_0(\widehat{\theta}))]$$

$$= E_{\widehat{\theta}}\left[E_{\theta_0|Y}\left\{-\frac{mn}{2}\log 2\pi - \frac{n}{2}\log\left|\frac{1}{n}\sum_{i=1}^n \mathbf{x}_i\mathbf{x}_i'\right| - \frac{1}{2}\sum_{i=1}^n \mathbf{y}_i'\widehat{\Sigma}^{-1}\mathbf{y}_i\right\}\right]$$

$$= E_{\widehat{\theta}}[\log L_0(\widehat{\theta})] + \frac{n}{2} - \frac{1}{2}E_{\theta_0|Y}\left\{\sum_{i=1}^n \mathbf{y}_i'\widehat{\Sigma}^{-1}\mathbf{y}_i\right\},$$

Table 3.5 Approximate Critical Values of SIC, $m = 2$

n/α	0.010	0.025	0.050	0.100
10	20.768	13.692	9.010	4.866
11	20.126	13.294	8.741	4.689
12	19.623	12.961	8.504	4.525
13	19.198	12.668	8.289	4.371
14	18.824	12.405	8.092	4.224
15	18.490	12.165	7.909	4.085
16	18.187	11.945	7.738	3.953
17	17.910	11.740	7.578	3.827
18	17.655	11.550	7.427	3.706
19	17.419	11.372	7.284	3.591
20	17.199	11.205	7.149	3.481
21	16.993	11.047	7.020	3.375
22	16.800	10.898	6.897	3.273
23	16.618	10.756	6.780	3.175
24	16.445	10.621	6.668	3.081
25	16.282	10.492	6.560	2.990
26	16.126	10.369	6.457	2.902
27	15.978	10.251	6.357	2.817
28	15.837	10.138	6.261	2.735
29	15.701	10.029	6.169	2.655
30	15.571	9.924	6.079	2.578
35	14.992	9.452	5.672	2.223
40	14.504	9.048	5.319	1.911
45	14.082	8.694	5.007	1.632
50	13.710	8.380	4.728	1.381
55	13.378	8.097	4.475	1.152
60	13.079	7.840	4.244	0.942
70	12.555	7.386	3.833	0.566
80	12.108	6.995	3.477	0.238
90	11.718	6.652	3.162	0.000
100	11.372	6.345	2.881	0.000
120	10.780	5.817	2.392	0.000
140	10.284	5.371	1.977	0.000
160	9.858	4.986	1.617	0.000
180	9.485	4.647	1.300	0.000
200	9.152	4.344	1.015	0.0000

where

$$\widehat{\Sigma} = \frac{1}{n} \sum_{i=1}^{n} \mathbf{x}_i \mathbf{x}_i',$$

and

$$L_0(\widehat{\theta}) = -\frac{mn}{2} \log 2\pi - \frac{n}{2} \log |\widehat{\Sigma}| - \frac{n}{2}.$$

Here, $\mathbf{x}_i$s are iid $N_m(0, \Sigma)$, $\mathbf{y}_i$s are iid $N_m(0, \Sigma)$, for $i = 1, 2, \ldots, n$ and X is independent of Y. Now let $T^2 = \mathbf{y}_i' \widehat{\Sigma}^{-1} \mathbf{y}_i$; then from Theorem 5.2.2 of Anderson (1984) we have:

$$\frac{T^2}{n}\frac{n-m+1}{m} \sim F_{m,n-m+1};$$

that is,

$$T^2 \sim \frac{mn}{n-m+1}F_{m,n-m+1}.$$

Hence,

$$E_{\theta_0|Y}\{\mathbf{y}_i'\widehat{\Sigma}^{-1}\mathbf{y}_i\} = \frac{mn}{n-m+1}\frac{n-m+1}{n-m-1} = \frac{mn}{n-m-1}.$$

Then, J is reduced to

$$J = E_{\widehat{\theta}}[\log L_0(\widehat{\theta})] + \frac{n}{2} - \frac{mn^2}{2(n-m-1)}$$

$$= E_{\widehat{\theta}}[\log L_0(\widehat{\theta})] - \frac{n(mn-n+m+1)}{2(n-m-1)}.$$

Table 3.6 Approximate Critical Values
of SIC, $m = 3$

n/α	0.01	0.025	0.05
10	14.077	6.986	2.284
11	13.344	6.465	1.871
12	12.735	6.001	1.488
13	12.192	5.574	1.128
14	11.697	5.178	0.788
15	11.242	4.807	0.466
16	10.819	4.458	0.161
17	10.425	4.130	0.000
18	10.056	3.819	0.000
19	9.708	3.525	0.000
20	9.380	3.245	0.000
21	9.069	2.978	0.000
22	8.773	2.723	0.000
23	8.492	2.479	0.000
24	8.224	2.245	0.000
25	7.967	2.020	0.000
26	7.721	1.804	0.000
27	7.485	1.596	0.000
28	7.257	1.395	0.000
29	7.039	1.201	0.000
30	6.828	1.013	0.000
35	5.873	0.159	0.000
40	5.052	0.000	0.000
45	4.331	0.000	0.000
50	3.688	0.000	0.000
55	3.108	0.000	0.000
60	2.581	0.000	0.000
70	1.648	0.000	0.000
80	0.843	0.000	0.000
90	0.135	0.000	0.000

Thus, $\log L_0(\widehat{\theta}) - ((n(mn - n + m + 1))/(2(n - m - 1)))$ is unbiased for J and we define, under H_0, the $u - \text{SIC}(n)$ by

$$u - \text{SIC}(n) = -2\log L_0(\widehat{\theta}) + \frac{n(mn - n + m + 1)}{n - m - 1}$$

$$= \text{SIC}(n) + \frac{n(mn - n + m + 1)}{n - m - 1} - \frac{m(m + 1)}{2}\log n,$$

where $\text{SIC}(n)$ is given in (3.11).

(2) Unbiased SIC Under $H_1 - u - \text{SIC}(k)$

Under H_1, let $Y = (\mathbf{y}_1, \mathbf{y}_2, \ldots, \mathbf{y}_n)$ be a sample of the same size and distribution as the Xs, $X = (\mathbf{x}_1, \mathbf{x}_2, \ldots, \mathbf{x}_n)$, and Y be independent of X. That is, $\mathbf{x}_1, \mathbf{x}_2, \ldots, \mathbf{x}_k, \mathbf{y}_1, \mathbf{y}_2, \ldots, \mathbf{y}_k$ are iid $N_m(0, \Sigma_1)$ random vectors, and $\mathbf{x}_{k+1}, \mathbf{x}_{k+2}, \ldots, \mathbf{x}_n, \mathbf{y}_{k+1}, \mathbf{y}_{k+2}, \ldots, \mathbf{y}_n$ are iid $N_m(0, \Sigma_n)$ random vectors.

$$J = E_{\widehat{\theta}}[E_{\theta_0|Y}(\log L_1(\widehat{\theta}))]$$

$$= E_{\widehat{\theta}}\left[E_{\theta_0|Y}\left\{-\frac{mn}{2}\log 2\pi - \frac{k}{2}\log\left|\frac{1}{k}\sum_{i=1}^{k}\mathbf{x}_i\mathbf{x}_i'\right|\right.\right.$$

$$\left.\left. -\frac{n-k}{2}\log\left|\frac{1}{n-k}\sum_{i=k+1}^{n}\mathbf{x}_i\mathbf{x}_i'\right| - \frac{1}{2}\sum_{i=1}^{k}\mathbf{y}_i'\widehat{\Sigma}^{-1}\mathbf{y}_i - \frac{1}{2}\sum_{i=k+1}^{n}\mathbf{y}_i'\widehat{\Sigma}_1^{-1}\mathbf{y}_i\right\}\right],$$

where

$$\widehat{\Sigma}_1 = \frac{1}{k}\sum_{i=1}^{k}\mathbf{x}_i\mathbf{x}_i' \quad \text{and} \quad \widehat{\Sigma}_n = \frac{1}{n-k}\sum_{i=k+1}^{n}\mathbf{x}_i\mathbf{x}_i'.$$

Because for $i = 1, 2, \ldots, k$,

$$\frac{y_i'\widehat{\Sigma}_1^{-1}y_i}{k}\frac{k - m + 1}{m} \sim F_{m,k-m+1},$$

and for $i = k + 1, \ldots, n$,

$$\frac{y_i'\widehat{\Sigma}_n^{-1}y_i}{n - k}\frac{n - k - k + 1}{m} \sim F_{m,n-k-m+1},$$

we have for $i = 1, \ldots, m$,

$$E[y_i'\widehat{\Sigma}_1^{-1}y_i] = \frac{km}{k - m + 1}\frac{k - m + 1}{k - m - 1} = \frac{km}{k - m - 1},$$

and for $i = 1, \ldots, m$,

$$E[y_i' \widehat{\Sigma}_n^{-1} y_i] = \frac{(n-k)m}{n-k-m+1} \frac{n-k-m+1}{n-k-m-1} = \frac{(n-k)m}{n-k-m+1}.$$

Then J is reduced to

$$J = E_{\widehat{\theta}}[\log L_1(\widehat{\theta})] + \frac{n}{2} - \frac{1}{2} \frac{k^2}{k-m-1} - \frac{1}{2} \frac{(n-k)^2 m}{n-k-m-1}$$

$$= E_{\widehat{\theta}}[\log L_1(\widehat{\theta})] - \frac{c}{2},$$

where

$$\log L_1(\widehat{\theta}) = -\frac{mn}{2} \log 2\pi - \frac{k}{2} \log |\widehat{\Sigma}_1| - \frac{n-k}{2} \log |\widehat{\Sigma}_n| - \frac{n}{2},$$

and

$$c = \frac{k^2 m(n-k-m-1) + (n-k)^2 m(k-m-1)}{(k-m-1)(n-k-m-1)}$$

$$- \frac{n(k-m-1)(n-k-m-1)}{(k-m-1)(n-k-m-1)}.$$

Therefore, we define the unbiased $SIC(k)$ as

$$u - SIC(k) = -2 \log L_1(\widehat{\theta}) + c$$

for $m < k < n - m$, or

$$u - SIC(k) = SIC(k) + c - m(m+1) \log n,$$

where $SIC(k)$ is given in (3.12).

3.2.3 Application to Multivariate Stock Market Data

Now, we illustrate an application of our test procedure. We collect the Friday closing prices from January, 1990 through December, 1991 for two stocks (Exxon and General Dynamics) from *Daily Stock Price Record: New York Stock Exchange*, published quarterly by Standard & Poor's. The prices are listed in Table 3.7, in which the digit following the dash denotes eights. The 3D scatterplot of the bivariate stock prices is given in Figure 3.1.

Table 3.7 Friday Closing Price, Jan. 1, 1990–Dec. 31, 1991

Obs.	Exxon	General Dynamics	Obs.	Exxon	General Dynamics
1	48–06	45–05	53	51–04	25–03
2	47–06	41–05	54	49–07	20–07
3	48–05	41–04	55	51–05	25–02
4	46–06	39–07	56	52–05	27–04
5	47–07	39–06	57	50–04	28–04
6	48–02	37	58	53–01	27–04
7	48	36–06	59	53–03	25–04
8	47–01	36–02	60	53–06	26–02
9	46–06	38	61	55–06	24–04
10	46–03	37–06	62	55–03	23–05
11	47–04	37–07	63	57–01	28–07
12	46–01	37–05	64	57–07	29–04
13	46–01	37–04	65	58–04	33–04
14	46–01	37–06	66	57–03	32
15	45–06	37–03	67	59–06	33–04
16	46–03	37–02	68	60–01	35–02
17	45	35–04	69	59–04	36–04
18	46–04	34–04	70	59–02	39
19	47–07	33–04	71	57–07	38–04
20	47–06	34–05	72	58	39
21	46–04	34–03	73	58–05	39
22	47–06	35	74	58–02	38–03
23	47	36	75	57–04	39–02
24	47–07	35–05	76	58–02	38–06
25	47–07	33–04	77	58–04	41–05
26	47–07	32	78	58–01	41–07
27	47–06	32–01	79	57–04	42–02
28	48–06	31–05	80	58–06	43–06
29	48–05	31–06	81	59–05	44–03
30	49–02	27–03	82	58–06	43–03
31	53–01	29	83	58–06	44
32	51–05	27–04	84	57–04	44–01
33	52–01	26–01	85	57–04	44–02
34	48–07	26–02	86	59	42–03
35	50	24–07	87	58–02	42–04
36	50–07	25–05	88	59	42–02
37	51–03	25–03	89	58–03	40–06
38	51–02	26–04	90	59	40–07
39	49	23–05	91	58–06	46–02
40	49–07	25–04	92	59–01	48–04
41	48–04	22–07	93	60–01	47–02
42	49–07	20–06	94	61–04	50–01
43	47–02	22–03	95	60–05	47–07
44	49–05	23–04	96	60–04	51–05
45	50–03	23–07	97	60–01	51–04
46	50–03	22–05	98	57–02	50–04
47	51–01	23	99	57–06	47–04
48	50–05	23–03	100	58–05	48–06
49	49–02	24–05	101	57–01	50
50	50–06	25–03	102	58	52–07
51	50–06	26	103	59–01	52–01
52	51–05	25	104	58–02	52–03

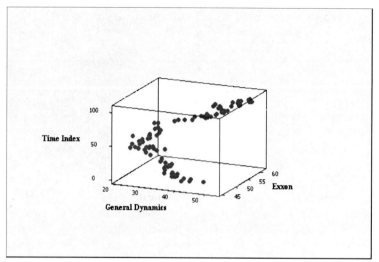

Fig. 3.1 The scatterplot of the original stock prices of the two stocks

The weekly rates of return for these two stocks are analyzed for any change points, where the weekly rates of return $R_t = (R_{t_1}, R_{t_2})$, R_{t_j} of stock j, $j = 1, 2$, is defined as (Johnson and Wichern, 1988)

$$R_{t_j} = \frac{Current\ Friday\ closing\ price - Previous\ Friday\ closing\ price}{Previous\ Friday\ closing\ price}.$$

The 3D scatterplot of the return series is given in Figure 3.2.

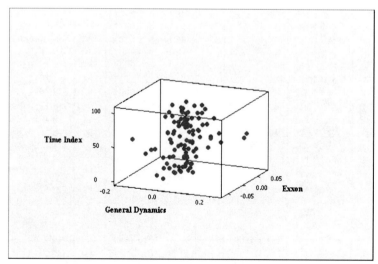

Fig. 3.2 The scatterplot of the return series of the two stocks

Assuming bivariate normal distribution with mean 0 for the R_t series, $t = 1, 2, \ldots, 103$, we test the following hypothesis,

$$H_0 : \Sigma_1 = \Sigma_2 = \cdots = \Sigma_{103} = \Sigma \text{ (unknown)},$$

versus the alternative:

$$H_1 : \Sigma_1 = \Sigma_{k_1} \neq \Sigma_{k_1+1} = \Sigma_{k_2} \neq \cdots \neq \Sigma_{k_q+1} = \cdots = \Sigma_{k_{q+1}} = \Sigma_{103},$$

where q, $m < q < n - m$, is the unknown number of changes and $k_1, k_2, \ldots, k_q$ are the unknown positions of the change points, and $n = 103$.

The SIC(k) for $3 \leq k \leq 100$, and SIC(103) are obtained and listed in Table 3.8. Here SIC(103) $= -871.6174 > \min_{3 \leq k \leq 103}$ SIC(k) $=$ SIC(66) $= -877.8032$. If we choose $\alpha = 0.025$, according to Table 3.5, we still have SIC(103) $> \min_{3 \leq k \leq 103}$ SIC(k) $+ c_\alpha$. Hence, the 66th observation is a change point for the return series. At the same α level, eight more change points are found; they are the 3rd, 23rd, 27th, 28th, 70th, 90th, 94th, and 95th observations of the R_t series.

Transferring back to the Friday closing price and the corresponding dates, we conclude that there are nine significant ($\alpha = 0.025$) changes in the stock

Table 3.8 SIC Values of Example 3.3 at the First Stage

k	SIC(k)	k	SIC(k)	k	SIC(k)	k	SIC(k)
3	−859.2419	28	−875.7469	53	−863.0627	78	−865.6730
4	−859.3577	29	−867.7354	54	−861.7981	79	−865.4149
5	−858.1211	30	−864.9223	55	−861.8698	80	−864.5139
6	−858.6054	31	−864.9908	56	−863.1338	81	−863.7539
7	−859.2951	32	−864.9012	57	−867.1126	82	−862.6420
8	−859.2241	33	−867.1411	58	−866.5134	83	−862.1341
9	−860.2009	34	−867.2714	59	−865.6375	84	−861.1433
10	−860.8116	35	−867.6294	60	−868.8383	85	−861.1606
11	−861.4430	36	−868.2560	61	−867.8304	86	−860.4344
12	−862.7146	37	−868.2560	62	−873.0662	87	−859.8044
13	−864.1533	38	−865.9297	63	−871.8757	88	−859.3132
14	−865.7002	39	−864.9329	64	−875.6297	89	−858.8113
15	−867.1564	40	−863.5814	65	−875.1426	90	−860.5685
16	−867.4550	41	−862.3220	66	−877.8032	91	−860.2412
17	−866.6043	42	−864.5585	67	−876.9840	92	−869.7628
18	−866.6930	43	−866.4932	68	−875.8014	93	−859.9674
19	−867.4225	44	−866.5224	69	−875.7610	94	−859.7997
20	−868.5823	45	−866.5224	70	−875.1666	95	−861.4867
21	−869.6860	46	−866.0638	71	−873.4096	96	−860.4341
22	−870.4244	47	−866.1460	72	−871.9115	97	−860.5178
23	−871.7620	48	−866.9320	73	−870.4108	98	−860.8406
24	−870.8104	49	−867.6632	74	−869.2914	99	−859.9722
25	−871.1048	50	−867.8991	75	−868.1249	100	−859.6739
26	−872.7699	51	−867.8991	76	−868.2891	103	−871.6174
27	−873.9515	52	−867.9753	77	−866.8725		

price for these two companies. These changes occurred during Jan. 29 to Feb. 11 of 1990, Jul. 23 to Aug. 5 of 1990, Apr. 8 to May 3 of 1991, Sep. 16 to Sep. 23 of 1991, and Oct. 13 to Oct. 28 of 1991. Historical materials showed that during these five periods several important economic, political, and social events happened, which might cause the changes in the price of the two stocks. In the first period, from Jan. 29 to Feb. 11 of 1990, interest rates around the world had suddenly surged, sending stock prices tumbling on exchanges from Tokyo to London and threatening to put the sickly U.S. economy into the intensive-care ward. In the second period, from Jul. 23 to Aug. 5 of 1990, Saddam Hussein called out the troops to frighten Kuwait, and his tactic raised oil prices. On Aug. 2, Iraqi troops invaded Kuwait. World stock and currency markets gyrated nervously in response to the news of Iraq's invasion. The sluggish American economy was in danger of recession. The third period, from Apr. 8 to May 3 of 1991, was called a period "from war fare to fare wars." After staying home through a bleak winter of conflict and recession, Americans were eager to bust loose, and travel companies were wooing them. In the fourth period, from Sep. 16 to Sep. 23 of 1990, mortgage rates had plunged sharply since 1984, and in September it was found that the pipeline to quick refinancing became more jammed. The number of qualified appraisers in many areas could barely keep up with demand. Despite such inconvenience, experts expected the rush to refinancing to continue, especially if interest rates dropped by another half percentage point. Finally, in the last period, from Oct. 13 to Oct. 28, the recovery of the recession was very slow. Even if President Bush continued to lower interest rates, the effects would not be felt until next spring. (Note, for historical evidence, see *Times* 1990 and 1991).

Furthermore, for the bivariate return series (R_{t_1}, R_{t_2}), we used Mardia's test (see Mardia, 1970) to check the multivariate normality assumption for every subsequence (separated by the change points). Our results show that the normality assumption is indeed valid.

3.3 Mean Vector and Covariance Change

In Chapter 2, Section 2.3, we investigated the change point problem of mean and variance for a sequence of normal random variables by using the information criterion, SIC, and studied some properties of the corresponding statistic. In this section, we study the problem of simultaneous changes in the mean vector and covariance matrix of a multivariate Gaussian model.

Suppose $\mathbf{x}_1, \mathbf{x}_2, \ldots, \mathbf{x}_n$ is a sequence of independent m-dimensional normal random vectors with parameters $(\mu_1, \Sigma_1), (\mu_2, \Sigma_2), \ldots, (\mu_n, \Sigma_n)$, respectively. We are interested in testing the following hypothesis,

$$H_0 : \mu_1 = \mu_2 = \cdots = \mu_n \quad \text{and} \quad \Sigma_1 = \Sigma_2 = \cdots = \Sigma_n (\mu, \Sigma \text{ unknown})$$
$$(3.14)$$

versus the alternative:

$$H_A : \mu_1 = \cdots = \mu_{k_1} \neq \mu_{k_1+1} = \cdots = \mu_{k_2} \neq \cdots \neq \mu_{k_q+1}$$
$$= \cdots = \mu_{k_{q+1}} = \mu_n$$

and

$$\Sigma_1 = \cdots = \Sigma_{k_1} \neq \Sigma_{k_1+1} = \cdots = \Sigma_{k_2} \neq \cdots \neq \Sigma_{k_q+1} = \cdots = \Sigma_{k_{q+1}} = \Sigma_n,$$

where $q, m < q < n-m$ are the unknown number of changes and $k_1, k_2, \ldots, k_q$ are the unknown positions of the change points. Similar to the previous discussion, by means of the binary segmentation procedure, we just need to test the single change point hypothesis and then repeat the procedure for each subsequence. Hence, we turn to test (3.14) against the alternative:

$$H_1 : \mu_1 = \cdots = \mu_k \neq \mu_{k+1} = \cdots = \mu_n$$
$$\text{and} \quad \Sigma_1 = \cdots = \Sigma_k \neq \Sigma_{k+1} = \cdots = \Sigma_n, \tag{3.15}$$

where $m < k < n - m$.

3.3.1 Likelihood-Ratio Procedure

Under H_0, the log likelihood function is given by

$$\log L_0(\theta) = -\frac{1}{2} mn \log 2\pi - \frac{n}{2} \log |\Sigma| - \frac{1}{2} \sum_{i=1}^{n} (\mathbf{x}_i - \mu)' \Sigma^{-1} (\mathbf{x}_i - \mu),$$

where $\theta = (\mu, \Sigma)$. Let $\widehat{\theta} = (\widehat{\mu}, \widehat{\Sigma})$ be the MLE of θ, then

$$\widehat{\mu} = \overline{\mathbf{x}} \quad \text{and} \quad \widehat{\Sigma} = \frac{1}{n} \sum_{i=1}^{n} (\mathbf{x}_i - \overline{\mathbf{x}})(\mathbf{x}_i - \overline{\mathbf{x}})'.$$

Hence, the maximum log likelihood is:

$$\log L_0(\widehat{\theta}) = -\frac{1}{2} mn \log 2\pi - \frac{n}{2} \log |\widehat{\Sigma}| - \frac{n}{2}.$$

Under H_1, the log likelihood function is:

$$\log L_1(\theta) = -\frac{1}{2}mn \log 2\pi - \frac{k}{2}\log|\Sigma_1| - \frac{n-k}{2}\log|\Sigma_n|$$

$$-\frac{1}{2}\sum_{i=1}^{k}(\mathbf{x}_i - \mu_1)'\Sigma_1^{-1}(\mathbf{x}_i - \mu_1)$$

$$-\frac{1}{2}\sum_{i=k+1}^{n}(\mathbf{x}_i - \mu_n)'\Sigma_n^{-1}(\mathbf{x}_i - \mu_n).$$

Here, $\theta = (\mu_1, \mu_n, \Sigma_1, \Sigma_n)$. Denote the MLE of θ by $\widehat{\theta} = (\widehat{\mu}_1, \widehat{\mu}_n, \widehat{\Sigma}_1, \widehat{\Sigma}_n)$; then

$$\widehat{\mu}_1 = \overline{\mathbf{x}}_k = \frac{1}{k}\sum_{i=1}^{k}\mathbf{x}_i, \quad \widehat{\Sigma}_1 = \frac{1}{k}\sum_{i=1}^{k}(\mathbf{x}_i - \overline{\mathbf{x}}_k)(\mathbf{x}_i - \overline{\mathbf{x}}_k)',$$

$$\widehat{\mu}_n = \overline{\mathbf{x}}_{n-k} = \frac{1}{n-k}\sum_{i=k+1}^{n}\mathbf{x}_i, \quad \widehat{\Sigma}_n = \frac{1}{n-k}\sum_{i=k+1}^{n}(\mathbf{x}_i - \overline{\mathbf{x}}_{n-k})(\mathbf{x}_i - \overline{\mathbf{x}}_{n-k})'.$$

Then the maximum log likelihood is:

$$\log L_1(\widehat{\theta}) = -\frac{1}{2}mn \log 2\pi - \frac{k}{2}\log|\widehat{\Sigma}_1| - \frac{n-k}{2}\log|\widehat{\Sigma}_n| - n/2.$$

Then the log likelihood procedure statistic is

$$\lambda_n = \left(\max_{m<k<n-m} \log \frac{|\widehat{\Sigma}|^n}{|\widehat{\Sigma}_1|^k|\widehat{\Sigma}|^{n-k}}\right)^{1/2}.$$

Notice that, in order to obtain the MLEs here, we can detect changes only when $m < k < n - m$. The position of the change point is estimated by $\widehat{k}$, where $\widehat{k}$ is the value of k such that λ_n attains its maximum.

(i) Asymptotic Null Distribution of the Test Statistic

Lemma 3.11 *Under H_0, when $n \to \infty$, $k \to \infty$ such that $(k/n) \to 0$, λ_n^2 is asymptotically distributed as*

$$\max_{1<k<n-1}\left\{\sum_{j=1}^{m}[n \log t_{n-1}^{(j)} - k \log t_{k-1}^{(j)} - (n-k)\log t_{n-k-1}^{(j)}]\right\},$$

where $nt_{n-1}^{(j)}, kt_{k-1}^{(j)}$, and $(n-k)t_{n-k-1}^{(j)}$ are distributed as chi-square random variables with $n-1, k-1$, and $n-k-1$ degrees of freedom, respectively, and $t^{(j)}_{k-1}$ and $t^{(j)}_{n-k-1}$ are independent. Furthermore, $\{t^{(j)}_{n-1}, t^{(j)}_{k-1}, t^{(j)}_{n-k-1}\}$, $j = 1, \ldots, m$, are also independent.

Proof. From Theorem 7.5.3 in Anderson (1984), under H_0, we obtain

$$\lambda^2_n \stackrel{D}{=} \max_{m < k < n-m} \left\{ n \log \chi^2_{n-1} - k \log \chi^2_{k-1} - (n-k) \log \chi^2_{n-k-1}] \right.$$
$$+ [n \log \chi^2_{n-2} - k \log \chi^2_{k-2} - (n-k) \log \chi^2_{n-k-2}]$$
$$\vdots$$
$$+ [n \log \chi^2_{n-m} - k \log \chi^2_{k-m} - (n-k) \log \chi^2_{n-k-m}]$$
$$\left. + \log \frac{k^{mk}(n-k)^{m(n-k)}}{n^{mn}} \right\},$$

where χ^2_{n-j}s, $j = 1, \ldots, m$, are independent, χ^2_{k-j}s, $j = 1, \ldots, m$, are independent, and χ^2_{n-k-j}s, $j = 1, \ldots, m$, are independent; and "$\stackrel{D}{=}$" means "distributed as." Because, $\chi^2_{n-l} \stackrel{AD}{=} \chi^2_{n-1}, \chi^2_{k-l} \stackrel{AD}{=} \chi^2_{k-1}$, and $\chi^2_{n-k-l} \stackrel{AD}{=} \chi^2_{n-k-1}$ for $l = 1, 2, \ldots, m$, where "$\stackrel{AD}{=}$" means "asymptotically distributed as," we have

$$\lambda^2_n \stackrel{AD}{=} \max_{1 < k < n} \left\{ \sum_{j=1}^m [n \log \chi^{2^{(j)}}_{n-1} - k \log \chi^{2^{(j)}}_{k-1} - (n-k) \log \chi^{2^{(j)}}_{n-k-1}] \right.$$
$$\left. + \log \frac{k^{km}(n-k)^{m(n-k)}}{n^{mn}} \right\}$$
$$\stackrel{D}{=} \max_{1 < k < n-1} \left\{ \sum_{j=1}^m [n \log t^{(j)}_{n-1} - k \log t^{(j)}_{k-1} - (n-k) \log t^{(j)}_{n-k-1}] \right\},$$

where $\{t^{(j)}_{n-1}, t^{(j)}_{k-1}, t^{(j)}_{n-k-1}\}$, $j = 1, \ldots, m$ are independent. $\square$

Let $\xi^{(j)}_k = n \log t^{(j)}_{n-1} - k \log t^{(j)}_{k-1} - (n-k) \log t^{(j)}_{n-k-1}$; then $\lambda^2_n = \max_{1 < k < n-1} \{\sum_{j=1}^m \xi^{(j)}_k\}$. Now for each $\xi^{(j)}_k, j = 1, 2, \ldots, m$, using the three-term Taylor expansion, we obtain:

$$\xi^{(j)}_k = n(t^{(j)}_{n-1} - 1) - \frac{n}{2}(t^{(j)}_{n-1} - 1)^2 + \frac{1}{3}(\theta^{(1)}_n)^{-3} n(t^{(j)}_{n-1} - 1)^3$$
$$- k(t^{(j)}_{k-1} - 1) + \frac{k}{2}(t^{(j)}_{k-1} - 1)^2 - \frac{1}{3}(\theta^{(2)}_k)^{-3} k(t^{(j)}_{k-1} - 1)^3$$

$$- (n-k)(t^{(j)}_{n-k-1} - 1) + \frac{n-k}{2}(t^{(j)}_{n-k-1} - 1)^2$$

$$- \frac{1}{3}(\theta^{(3)}_{n-k})^{-3}(n-k)(t^{(j)}_{n-k-1} - 1)^3,$$

where $|\theta^{(1)}_n - 1| < |t^{(j)}_{n-1} - 1|, |\theta^{(2)}_k - 1| < |t^{(j)}_{k-1} - 1|$, and $|\theta^{(3)}_{n-k} - 1| < |t^{(j)}_{n-k-1} - 1|$.
Next denote

$$t^{(j)}_{n-1} = \frac{1}{n}\sum_{i=1}^{n}(z^{(j)}_i - \bar{z}^{(j)}_k)^2, \qquad \bar{z}^{(j)}_n = \frac{1}{n}\sum_{i=1}^{n}z^{(i)}_i$$

$$t^{(j)}_{k-1} = \frac{1}{k}\sum_{i=1}^{k}(z^{(j)}_i - \bar{z}^{(j)}_k)^2, \qquad \bar{z}^{(j)}_k = \sum_{i=1}^{k}z^{(j)}_i$$

$$t^{(j)}_{n-k-1} = \frac{1}{n-k}\sum_{i=k+1}^{n}(z^{(j)}_i - \bar{z}^{(j)}_{n-k})^2, \qquad \bar{z}^{(j)}_{n-k} = \frac{1}{n-k}\sum_{i=k+1}^{n}z^{(j)}_i,$$

where $z^{(j)}_i$s are iid $N(0,1)$ random variables, $i = 1,\dots,n, j = 1,\dots,m$. Then

$$\xi^{(j)}_k = jW^{(1)}_k + jW^{(2)}_k + jQ^{(1)}_k + jQ^{(2)}_k,$$

where

$$jW^{(1)}_k = k\bar{z}^{(j)^2}_k + (n-k)\bar{z}^{(j)^2}_{n-k} - n\bar{z}^{(j)^2}_n,$$

$$jW^{(2)}_k = -\frac{1}{2n}\left[\sum_{i=1}^{n}(z^{(j)^2}_i - 1)\right]^2 + \frac{1}{2k}\left[\sum_{i=1}^{k}(z^{(j)^2}_i - 1)\right]^2$$

$$+ \frac{1}{2(n-k)}\left[\sum_{i=k+1}^{n}(z^{(j)}_i - 1)\right]^2,$$

$$jQ^{(1)}_k = \frac{n}{3}(\theta^{(1)}_n)^{-3}(t^{(j)}_{n-1} - 1)^3 - \frac{k}{3}(\theta^{(2)}_k)^{-3}(t^{(j)}_{k-1} - 1)^3 + \bar{z}^{(j)^2}_n\sum_{i=1}^{n}(z^{(j)^2}_i - 1)$$

$$+ \frac{n}{2}\bar{z}^{(j)^4}_n - \bar{z}^{(j)^2}_k\sum_{i=1}^{k}(z^{(j)^2}_i - 1) + \frac{k}{2}\bar{z}^{(j)^4}_k,$$

$$jQ^{(2)}_k = \frac{(n-k)}{3}(\theta^{(3)}_{n-k})^{-3}(t^{(j)}_{n-k-1} - 1)^3 + \frac{(n-k)}{2}\bar{z}^{(j)^4}_{n-k}$$

$$- \bar{z}^{(j)^2}_{n-k}\sum_{i=k+1}^{n}(z^{(j)^2}_i - 1).$$

Let $\xi_k = \sum_{j=1}^m \xi_k^{(j)}$, $W_k^{(1)} = \sum_{j=1}^m j W_k^{(1)}$, $W_k^{(2)} = \sum_{j=1}^m W_k^{(2)}$, $Q_k^{(1)} = \sum_{j=1}^m j Q_k^{(1)}$, and $Q_k^{(2)} = \sum_{j=1}^m j Q_k^{(2)}$; then $\xi_k = W_k^{(1)} + W_k^{(2)} + Q_k^{(1)} + Q_k^{(2)}$ and $\lambda_n^2 = \max_{1 < k < n-1} \xi_k$.

Lemma 3.12 $\max_{1 \le k \le n} k^{1/2} (\log \log k)^{-(3/2)} |Q_k^{(1)}| = O_p(1)$.

Proof. From the law of iterated logarithm, one can easily establish that:

$$\max_{1<k<n} k^{1/2} (\log \log k)^{-(3/2)} |(\theta_k^{(2)})^{-3} k (t_{k-1}^{(j)} - 1)^3| = O_p(1), \tag{3.16}$$

$$\max_{1<k<n} k^{1/2} (\log \log k)^{-(3/2)} \bar{z}_k^{(j)^2} \sum_{i=1}^k (z_k^{(j)^2} - 1) = O_p(1), \tag{3.17}$$

$$\max_{1<k<n} k^{1/2} (\log \log k)^{-(3/2)} k \bar{z}_k^{(j)^4} = O_p(1), \tag{3.18}$$

$$\max_{1<k<n} k^{1/2} (\log \log k)^{-(3/2)} |(\theta_n^{(1)})^{-3} n (t_{n-1}^{(j)} - 1)^3| = O_p(1), \tag{3.19}$$

$$\max_{1<k<n} k^{1/2} (\log \log k)^{-(3/2)} \bar{z}_n^{(j)^2} \sum_{i=1}^n (z_i^{(j)^2} - 1) = O_p(1), \tag{3.20}$$

$$\max_{1<k<n} k^{1/2} (\log \log k)^{-(3/2)} n \bar{z}_n^{(j)^4} = O_p(1). \tag{3.21}$$

Hence,

$$\max_{1<k<n} k^{1/2} (\log \log k)^{-(3/2)} |j Q_k^{(1)}| = O_p(1).$$

Therefore, for m fixed,

$$\max_{1<k<n} k^{1/2} (\log \log k)^{-(3/2)} |Q_k^{(1)}| = O_p(1). \qquad \square$$

Lemma 3.13 $\max_{1<k<n} (n-k)^{1/2} (\log \log (n-k))^{-(3/2)} |Q_k^{(2)}| = O_p(1)$.

Proof. Proceeding as in Lemma 3.12, we obtain

$$\max_{1<k<n} (n-k)^{1/2} (\log \log (n-k)^{-(3/2)}) |(\theta_{n-k}^{(3)})^{-3}| (n-k) (t_{n-k-1}^{(j)} - 1)^3|$$
$$= O_p(1), \tag{3.22}$$

$$\max_{1<k<n} (n-k)^{1/2} (\log \log (n-k)^{-(3/2)}) \bar{z}_{n-k}^{(j)^2} \sum_{i=k+1}^n (z_i^{(j)^2} - 1) = O_p(1), \tag{3.23}$$

$$\max_{1<k<n} (n-k)^{1/2} (\log \log (n-k)^{-(3/2)}) (n-k) \bar{z}_{n-k}^{(j)^4} = O_p(1). \tag{3.24}$$

Hence,

$$\max_{1<k<n} (n-k)^{1/2} (\log\log(n-k)^{-(3/2)}) |Q_k^{(2)}| = O_p(1).$$

Q.E.D.   □

Lemma 3.14 $a^2(\log n) \max_{1<k<\log n} (W_k^{(1)} + W_k^{(2)}) - (x + b_{2m}(\log n))^2 \xrightarrow{P} \infty$ as $n \to \infty$, where $x \in R$,

$$a(\log n) = (2\log\log n)^{1/2}, \tag{3.25}$$

and

$$b_{2m}(\log n) = 2\log\log n + m\log\log\log n - \log \Gamma(m). \tag{3.26}$$

Proof. Convergence properties imply that as $n \to \infty$,

$$a^2(\log n) \max_{1<k<\log n} k\bar{z}_k^{(j)^2} + (x + b_{2m}(\log n))^2 \xrightarrow{P} -\infty, \tag{3.27}$$

and

$$a^2(\log n) \max_{1<k<\log n} (n-k)\bar{z}_{n-k}^{(j)^2} - (x + b_{2m}(\log n))^2 \xrightarrow{P} -\infty. \tag{3.28}$$

Moreover,

$$a^2(\log n) \max_{1<k<\log n} (-n\bar{z}_n^{(j)^2}) - (x + b_{2m}(\log n))^2 \xrightarrow{P} -\infty. \tag{3.29}$$

Hence,

$$a^2(\log n) \max_{1<k<\log n} W_k^{(1)} - (x + b_{2m}(\log n))^2 \xrightarrow{P} -\infty. \tag{3.30}$$

Similarly,

$$a^2(\log n) \max_{1<k<\log n} W_k^{(2)} - (x + b_{2m}(\log n))^2 \xrightarrow{P} -\infty. \tag{3.31}$$

Therefore, as $n \to \infty$

$$a^2(\log n) \max_{1<k<\log n} (W_k^{(1)} + W_k^{(2)}) - (x + b_{2m}(\log n))^2 \xrightarrow{P} -\infty. \qquad □$$

Lemma 3.15 For all $x \in R$, as $n \to \infty$, the following hold.

(i) $a^2(\log n) \max_{1<k<\log n} \xi_k - (x + b_{2m}(\log n))^2 \xrightarrow{P} -\infty$;

(ii) $a^2(\log n) \max_{n-\log n<k<n} (W_k^{(1)} + W_k^{(2)}) - (x + b_{2m}(\log n))^2 \xrightarrow{P} -\infty$;

(iii) $a^2(\log n) \max_{n-\log n<k<n} \xi_k - (x + b_{2m}(\log n))^2 \xrightarrow{P} -\infty$.

Proof. From Lemma 3.13 and Lemma 3.14, one can obtain (i).

Proceeding in the same manner as in the proof of Lemma 3.14, one can immediately obtain (ii).

Now from Lemma 3.13, Lemma 3.14, and the law of iterated logarithm, one can obtain:

$$a^2(\log n) \max_{n-\log n < k < n} Q_k^{(1)} - (x + b_{2m}(\log n))^2 \xrightarrow{P} -\infty, \qquad (3.32)$$

and

$$a^2(\log n) \max_{n-\log n < k < n} Q_k^{(2)} - (x + b_{2m}(\log n))^2 \xrightarrow{P} -\infty, \qquad (3.33)$$

as $n \to \infty$, for all $x \in R$. In view of (ii) above, we thus conclude that (iii) holds. □

Lemma 3.16 *We have the following results.*

(i) $a^2(\log n) \max_{\log n \leq k \leq n - \log n} |\xi_k - W_k^{(1)} + W_k^{(2)}| = O_p(1)$;

(ii) $a^2(\log n) \max_{1 \leq k < \frac{n}{\log n}} |(n-k)\bar{z}_{n-k}^{(j)^2} - n\bar{z}_n^{(j)^2}| = O_p(1)$, $j = 1, 2, \ldots, m$;

(iii) for $j = 1, 2, \ldots, m$,

$$a^2(\log n) \max_{1 \leq k < (n/\log n)} \left| \frac{1}{n-k} \left[\sum_{i=k+1}^{n} (\bar{z}_i^{(j)^2} - 1) \right]^2 - \frac{1}{n} \left[\sum_{i=1}^{n} (\bar{z}_i^{(j)^2} - 1) \right]^2 \right| =$$

$$O_p(1).$$

Proof. From Lemmas 3.14 and 3.15, we have

$$\max_{1 \leq k < (n/\log n)} k^{1/2} (\log \log k)^{-(3/2)} |\xi_k - (W_k^{(1)} + W_k^{(2)})| = O_p(1). \qquad (3.34)$$

Hence, after some calculations, we obtain (i).

Starting with the identity

$$(n-k)\bar{z}_{n-k}^{(j)^2} - n\bar{z}_n^{(j)^2} = \frac{k}{n(n-k)} \left(\sum_{i=1}^{n} z_i^{(j)} \right)^2 - \frac{2}{n-k} \left(\sum_{i=1}^{n} z_i^{(j)} \right) \left(\sum_{i=1}^{k} z_i^{(j)} \right)$$

$$+ \frac{1}{n-k} \left(\sum_{i=1}^{k} z_i^{(j)} \right)^2, \qquad j = 1, 2, \ldots, m,$$

and combining it with the law of iterated logarithm, we can obtain (ii). Similar to (ii), we can easily prove (iii). □

Lemma 3.17 *As* $n \to \infty$, *for all* $x \in R$, *we have*

$$a^2(\log n) \max_{n/\log n < k < n - n/\log n} (W_k^{(1)} + W_k^{(2)}) - (x + b_{2m}(\log n))^2 \xrightarrow{P} -\infty.$$

Proof. From Theorem 2 of Darling and Erdös (1956), we obtain:

$$a^2(\log n) \max_{n/\log n < k < n - n/\log n} |W_k^{(1)} + W_k^{(2)}| = O_p((\log \log \log n)^2). \qquad (3.35)$$

Therefore, one can conclude that

$$a^2(\log n) \max_{n/\log n < k < n - n/\log n} (W_k^{(1)} + W_k^{(2)}) - (x + b_{2m}(\log n))^2 \xrightarrow{P} -\infty$$

as $n \to \infty$ for $x \in R$. $\qquad\qquad\square$

Similar to Lemma 3.16(ii) and (iii), we obtain the following result.

Lemma 3.18

(i) $a^2(\log n) \max_{n-n/\log n \le k < n} |k\bar{z}_k^{(j)^2} - n\bar{z}_n^{(j)^2}| = O_p(1)$, $j = 1, \ldots, m$.

(ii) $a^2(\log n) \max_{n-n/\log n \le k < n} |1/k[\sum_{i=1}^k (z_i^{(j)^2} - 1)]^2 - 1/n[\sum_{i=1}^n (z_i^{(j)^2} - 1)]^2| = O_p(1)$.
$j = 1, \ldots, m$,

Now we are in a position to state and prove our main result.

Theorem 3.19 *Under the null hypothesis H_0, when $n \to \infty$, $k \to \infty$ such that $(k/n) \to 0$, then*

$$\lim_{n\to\infty} P\{a(\log n)\lambda_n - b_{2m}(\log n) \le x\} = \exp\{-2e^{-x}\}$$

for $x \in R$, where $a(\log n)$ and $b_{2m}(\log n)$ are defined in (3.25) and (3.26), respectively.

Proof. Because of the above Lemmas 3.13–3.18, it suffices to show that

$$\lim_{n\to\infty} P\left\{a(\log n) \max_{\log n \le k < n - \log n} (W_k^{(1)} + W_k^{(2)})^{1/2} - b_{2m}(\log n) \le x\right\} = e^{-2e^{-x}}.$$

$$(3.36)$$

The left-hand side of (3.36) is equivalent to

$$\lim_{n\to\infty} P\left\{a^2(\log n) \max_{\log n \le k < n - \log n} (W_k^{(1)} + W_k^{(2)}) \le (x + b_{2m}(\log n))^2\right\}$$

$$= \lim_{n\to\infty} P\left\{a^2(\log n) \left[\max_{1 \le k < n - \log n} \left(\sum_{j=1}^m \left[k\bar{z}_k^{(j)^2} + \frac{1}{2k}\left(\sum_{i=1}^k (z_i^{(j)^2} - 1)^2\right)\right]\right),\right.\right.$$

$$\max_{n-n/\log n \le k < n} \left(\sum_{j=1}^m \left[(n-k)\bar{z}_{n-k}^{(j)^2} + \frac{1}{2(n-k)}\left(\sum_{i=k+1}^n (z_i^{(j)^2} - 1)^2\right)\right]\right)\right]$$

$$\left.\left. \le [x + b_{2m}(\log n)]^2\right\}.$$

$$(3.37)$$

Because $\{z_i^{(j)}, 1 \leq i < (n/\log n)\}$ and $\{z_i^{(j)}, n - (n/\log n) \leq i \leq n\}$ are independent, (3.37) reduces to

$$\lim_{n\to\infty} P\left\{ a^2(\log n) \max_{1 \leq k < n/\log n} \left[\sum_{j=1}^{m} \left(k\bar{z}_k^{(j)^2} + \frac{1}{2k}\left(\sum_{i=1}^{k}(z_i^{(j)^2} - 1)^2\right)\right) \right] \right.$$

$$\left. \leq (x + b_{2m}(\log n))^2 \right\}$$

$$\cdot \lim_{n\to\infty} P\left\{ a^2(\log n) \max_{n-n/\log n \leq k < n} \left[\sum_{j=1}^{m}(n-k)\bar{z}_{n-k}^{(j)^2} \right.\right.$$

$$\left.\left. + \frac{1}{2(n-k)}\sum_{i=k+1}^{n}(z_i^{(j)^2} - 1)^2 \right] \leq (x + b_{2m}(\log n))^2 \right\}$$

$$= \lim_{n\to\infty} P\left\{ a(\log n) \max_{1 \leq k < n/\log n} \left[\sum_{j=1}^{m}(k\bar{z}_k^{(j)^2} \right.\right.$$

$$\left.\left. + \frac{1}{2k}\left(\sum_{i=1}^{k}(z_i^{(j)^2} - 1)^2\right)\right]^{1/2} - b_{2m}(\log n) \leq x \right\}$$

$$\cdot \lim_{n\to\infty} P\left\{ a(\log n) \max_{n-n/\log n \leq k < n} \left[\sum_{j=1}^{m}(n-k)\bar{z}_{n-k}^{(j)^2} \right.\right.$$

$$\left.\left. + \frac{1}{2(n-k)}\left(\sum_{i=k+1}^{n}(z_i^{(j)^2} - 1)^2\right)\right]^{1/2} - b_{2m}(\log n) \leq x \right\}.$$

$$(3.38)$$

On the R.H.S. of (3.38) denote the first term by (e), and the second by (f). Let us consider (e) first

$$k\bar{z}_k^{(j)^2} + \frac{1}{2k}\left(\sum_{i=1}^{k}(z_i^{(j)^2} - 1)^2\right) = \left(\sum_{i=1}^{k}\frac{z_i^{(j)}}{\sqrt{k}}\right)^2 + \left(\sum_{i=1}^{k}\frac{z_i^{(j)^2} - 1}{\sqrt{2k}}\right)^2.$$

Let

$$\mathbf{v}_i = \left(z_i^{(1)}, \ldots, z_i^{(m)}, \frac{z_i^{(1)^2} - 1}{\sqrt{2}}, \ldots, \frac{z_i^{(m)^2} - 1}{\sqrt{2}} \right), \qquad 1 \leq i < \infty;$$

then $\{\mathbf{v}_i, 1 \leq i < \infty\}$ is a sequence of iid d-dimensional random vectors with $d = 2m$, and

$$v_i^{(1)} = z_i^{(1)}, \ldots, v_i^{(m)} = z_i^{(m)},$$

$$\mathbf{v_i^{(m+1)}} = \frac{\mathbf{z_i}^{(1)^2} - 1}{\sqrt{2}}, \ldots, \mathbf{v_i^{(2m)}} = \frac{\mathbf{z_i}^{(m)^2} - 1}{\sqrt{2}}.$$

After some computations, we have $E[v_i^{(j)}] = 0, \text{Var}[v_i^{(j)}] = 1$ for $1 \leq j \leq 2m$, and $\text{cov}(v_i^{(j)}, v_i^{(1)}) = 1$ for $j \neq 1, 1 < j, 1 < 2m$. Hence, the covariance matrix of $\mathbf{v_i}$ is the identity matrix. And clearly, $E|v_i^{(j)}|^r < \infty$ for all j and some $r > 2$.

Let $S_i^{(j)} = \sum_{i=1}^{k} v_i^{(j)}$ for $1 \leq j \leq 2m$; then

$$\sum_{j=1}^{2m} \left(\frac{S_i^{(j)}}{\sqrt{k}} \right)^2 = k z_k^{(j)^2} + \frac{1}{2k} \sum_{i=1}^{k} (z_i^{(j)^2} - 1)^2.$$

Hence, in view of Lemma 2.18, we have

$$(e) = \exp\{-e^{-x}\}.$$

Similarly, we can show that $(f) = \exp\{-e^{-x}\}$. This completes the proof of the theorem. $\qquad\square$

3.3.2 Informational Approach

(i) Derivation of the SICs

Under H_0, the log likelihood function is given by

$$\log L_0(\theta) = -\frac{1}{2} mn \log 2\pi - \frac{n}{2} \log |\Sigma| - \frac{1}{2} \sum_{i=1}^{n} (\mathbf{x}_i - \mu)' \Sigma^{-1} (\mathbf{x}_i - \mu),$$

where $\theta = (\mu, \Sigma)$. Let $\widehat{\theta} = (\widehat{\mu}, \widehat{\Sigma})$ be the MLE of θ; then $\widehat{\mu} = \bar{\mathbf{x}}$ and $\widehat{\Sigma} = (1/n) \sum_{i=1}^{n} (\mathbf{x}_i - \bar{\mathbf{x}})(\mathbf{x}_i - \bar{\mathbf{x}})'$. Hence, the maximum log likelihood is

$$\log L_0(\widehat{\theta}) = -\frac{1}{2} mn \log 2\pi - \frac{n}{2} \log |\widehat{\Sigma}| - \frac{n}{2},$$

and the SIC under H_0 is

$$\text{SIC}(n) = mn \log 2\pi + n \log |\widehat{\Sigma}| + n + \frac{m(m+3)}{2} \log n. \tag{3.39}$$

Under H_1, the log likelihood function is

$$\log L_1(\theta) = -\frac{1}{2}mn \log 2\pi - \frac{k}{2}\log|\Sigma_1| - \frac{n-k}{2}\log|\Sigma_n|$$

$$-\frac{1}{2}\sum_{i=1}^{k}(\mathbf{x}_i-\mu_1)'\Sigma_1^{-1}(\mathbf{x}_i-\mu_1) - \frac{1}{2}\sum_{i=k+1}^{n}(\mathbf{x}_i-\mu_n)'\Sigma_1^{-1}(\mathbf{x}_i-\mu_n).$$

Here, $\theta = (\mu_1, \mu_n, \Sigma_1, \Sigma_n)$. Denote the MLE of θ by $\widehat{\theta} = (\widehat{\mu}_1, \widehat{\mu}_n, \widehat{\Sigma}_1, \widehat{\Sigma}_n)$; then

$$\widehat{\mu}_1 = \overline{\mathbf{x}}_k = \frac{1}{k}\sum_{i=1}^{k}\mathbf{x}_i, \qquad \widehat{\mu}_n = \overline{\mathbf{x}}_{n-k} = \frac{1}{n-k}\sum_{i=k+1}^{n}\mathbf{x}_i,$$

$$\widehat{\Sigma}_1 = \frac{1}{k}\sum_{i=1}^{k}(\mathbf{x}_i-\overline{\mathbf{x}}_k)(\mathbf{x}_i-\overline{\mathbf{x}}_k)', \quad \text{and}$$

$$\widehat{\Sigma}_n = \frac{1}{n-k}\sum_{i=k+1}^{n}(\mathbf{x}_i-\overline{\mathbf{x}}_{n-k})(\mathbf{x}_i-\overline{\mathbf{x}}_{n-k})'.$$

Then the maximum log likelihood is

$$\log L_1(\widehat{\theta}) = -\frac{1}{2}mn \log 2\pi - \frac{k}{2}\log|\widehat{\Sigma}_1| - \frac{n-k}{2}\log|\widehat{\Sigma}_n| + n;$$

hence, the SICs under H_1 are:

$$\text{SIC}(k) = mn\log 2\pi + k\log|\widehat{\Sigma}_1| + (n-k)\log|\widehat{\Sigma}_n| + n + m(m+3)\log n, \quad (3.40)$$

where $m < k < n - m$.

Notice that, in order to obtain the MLEs here, we can detect changes only when $m < K < n - m$. According to the principle of information criterion, we accept H_0 if

$$\text{SIC}(n) < \min_{m<k<n-m} \text{SIC}(k),$$

and we accept H_1 if

$$\text{SIC}(n) > \text{SIC}(k)$$

for some k and estimate the position of the change point by $\widehat{k}$ such that

$$\text{SIC}(\widehat{k}) = \min_{m<k<n-m} \text{SIC}(k). \qquad (3.41)$$

(ii) Asymptotic property of the SIC

In this section, we provide an asymptotic property of the SIC derived in (i). Because $\lambda_n = [(m(m+3))/2 \log n - \Delta_n]^{1/2}$, we obtain the following theorem as a Corollary of Theorem 3.20.

Theorem 3.20 Let $\Delta_n = \min_{m<k<n-m}[\text{SIC}(k) - \text{SIC}(n)]$, where $\text{SIC}(k)$ is given by (3.40), and $\text{SIC}(n)$ by (3.39). Under H_0, then

$$\lim_{n\to\infty} P\left[a(\log n)\left[\frac{m(m+3)}{2}\log n - \Delta_n\right]^{1/2} - b_{2m}(\log n) \le x\right]$$

$$= \exp\{-2e^{-x}\}, \tag{3.42}$$

where $a(\log n)$ and $b_{2m}(\log n)$ are defined in (3.25) and (3.26).

(iii) Approximate Critical Values c_α

As in Chapter 2, we modify the minimum SIC procedure by introducing the significant level α and its associate critical value c_α. Then we accept H_0 if $\text{SIC}(n) < \min_{m<k<n-m}\text{SIC}(k) + c_\alpha$, where c_α is determined by

$$1 - \alpha = P\left(\text{SIC}(n) < \min_{m<k<n-m}\text{SIC}(k) + c_\alpha|H_0\right).$$

Using Theorem 3.21,

$$1 - \alpha = P(\Delta_n > -c_\alpha|H_0)$$

$$= P\left(\frac{m(m+3)}{2}\log n - \Delta_n < \frac{m(m+3)}{2}\log n + c_\alpha|H_0\right)$$

$$= P\left(\lambda_n^2 < \left[\frac{m(m+3)}{2}\log n + c_\alpha\right]^{1/2}|H_0\right)$$

$$= P\left(-b_{2m}(\log n) < a(\log n)\lambda_n - b_{2m} < a(\log n)\left[\frac{m(m+3)}{2}\log n + c_\alpha\right]^{1/2}\right.$$

$$\left. - b_{2m}(\log n)\right)$$

$$\cong \exp\left(-2\exp\left[-a(\log n)\left\{\frac{m(m+3)}{2}\log n + c_\alpha\right\}^{1/2} + b_{2m}(\log n)\right]\right)$$

$$- \exp(-2\exp[b_{2m}(\log n)]).$$

Solving for c_α, we get

$$c_\alpha = \left\{ -\frac{1}{a(\log n)} \log \log[1 - \alpha + \exp(-2\exp(b_{2m}(\log n)))]^{-(1/2)} + \frac{b_{2m}(\log n)}{a(\log n)} \right\}$$
$$- \frac{m(m+3)}{2} \log n.$$

For a different α and sample size n, we computed c_α and listed them in Table 3.7 for $m = 2$.

(iv) Unbiased Version of SICs

As in Chapter 2, for the hypotheses (3.14) and (3.15), and finite sample size n, we derived the unbiased version $u-\mathrm{SIC}(n)$ and $u-\mathrm{SIC}(k)$, $m < k < n-m$, respectively.

(1) $u - \mathrm{SIC}(n)$

Let $Y = (\mathbf{y}_1, \mathbf{y}_2, \ldots, \mathbf{y}_n)$ be a sample of the same size and distribution as the Xs, where $X = (\mathbf{x}_1, \mathbf{x}_2, \ldots, \mathbf{x}_n)$, and Y be independent of X.

Under H_0 given by (3.14),

$$J = E_{\widehat{\theta}}[E_{\theta_0|Y}(\log L_0(\widehat{\theta}))]$$

$$= E_{\widehat{\theta}}\left[E_{\theta_0|Y} \left\{ -\frac{1}{2}mn \log 2\pi - \frac{n}{2} \log |\widehat{\Sigma}| - \frac{1}{2} \sum_{i=1}^{n} (\mathbf{y}_i - \widehat{\mu})' \widehat{\Sigma}^{-1} (\mathbf{y}_i - \widehat{\mu}) \right\} \right]$$

$$= E_{\widehat{\theta}}\left[-\frac{1}{2}mn \log 2\pi - \frac{n}{2} \log |\widehat{\Sigma}| - \frac{n}{2} + \frac{n}{2} \right.$$
$$\left. -\frac{1}{2} E_{\theta_0|Y} \left\{ \sum_{i=1}^{n} (\mathbf{y}_i - \widehat{\mu})' \widehat{\Sigma}^{-1} (\mathbf{y}_i - \widehat{\mu}) \right\} \right]$$

$$= E_{\widehat{\theta}}\left[\log L_0(\widehat{\theta}) + \frac{n}{2} - \frac{1}{2} E_{\theta_0|Y} \left\{ \sum_{i=1}^{n} (\mathbf{y}_i - \widehat{\mu})' \widehat{\Sigma}^{-1} (\mathbf{y}_i - \widehat{\mu}) \right\} \right],$$

where

$$\widehat{\Sigma} = \frac{1}{n} \sum_{i=1}^{n} (\mathbf{x}_i - \overline{\mathbf{x}})(\mathbf{x}_i - \overline{\mathbf{x}})', \qquad \widehat{\mu} = \overline{\mathbf{x}},$$

and

$$\log L_0(\widehat{\theta}) = -\frac{1}{2}mn \log 2\pi - \frac{n}{2} \log |\widehat{\Sigma}| - \frac{n}{2}.$$

Here $\mathbf{x}_i$s are iid $N_m(\mu, \Sigma)$, and $\mathbf{y}_i$s are iid $N_m(\mu, \Sigma)$ for $i = 1, 2, \ldots, n$, and X is independent of Y, thus we have $\mathbf{y}_i - \bar{\mathbf{x}} \sim N_m(0, ((n+1)/n)\Sigma)$ and

$$\sqrt{\frac{n+1}{n}}(\mathbf{y}_i - \bar{\mathbf{x}}) \sim N_m(0, \Sigma).$$

Now,

$$(n-1)S = \sum_{i=1}^{n}(\mathbf{x}_i - \bar{\mathbf{x}})(\mathbf{x}_i - \bar{\mathbf{x}})' = n\widehat{\Sigma}$$

is distributed as $\sum_{i=1}^{n-1}\mathbf{z}_i\mathbf{z}_i'$, where $\mathbf{z}_i$s are iid $N_m(0, \Sigma)$ random vectors, and S is independent of $\bar{\mathbf{x}}$. But S is also independent of Y, therefore, $\widehat{\Sigma} = ((n-1)/n)S$ is independent of $\mathbf{y}_i - \bar{\mathbf{x}}$ for $i = 1, 2, \ldots, n$. From Theorem 5.2.2 in Anderson (1984), by letting

$$T^2 = \frac{n}{n+1}(\mathbf{y}_i - \bar{\mathbf{x}})'S^{-1}(\mathbf{y}_i - \bar{\mathbf{x}}),$$

then we have

$$\frac{T^2}{n-1}\frac{n-m}{m} \sim F_{m,n-m}.$$

That is,

$$\frac{\frac{n}{n+1}(\mathbf{y}_i - \bar{\mathbf{x}})'S^{-1}(\mathbf{y}_i - \bar{\mathbf{x}})}{n-1}\frac{n-m}{m} \sim F_{m,n-m},$$

or

$$\frac{n-m}{(n+1)m}(\mathbf{y}_i - \bar{\mathbf{x}})'\widehat{\Sigma}^{-1}(\mathbf{y}_i - \bar{\mathbf{x}}) \sim F_{m,n-m}.$$

Hence,

$$E_{\theta_0|Y}[(\mathbf{y}_i - \bar{\mathbf{x}})'\widehat{\Sigma}^{-1}(\mathbf{y}_i - \bar{\mathbf{x}})] = \frac{(n+1)m}{n-m-2}.$$

Then, J is simplified to

$$J = E_{\widehat{\theta}}\left[\log L_0(\widehat{\theta}) + \frac{n}{2} - \frac{mn(n+1)}{2(n-m-2)}\right]$$

$$= E_{\widehat{\theta}}[\log L_0(\widehat{\theta})] - \frac{n(mn+2m+2-n)}{2(n-m-2)}.$$

Thus, $\log L_0(\widehat{\theta}) - (n(mn+2m+2-n))/(2(n-m-2))$ is unbiased for J and we define the $u - \text{SIC}(n)$ as

$$u - \text{SIC}(n) = -2\log L_0(\widehat{\theta}) + \frac{n(mn+2m+2-n)}{n-m-2}$$

$$= \text{SIC}(n) + \frac{n(mn+2m+2-n)}{n-m-2} - \frac{m(m+3)}{2}\log n,$$

where $\text{SIC}(n)$ is defined in (3.39).

(2) $u - \mathrm{SIC}(k), m < k < n - m$

Let $Y = (\mathbf{y}_1, \mathbf{y}_2, \ldots, \mathbf{y}_n)$ be a sample of the same size and distribution as the Xs, where $X = (\mathbf{x}_1, \mathbf{x}_2, \ldots, \mathbf{x}_n)$, and Y be independent of X. That is, under $H_1, \mathbf{x}_1, \mathbf{x}_2, \ldots, \mathbf{x}_k, \ \mathbf{y}_1, \mathbf{y}_2, \ldots, \mathbf{y}_k$ are iid $N_m(\mu_1, \Sigma_1)$ random vectors, and $\mathbf{x}_{k+1}, \mathbf{x}_{k+2}, \ldots, \mathbf{x}_n, \mathbf{y}_{k+1}, \mathbf{y}_{k+2}, \ldots, \mathbf{y}_n$ are iid $N_m(\mu_n, \Sigma_n)$ random vectors.

$$
\begin{aligned}
J &= E_{\widehat{\theta}}[E_{\theta_0|Y}(\log L_1(\widehat{\theta}))] \\
&= E_{\widehat{\theta}}\left\{ -\frac{1}{2}mn \log 2\pi - \frac{k}{2}\log|\widehat{\Sigma}_1| - \frac{n-k}{2}\log|\widehat{\Sigma}_n| \right. \\
&\qquad \left. -\frac{1}{2}\sum_{i=1}^{n}(\mathbf{y}_i - \overline{\mathbf{x}}_k)'\widehat{\Sigma}_1^{-1}(\mathbf{y}_i - \overline{\mathbf{x}}_k) - \frac{1}{2}\sum_{i=k+1}^{n}(\mathbf{y}_i - \overline{\mathbf{x}}_{n-k})'\widehat{\Sigma}_1^{-1}(\mathbf{y}_i - \overline{\mathbf{x}}_{n-k}) \right] \right\} \\
&= E_{\widehat{\theta}}\left\{ \log L_1(\widehat{\theta}) + \frac{n}{2} + E_{\theta_0|Y}\left[-\frac{1}{2}\sum_{i=1}^{n}(\mathbf{y}_i - \overline{\mathbf{x}}_k)'\widehat{\Sigma}_1^{-1}(\mathbf{y}_i - \overline{\mathbf{x}}_k) \right. \right. \\
&\qquad \left. \left. -\frac{1}{2}\sum_{i=k+1}^{n}(\mathbf{y}_i - \overline{\mathbf{x}}_{n-k})'\widehat{\Sigma}_1^{-1}(\mathbf{y}_i - \overline{\mathbf{x}}_{n-k}) \right] \right\}.
\end{aligned}
$$

Because

$$
y_i - \overline{\mathbf{x}}_k \sim \mathrm{iid}\,N_m\left(0, \frac{k+1}{k}\Sigma_1\right), \qquad i = 1, 2, \ldots, k,
$$

and

$$
(k-1)S_1 = \sum_{i=1}^{k}(\mathbf{x}_1 - \overline{\mathbf{x}}_k)(\mathbf{x}_1 - \overline{\mathbf{x}}_k)' = k\widehat{\Sigma}_1 \sim \sum_{i=1}^{k-1}\mathbf{z}_i\mathbf{z}_i',
$$

where $\mathbf{z}_i$s are iid $N_m(0, \Sigma_1)$ for $i = 1, 2, \ldots, k-1$. Furthermore, S_1 and $\overline{\mathbf{x}}_k$ are independent, and $\widehat{\Sigma}_1$ is independent of $\mathbf{y}_i - \overline{\mathbf{x}}_k$ for $i = 1, 2, \ldots, k$. From Theorem 5.2.2 in Anderson (1984), by letting

$$
T^2 = \frac{k}{k+1}(\mathbf{y}_i - \overline{\mathbf{x}}_k)'S_1^{-1}(\mathbf{y}_i - \overline{\mathbf{x}}_k),
$$

we have

$$
\frac{T^2}{k-1}\frac{k-m}{m} \sim F_{m,k-m}.
$$

Therefore,

$$
\frac{\frac{k}{k+1}(\mathbf{y}_i - \overline{\mathbf{x}}_k)'S_1^{-1}(\mathbf{y}_i - \overline{\mathbf{x}}_k)}{k-1}\frac{k-m}{m} \sim F_{m,k-m}.
$$

Then,

$$
E_{\theta_0|Y}\left[\sum_{i=1}^{n}(\mathbf{y}_i - \overline{\mathbf{x}}_k)'\widehat{\Sigma}_1^{-1}(\mathbf{y}_i - \overline{\mathbf{x}}_k)\right] = \frac{m(k+1)}{k-m-2} \quad \text{for } i = 1, 2, \ldots, k.
$$

Similarly, we obtain:

$$
E_{\theta_0|Y} \left[\sum_{i=k+1}^{n} (\mathbf{y}_i - \overline{\mathbf{x}}_{n-k})' \widehat{\Sigma}_1^{-1} (\mathbf{y}_i - \overline{\mathbf{x}}_{n-k}) \right]
$$

$$
= \frac{m(n-k-1)}{n-k-m-2} \quad \text{for } i = k+1, \ldots, n.
$$

Therefore, J reduces to:

$$
J = E_{\widehat{\theta}}\{\log L_1(\widehat{\theta})\} + \frac{n}{2} - \frac{km(k+1)}{2(k-m-2)} - \frac{m(n-k)(n-k+1)}{2(n-k-m-2)}
$$

$$
= E_{\widehat{\theta}}\{\log L_1(\widehat{\theta})\} - \frac{c}{2},
$$

where $\log L_1(\widehat{\theta})$ is defined as before, and

$$
c = \frac{m[k(k+1)(n-k-m-2)+(n-k)(n-k+1)(k-m-2)] - n(k-m-2)(n-k-m-2)}{(k-m-2)(n-k-m-2)}.
$$

Because $\log L_1(\widehat{\theta}) - (c/2)$ is unbiased for J, we define:

$$
u - \text{SIC}(k) = -2\log L_1(\widehat{\theta}) + c, \quad \text{for } m < k < n-m
$$

or

$$
u - \text{SIC}(k) = \text{SIC}(k) + c - m(m+3)\log n, \quad \text{for } m < k < n-m.
$$

3.3.3 Examples

In this section, we carry out the data analysis of change points for the following examples by using the SIC procedure.

Example 3.3 As in Anderson (1984), two-dimensional data from Shewhart (1931) are analyzed for testing any change in mean vector and covariance matrix. These are the tensile strength data we analyzed in Chapter 2. But here, in addition to the tensile strength, the hardness of the tensile is also taken into consideration. We test (3.14) against (3.15). The SIC$(k), 3 \le k \le 57$, at the first stage, along with SIC(60), are listed in Table 3.8.

From Table 3.8, we obtain that SIC(60) $= 1553.3, \min_{3 \le k \le 57} \text{SIC}(k) = \text{SIC}(47) = 1556.9$. SIC(60) $< \min_{3 \le k \le 57} \text{SIC}(k) + c_\alpha$ for any α, therefore we fail to reject H_0, and conclude that there is no change in both mean vectors and covariance matrix among the 60 tensile strength specimen. Clearly, our conclusion agrees with that of Anderson.

Example 3.4 As a second example in this chapter, we take the Indian agricultural data analyzed in Giri (1977). There were 27 randomly selected plants of Sonolika, a late-sown variety of wheat, which are observed on six different characters in two consecutive years (1971, 1972). These six different characters are:

x_1: plant height at harvesting (cm)
x_2: number of effective tillers
x_3: length of ear (cm)

Table 3.9 Approximate Critical Values of SIC, $m = 2$

n/α	0.010	0.025	0.050
10	15.105	7.966	3.363
11	14.367	7.620	3.159
12	13.951	7.372	2.985
13	13.621	7.148	2.814
14	13.324	6.933	2.642
15	13.046	6.725	2.472
16	12.783	6.523	2.303
17	12.534	6.328	2.138
18	12.296	6.139	1.976
19	12.069	5.956	1.818
20	11.852	5.780	1.664
21	11.645	5.610	1.514
22	11.446	5.445	1.369
23	11.255	5.286	1.227
24	11.072	5.131	1.089
25	10.895	4.982	0.955
26	10.724	4.837	0.825
27	10.560	4.696	0.698
28	10.401	4.560	0.574
29	10.247	4.427	0.454
30	10.098	4.299	0.337
35	9.416	3.703	0.000
40	8.821	3.176	0.000
45	8.293	2.704	0.000
50	7.818	2.277	0.000
55	7.386	1.887	0.000
60	6.992	1.527	0.000
70	6.289	0.884	0.000
80	5.678	0.321	0.000
90	5.137	0.000	0.000
100	4.562	0.000	0.000
120	3.810	0.000	0.000
140	3.096	0.000	0.000
160	2.476	0.000	0.000
180	1.928	0.000	0.000
200	1.436	0.000	0.000

x_4: number of fertile spikelets per 10 ears
x_5: number of grains per 10 ears
x_6: weight of grains per 10 ears.

We view these observations as taken in a sequence from 1971 to 1972. Then we have a sequence of 54 observations where the 28th through 54th observations correspond to the observations obtained in 1972. Assuming that the observations form a sample from a six-dimensional normal distribution, we test:

$$H_0 : \mu_1 = \mu_2 = \cdots = \mu_{54} = \mu \text{ (unknown)}$$

$$\text{and} \quad \Sigma_1 = \Sigma_2 = \cdots = \Sigma_{54} = \Sigma \text{ (unknown)}$$

versus the alternative:

$$H_1 : \mu_1 = \cdots = \mu_{k_1} \neq \mu_{k_1+1} = \cdots = \mu_{k_2} \neq \cdots \neq \mu_{k_q+1} = \cdots = \mu_{k_{q+1}} = \mu_{54}$$

and

$$\Sigma_1 = \cdots = \Sigma_{k_1} \neq \Sigma_{k_1+1} = \cdots = \Sigma_{k_2} \neq \cdots \neq \Sigma_{k_q+1} = \cdots = \Sigma_{k_{q+1}} = \Sigma_{54},$$

According to the binary segmentation procedure, at the first stage we detect a single change between the 7th and 47th observations. The $\mathrm{SIC}(k), 7 \leq k \leq 47$, along with $\mathrm{SIC}(54)$ are listed in Table 3.9.

From Table 3.9, we obtain that $\mathrm{SIC}(54) = 1474.0$, $\min_{7 \leq k \leq 47} \mathrm{SIC}(k)$, we reject H_0, and conclude that at the first stage, there is one change point which is located at the 27th observation. Next, two subsequences, from the 1st observation through the 27th observation, and from the 28th observation through the 54th observation, were tested for change but no change point was found.

Table 3.10 SIC Values of Example 3.3 at the First Stage

k	SIC(k)	k	SIC(k)	k	SIC(k)	k	SIC(K)
3	1566.3	17	1572.5	31	1566.8	45	1566.4
4	1562.4	18	1572.6	32	1565.4	46	1564.1
5	1564.5	19	1572.5	33	1565.5	47	1556.9
6	1563.3	20	1572.1	34	1563.9	48	1557.0
7	1562.8	21	1570.6	35	1564.2	49	1559.9
8	1560.4	22	1568.1	36	1563.8	50	1561.7
9	1565.6	23	1567.8	37	1564.0	51	1560.9
10	1563.7	24	1568.1	38	1565.9	52	1559.3
11	1564.5	25	1567.6	39	1567.0	53	1562.2
12	1562.8	26	1568.7	40	1566.7	54	1561.3
13	1568.1	27	1569.2	41	1565.1	55	1563.5
14	1569.1	28	1568.5	42	1565.2	56	1557.5
15	1568.6	29	1567.6	43	1564.2	57	1558.7
16	1569.6	30	1567.5	44	1564.5	60	1553.3

Table 3.11 SIC Values of Example 3.4 at the First Stage

k	SIC(k)	k	SIC(k)	k	SIC(k)
7	1527.7	21	1496.5	35	1480.7
8	1529.1	22	1489.8	36	1480.2
9	1517.2	23	1465.4	37	1484.5
10	1515.3	24	1464.0	38	1488.9
11	1518.7	25	1456.5	39	1501.4
12	1512.8	26	1445.9	40	1506.7
13	1509.0	27	1433.3	41	1498.7
14	1507.1	28	1452.2	42	1492.0
15	1503.1	29	1466.6	43	1484.9
16	1499.7	30	1479.3	44	1491.2
17	1492.3	31	1485.3	45	1504.0
18	1486.9	32	1475.3	46	1509.2
19	1493.1	33	1472.8	47	1496.0
20	1496.3	34	1476.8	54	1474.0

In other words, our conclusion simply says that the observations made in 1971 are different from those in 1972. This conclusion agrees with Giri's conclusion in his analysis.

Chapter 4
Regression Models

4.1 Literature Review

Regression analysis is an important statistical application employed in many disciplines. Before the introduction of a change point hypothesis into the regression study, the statistician faced problems of being unable to establish a regression model for some observed datasets. If the data structure has changed after a certain point of time, then using one regression model to study the data obviously leaves the data unfitted or leaves them poorly explained by a regression model. Ever since the change point hypothesis was introduced into statistical analyses, the study of switching regression models has taken place in regression analysis. This made some previously poorly fitted regression models better fitted to some datasets after the change point was been located in the regression models.

In the literature, many authors have studied the change point problem associated with regression models. Quandt (1958, 1960) derived a likelihood ratio-based test for testing and estimation about a linear regression model obeying two separate regimes. Ferreira (1975) studied a switching regression model from the Bayesian point of view with the assumption of a known number of regimes. Brown, Durbin, and Evans (1975) introduced a method of recursive residuals to test change points in multiple regression models. Hawkins (1989) used a union and intersection approach to test changes in a linear regression model. Kim (1994) considered a test for a change point in linear regression by using the likelihood ratio statistic, and studied the asymptotic behavior of the LRT statistic.

In this chapter, we present the change point problem in regression models by combining the work of Quandt (1958, 1960), Ferreira (1975), Hawkins (1989), Brown, Durbin, and Evans (1975), Kim (1994), and Chen (1998). Specifically, we discuss the change point problem for the simple linear regression model, as well as for the multiple linear regression model mainly by using the Schwarz information criterion, and by a Bayesian approach.

4.2 Simple Linear Regression Model

Let $(x_1, y_1), (x_2, y_2), \ldots, (x_n, y_n)$ be a sequence of observations obtained in a practical situation. The researcher currently might be interested in fitting a linear regression model to these data,

$$y_i = \beta_0 + \beta_1 x_i + \varepsilon_i, \qquad i = 1, \ldots, n,$$

where x_i, $i = 1, \ldots, n$, is a nonstochastic variable, β_0 and β_1 are unknown regression parameters, $\varepsilon_i, i = 1, \ldots, n$, is a random error distributed as $N(0, \sigma^2)$, with σ^2 unknown, and ε_is are uncorrelated from observation to observation. That is, y_i, $i = 1, \ldots, n$, is a random variable distributed as $N(\beta_0 + \beta_1 x_i, \sigma^2)$. The nature of the data might cause the researcher to ponder that the regression coefficients have changed after a certain point, say k, of the observations, therefore, the following hypothesis testing might be interesting to work on. That is, we want to test the null hypothesis:

$$H_0 : \mu_{y_i} = \beta_0 + \beta_1 x_i, \quad \text{for } i = 1, \ldots, n$$

versus the alternative hypothesis:

$$H_0 : \mu_{y_i} = \beta_0^1 + \beta_1^1 x_i, \quad \text{for } i = 1, \ldots, k$$
$$\text{and} \quad \mu_{y_i} = \beta_0^* + \beta_1^* x_i, \quad \text{for } i = k+1, \ldots, n,$$

where k, $k = 2, \ldots, n-2$, is the location of the change point; $\beta_0, \beta_1, \beta_0^1, \beta_1^1, \beta_0^*$, and β_1^* are unknown regression parameters. In the following sections we study several methods of locating the change point k.

4.2.1 Informational Approach

Again, if we use an information criterion such as SIC, we can locate the change point position by using the minimum SIC principle.

Under H_0, the likelihood function is

$$L_0(\beta_0, \beta_1, \sigma^2) = \prod_{i=1}^{n} f_{Y_i}(y_i; \beta_0, \beta_1, \sigma^2)$$

$$= \prod_{i=1}^{n} \frac{1}{\sqrt{2\pi\sigma^2}} e^{-(((y_i - \beta_0 - \beta_1 x_i)^2)/2\sigma^2)}$$

$$= \frac{1}{(\sqrt{2\pi\sigma^2})^n} \exp\left[-\sum_{i=1}^{n} (y_i - \beta_0 - \beta_1 x_i)^2/2\sigma^2 \right],$$

and the maximum likelihood estimates of β_0, β_1, and σ^2 are, respectively,

$$b_1 \triangleq \widehat{\beta}_1 = \frac{S_{xy}}{S_x},$$

$$b_0 \triangleq \widehat{\beta}_0 = \overline{y} - b_1 \overline{x},$$

$$\widehat{\sigma}^2 = \frac{1}{n} \sum_{i=1}^{n} (y_i - b_0 - b_1 x_i)^2,$$

where

$$\overline{x} = \frac{1}{n} \sum_{i=1}^{n} x_i,$$

$$\overline{y} = \frac{1}{n} \sum_{i=1}^{n} y_i,$$

$$S_x = \sum_{i=1}^{n} (x_i - \overline{x})^2,$$

$$S_{xy} = \sum_{i=1}^{n} (x_i - \overline{x})(y_i - \overline{y}).$$

It is clear to see that the MLEs obtained above coincide with least square estimates of β_0, β_1, and σ^2.

Therefore, the maximum likelihood under H_0 is

$$\sup L_0(\beta_0, \beta_1, \sigma^2) = L_0(\widehat{\beta}_0, \widehat{\beta}_1, \widehat{\sigma}^2)$$

$$= \frac{n^{n/2} e^{-n/2}}{\sqrt{2\pi}^{-n} \left[\sum\limits_{i=1}^{n} (y_i - b_0 - b_1 x_i)^2 \right]^{n/2}},$$

and the SIC under H_0, denoted by SIC(n), is obtained as

$$\text{SIC}(n) = -2 \log L_0(\widehat{\beta}_0, \widehat{\beta}_1, \widehat{\sigma}^2) + 3 \log n$$

$$= n \log 2\pi + n \log \left(\sum_{i=1}^{n} (y_i - b_0 - b_1 x_i)^2 \right) + n + 3 \log n - n \log n.$$

Under H_1, the likelihood function is

$$L_1(\beta_0^1, \beta_1^1, \beta_0^*, \beta_1^*, \sigma^2)$$

$$= \prod_{i=1}^{n} f_{Y_i}(y_i; \beta_0^1, \beta_1^1, \beta_0^*, \beta_1^*, \sigma^2)$$

$$= \prod_{i=1}^{k} \frac{1}{\sqrt{2\pi\sigma^2}} e^{-(((y_i - \beta_0^1 - \beta_1^1 x_i)^2)/2\sigma^2)} \prod_{i=k+1}^{n} \frac{1}{\sqrt{2\pi\sigma^2}} e^{-(((y_i - \beta_0^* - \beta_1^* x_i)^2)/2\sigma^2)}$$

$$= \frac{1}{(\sqrt{2\pi\sigma^2})^n} \exp\left[-\sum_{i=1}^{k} (y_i - \beta_0^1 - \beta_1^1 x_i)^2/2\sigma^2 \right]$$

$$\cdot \exp\left[-\sum_{i=k+1}^{n} (y_i - \beta_0^* - \beta_1^* x_i)^2/2\sigma^2 \right].$$

Similar calculations give the MLEs of $\beta_0^1, \beta_1^1, \beta_0^*, \beta_1^*$, and σ^2 as the following, respectively.

$$b_1^1 \triangleq \widehat{\beta}_1^1 = \frac{{}_k S_{xy}}{{}_k S_x},$$

$$b_0^1 \triangleq \widehat{\beta}_0^1 = \overline{y}_k - b_1^1 \overline{x}_k,$$

$$b_1^* \triangleq \widehat{\beta}_1^* = \frac{{}_{n-k} S_{xy}}{{}_{n-k} S_x},$$

$$b_0^* \triangleq \widehat{\beta}_0^* = \overline{y}_{n-k} - b_1^* \overline{x}_{n-k},$$

$$\widehat{\sigma}^2 = \frac{1}{n} \left[\sum_{i=1}^{k} (y_i - b_0^1 - b_1^1 x_i)^2 + \sum_{i=k+1}^{n} (y_i - b_0^* - b_1^* x_i)^2 \right],$$

where

$$\overline{x}_k = \frac{1}{k} \sum_{i=1}^{k} x_i,$$

$$\overline{y}_k = \frac{1}{k} \sum_{i=1}^{k} y_i,$$

$$\overline{x}_{n-k} = \frac{1}{n-k} \sum_{i=k+1}^{n} x_i,$$

$$\overline{y}_{n-k} = \frac{1}{n-k} \sum_{i=k+1}^{n} y_i,$$

$${}_k S_x = \sum_{i=1}^{k} (x_i - \overline{x}_k)^2,$$

$${}_k S_{xy} = \sum_{i=1}^{k} (x_i - \overline{x}_k)(y_i - \overline{y}_k),$$

$$n-k S_x = \sum_{i=k+1}^{n} (x_i - \bar{x}_{n-k})^2,$$

$$n-k S_{xy} = \sum_{i=k+1}^{n} (x_i - \bar{x}_{n-k})(y_i - \bar{y}_{n-k}).$$

Hence, we obtain the SIC under H_1, denoted by $SIC(k)$, for $k = 2, \ldots, n-2$, as follows.

$$SIC(k) = -2 \log L_1(\widehat{\beta}_0^1, \widehat{\beta}_1^1, \widehat{\beta}_0^*, \widehat{\beta}_1^*, \widehat{\sigma}^2) + 5 \log n$$

$$= n \log 2\pi + n \log \left[\sum_{i=1}^{k} (y_i - b_0^1 - b_1^1 x_i)^2 + \sum_{i=k+1}^{n} (y_i - b_0^* - b_1^* x_i)^2 \right]$$

$$+ n + 5 \log n - n \log n.$$

With the implementation of this information criterion, SIC, we transformed our task of hypothesis testing into a model selection process. The null hypothesis H_0 corresponds to a regression model with no change in the parameters, and the alternative hypothesis H_1 is represented by $n - 1$ regression models with a change point at position 2, or 3, $\ldots$, or $n - 2$. Therefore the decision rule for selecting one of the n regression models is: select the model with no change (or accept H_0) if $SIC(n) < SIC(k)$, for all k; select a model with a change at $\widehat{k}$ if $SIC(\widehat{k}) = \min_{1 < k < n-1} SIC(k) < SIC(n)$, where $\widehat{k} = 2, \ldots, n - 2$.

4.2.2 Bayesian Approach

In addition to the classical and informational approaches to the switching simple linear regression model, several authors proposed the solution of such problems from a Bayesian point of view. Chin Choy and Broemeling (1980) studied Bayesian inference for a switching linear model. Holbert (1982) investigated the switching simple linear regression model and multiple regression model by employing Bayesian methodology. We give a detailed presentation of the Bayesian approach on the basis of Holbert's work.

In the Bayesian settings, a change in the regression model is assumed to have taken place. The problem is to find where this change point is located. It is more appropriate to present the problem in the following setting.

$$\mu_{y_i} = \beta_0 + \beta_1 x_i, \quad \text{for } i = 1, \ldots, k \quad \text{and}$$

$$\mu_{y_i} = \beta_0^* + \beta_1^* x_i; \quad \text{for } i = k + 1, \ldots, n,$$

for $k = 2, \ldots, n - 2$, where $\beta_0, \beta_1, \beta_0^*$, and β_1^* are unknown regression parameters. Our goal here is to find the value of k, or an estimated value of k according to the data information.

The following general vague prior probability densities $\pi_0(\cdot)$ are assigned to the parameters.

$$\pi_0(\beta_0, \beta_1, \beta_0^*, \beta_1^* | k, \sigma^2) \propto \text{constant}, \quad -\infty < \beta_0, \beta_1, \beta_0^*, \beta_1^* < \infty,$$

$$\pi_0(k) = \begin{cases} \frac{1}{n-3}, & k = 2, \ldots, n-2 \\ 0, & \text{otherwise} \end{cases},$$

$$\pi_0(\sigma^2 | k) \propto \begin{cases} \frac{1}{\sigma^2}, & 0 < \sigma^2 < \infty \\ 0, & \text{otherwise} \end{cases}.$$

Because $Y_i, i = 1, \ldots, k, \sim$ iid $N(\beta_0 + \beta_1 x_i, \sigma^2), Y_j, j = k+1, \ldots, n,$ $\sim$ iid $N(\beta_0^* + \beta_1^* x_j, \sigma^2)$, the joint density function (likelihood function) of the data given the parameters is

$$L(\beta_0, \beta_1, \beta_0^*, \beta_1^*, k, \sigma^2) = f(y_1, \ldots, y_n | \beta_0, \beta_1, \beta_0^*, \beta_1^*, k, \sigma^2)$$

$$= \frac{1}{(\sqrt{2\pi\sigma^2})^n} \exp\left[-\sum_{i=1}^{k} (y_i - \beta_0 - \beta_1 x_i)^2 / 2\sigma^2 \right]$$

$$\cdot \exp\left[-\sum_{i=k+1}^{n} (y_i - \beta_0^* - \beta_1^* x_i)^2 / 2\sigma^2 \right].$$

Then, the joint posterior density of all the parameters is

$$\pi_1(k, \beta_0, \beta_1, \beta_0^*, \beta_1^*, \sigma^2)$$

$$= f(k, \beta_0, \beta_1, \beta_0^*, \beta_1^*, \sigma^2 | y_1, \ldots, y_n)$$

$$= \frac{f(y_1, \ldots, y_n | \beta_0, \beta_1, \beta_0^*, \beta_1^*, k, \sigma^2) \cdot f(\beta_0, \beta_1, \beta_0^*, \beta_1^*, k, \sigma^2)}{f(y_1, \ldots, y_n)}$$

$$\propto f(y_1, \ldots, y_n | \beta_0, \beta_1, \beta_0^*, \beta_1^*, k, \sigma^2) \cdot f(\beta_0, \beta_1, \beta_0^*, \beta_1^*, k, \sigma^2)$$

$$= \pi_0(k) \cdot \pi_0(\sigma^2 | k) \cdot \pi_0(\beta_0, \beta_1, \beta_0^*, \beta_1^* | k, \sigma^2) \cdot L(\beta_0, \beta_1, \beta_0^*, \beta_1^*, k, \sigma^2)$$

$$\propto \frac{1}{n-3} \cdot \frac{1}{\sigma^2} \cdot \frac{1}{(\sqrt{2\pi\sigma^2})^n} \cdot \exp\left[-\sum_{i=1}^{k} (y_i - \beta_0 - \beta_1 x_i)^2 / 2\sigma^2 \right]$$

$$\cdot \exp\left[-\sum_{i=k+1}^{n} (y_i - \beta_0^* - \beta_1^* x_i)^2 / 2\sigma^2 \right]$$

$$\propto \left(\frac{1}{\sigma^2} \right)^{n/2+1} \cdot \exp\left[-\sum_{i=1}^{k} (y_i - \beta_0 - \beta_1 x_i)^2 / 2\sigma^2 \right]$$

$$\cdot \exp\left[-\sum_{i=k+1}^{n} (y_i - \beta_0^* - \beta_1^* x_i)^2 / 2\sigma^2 \right].$$

Now, integrating $\pi_1(k, \beta_0, \beta_1, \beta_0^*, \beta_1^*, \sigma^2)$ with respect to $\beta_0, \beta_1, \beta_0^*, \beta_1^*$, and σ^2, we obtain the posterior density of the change point location k as

$$\pi_1(k)$$

$$= f(k|y_1, \ldots, y_n)$$

$$= \int_{-\infty}^{\infty} \int_{-\infty}^{\infty} \int_{-\infty}^{\infty} \int_{-\infty}^{\infty} \int_{-\infty}^{\infty} \Pi_1(k, \beta_0, \beta_1, \beta_0^*, \beta_1^*, \sigma^2) d\beta_0 d\beta_1 d\beta_0^* d\beta_1^* d\sigma^2$$

$$\propto \int_0^{\infty} \int_{-\infty}^{\infty} \int_{-\infty}^{\infty} \int_{-\infty}^{\infty} \int_{-\infty}^{\infty} \left(\frac{1}{\sigma^2} \right)^{n/2+1}$$

$$\cdot \exp\left[-\sum_{i=1}^{k} (y_i - \beta_0 - \beta_1 x_i)^2 / 2\sigma^2 \right]$$

$$\cdot \exp\left[-\sum_{i=k+1}^{n} (y_i - \beta_0^* - \beta_1^* x_i)^2 / 2\sigma^2 \right] d\beta_0 d\beta_1 d\beta_0^* d\beta_1^* d\sigma^2$$

$$\triangleq I.$$

To simplify the expression I, we calculate the following.

$$\int_{-\infty}^{\infty} \frac{1}{\sqrt{\sigma^2}} \exp\left[-\sum_{i=1}^{k} (y_i - \beta_0 - \beta_1 x_i)^2 / 2\sigma^2 \right] d\beta_0$$

$$= \int_{-\infty}^{\infty} \frac{1}{\sqrt{\sigma^2}} \exp\left\{ -\left[\sqrt{k}\beta_0 - \left(\sum_{i=1}^{k} y_i - \beta_1 \sum_{i=1}^{k} x_i \right) \frac{1}{\sqrt{k}} \right]^2 \bigg/ 2\sigma^2 \right\}$$

$$\cdot \exp\{ -[(\sqrt{{}_kS_x}\beta_1 - \sqrt{{}_kS_x}\widehat{\beta}_1)^2 - \widehat{\beta}_{1k}^2 {}_kS_x + {}_kS_y]/2\sigma^2 \} d\beta_0$$

$$= \sqrt{\frac{2\pi}{k}} \cdot \exp\{ -[(\sqrt{{}_kS_x}\beta_1 - \sqrt{{}_kS_x}\widehat{\beta}_1)^2 - \widehat{\beta}_{1k}^2 {}_kS_x + {}_kS_y]/2\sigma^2 \},$$

and

$$\int_{-\infty}^{\infty} \frac{1}{\sqrt{\sigma^2}} \sqrt{\frac{2\pi}{k}} \cdot \exp\{ -[(\sqrt{{}_kS_x}\beta_1 - \sqrt{{}_kS_x}\widehat{\beta}_1)^2$$

$$- \widehat{\beta}_{1k}^2 {}_kS_x + {}_kS_y]/2\sigma^2 \} d\beta_1$$

$$= \frac{2\pi}{\sqrt{k {}_kS_x}} \exp\{ -({}_kS_y - \widehat{\beta}_{1k}^2 {}_kS_x)/2\sigma^2 \}.$$

Moreover,

$$\int_{-\infty}^{\infty} \frac{1}{\sqrt{\sigma^2}} \exp\left[-\sum_{i=k+1}^{n} (y_i - \beta_0^* - \beta_1^* x_i)^2 / 2\sigma^2 \right] d\beta_0^*$$

$$= \sqrt{\frac{2\pi}{n-k}} \cdot \exp\{-[(\sqrt{_{n-k}S_x}\beta_1^* - \sqrt{_{n-k}S_x}\widehat{\beta}_1^*)^2$$

$$-_{n-k} S_x(\widehat{\beta}_1^{*2}) +_{n-k} S_y]/2\sigma^2\},$$

and

$$\int_{-\infty}^{\infty} \frac{1}{\sqrt{\sigma^2}} \exp\{-[(\sqrt{_{n-k}S_x}\beta_1^* - \sqrt{_{n-k}S_x}\widehat{\beta}_1^*)^2$$

$$-_{n-k} S_x(\widehat{\beta}_1^{*2}) +_{n-k} S_y]/2\sigma^2\}d\beta_1^*$$

$$= \sqrt{\frac{2\pi}{_{n-k}S_x}} \exp\{-[_{n-k}S_x -_{n-k} S_x(\widehat{\beta}_1^{*2})]/2\sigma^2\},$$

where $_kS_x$, and $_{n-k}S_x$ were given in the previous section, and

$$_kS_y = \sum_{i=1}^{k} (y_i - \bar{y}_k)^2, \qquad _{n-k}S_y = \sum_{i=k+1}^{n} (y_i - \bar{y}_{n-k})^2.$$

Then, I reduces to

$$I = \frac{(2\pi)^2}{\sqrt{k(n-k)_kS_{xn-k}S_x}}$$

$$\cdot \int_0^{\infty} \left(\frac{1}{\sigma^2}\right)^{n/2+1} \exp\{-[_kS_y - \widehat{\beta}_{1k}^2 S_x +_{n-k} S_y$$

$$-_{n-k} S_x\widehat{\beta}_1^{*2}]/2\sigma^2\}d\sigma^2.$$

After some algebraic simplifications, we obtain

$$D \overset{\triangle}{=}_k S_y - \widehat{\beta}_{1k}^2 S_x +_{n-k} S_y -_{n-k} S_x(\widehat{\beta}_1^{*2})$$

$$= \sum_{i=1}^{k} (y_i - \widehat{y}_{i(1,k)})^2 + \sum_{i=k+1}^{n} (y_i - \widehat{y}_{i(k+1,n)})^2,$$

where

$$\widehat{y}_{i(1,k)} = \widehat{\beta}_0 + \widehat{\beta}_1 x_i, \quad \text{for } i = 1, \ldots, k,$$

and

$$\widehat{y}_{i(k+1,n)} = \widehat{\beta}_0^* + \widehat{\beta}_1^* x_i, \quad \text{for } i = k+1, \ldots, n.$$

Therefore,

$$I = \frac{(2\pi)^2}{\sqrt{k(n-k)_kS_x \cdot_{n-k} S_x}} \int_0^{\infty} \left(\frac{1}{\sigma^2}\right)^{n/2-1} \exp\{-D/2\sigma^2\}d\sigma^2.$$

Note that to be able to build two regression models, it is required that both $n > 2$, and $n - 2 > 2$; hence, we have $n \geq 5$.

For $n = 2m$, with $m = 3, 4, \ldots,$

$$\int_0^\infty \left(\frac{1}{\sigma^2}\right)^{n/2-1} \exp\{-D/2\sigma^2\} d\sigma^2$$

$$= \frac{(m-3)!}{(D/2)^{m-2}}$$

$$= \frac{(n/2-3)!}{(D/2)^{(n-4)/2}}$$

$$\propto D^{-((n-4)/2)},$$

For $n = 2m - 1$, with $m = 3, 4, \ldots,$

$$\int_0^\infty \left(\frac{1}{\sigma^2}\right)^{n/2-1} \exp\{-D/2\sigma^2\} d\sigma^2 = \frac{\sqrt{2\pi}(2m-7)!!}{D^{m-5/2}}$$

$$\propto D^{-((n-4)/2)}.$$

Finally, we obtain:

$$I \propto [k(n-k)_k S_x \cdot_{n-k} S_x]^{-1/2} D^{-((n-4)/2)}, \quad \text{for } k = 2, \ldots, n - 2;$$

that is,

$$\pi_1(k) \propto [k(n-k)_k S_x \cdot_{n-k} S_x]^{-1/2} D^{-((n-4)/2)}, \quad \text{for } k = 2, \ldots, n - 2.$$

From the values of $\pi_1(k)$, a change point is located at $\widehat{k}$ if $\pi_1(\widehat{k}) = \max_{2 \leq k \leq n-2} \pi_1(k)$.

4.2.3 Application to Stock Market Data

Holbert (1982) studied the switching simple linear regression models and switching linear model from a Bayesian point of view. He assigned some vague prior densities to the unknown position of the change point and to the unknown parameters of the model, and obtained the posterior density of the change point. He analyzed the dataset on stock market sales volumes to illustrate the estimation of the change point in two-phase regression by calculating the posterior density of the change point. He found out that the maximum posterior density occurred at position 24, which corresponded to the calendar month of December of 1968, and concluded that it is a change point caused by the abolition of give-ups (commission splitting) in December of 1968.

We take the same data that Holbert used to illustrate the SIC method for locating the switching change point in linear regression. The monthly dollar volume of sales (in millions) on the Boston Stock Exchange (BSE) is considered as the response variable, and the combined New York American Stock Exchange (NYAMSE) is considered as the regressor. The computed SIC values are listed in Table 4.1 along with the original BSE and NYAMSE values given in Holbert (1982). The starred SIC value in this table is the minimum SIC value, which corresponds to time point 23, hence the regression model change starts at the time point 24, which is December of 1968. This conclusion coincides with the one drawn by Holbert using his method. As the reader may notice, the minimum SIC principle leads us firmly to the conclusion on the change point. Although Holbert (1982) found the same change

Table 4.1 NYAMSE and BSE Values, Computed SIC Values

Time Point	Calendar Month	NYAMSE	BSE	SIC
1	Jan. 1967	10581.6	78.8	—
2	Feb. 1967	10234.3	69.1	368.5736
3	Mar. 1967	13299.5	87.6	368.0028
4	Apr. 1967	10746.5	72.8	367.9975
5	May 1967	13310.7	79.4	366.8166
6	Jun. 1967	12835.5	85.6	366.1827
7	Jul. 1967	12194.2	75.0	365.3197
8	Aug. 1967	12860.4	85.3	364.4143
9	Set. 1967	11955.6	86.9	364.0418
10	Oct. 1967	13351.5	107.8	364.0670
11	Nov. 1967	13285.9	128.7	365.1320
12	Dec. 1967	13784.4	134.5	365.8783
13	Jan. 1968	16336.7	148.7	365.7791
14	Feb. 1968	11040.5	94.2	366.0318
15	Mar. 1968	11525.3	128.1	367.1252
16	Apr. 1968	16056.4	154.1	367.2805
17	May 1968	18464.3	191.3	367.4632
18	Jun. 1968	17092.2	191.9	367.6615
19	Jul. 1968	15178.8	159.6	367.8082
20	Aug. 1968	12774.8	185.5	368.8873
21	Sep. 1968	12377.8	178.0	368.9790
22	Oct. 1968	16856.3	271.8	364.2126
23	Nov. 1968	14635.3	212.3	359.3774*
24	Dec. 1968	17436.9	139.4	362.7803
25	Jan. 1969	16482.2	106.0	366.7591
26	Feb. 1969	13905.4	112.1	367.4118
27	Mar. 1969	11973.7	103.5	367.6757
28	Apr. 1969	12573.6	92.5	368.4138
29	May 1969	16566.8	116.9	370.6948
30	Jun. 1969	13558.7	78.9	372.0000
31	Jul. 1969	11530.9	57.4	372.1517
32	Aug. 1969	11278.0	75.9	371.8513
33	Sep. 1969	11263.7	109.8	372.1726
34	Oct. 1969	15649.5	129.2	—
35	Nov. 1969	12197.1	115.1	361.4956

point as here, his conclusion is less affirmative. As he pointed out, there is a tendency of a relative maximum at the endpoints using the Bayesian posterior density.

4.3 Multiple Linear Regression Model

As an analogue to the simple linear regression model, we consider the switching multiple linear regression model in this section. The model we discuss is

$$y_i = \mathbf{x}_i'\beta + \varepsilon_i, \qquad i = 1, \ldots, n,$$

where $\mathbf{x}_i \ i = 1, \ldots, n$ is a nonstochastic $(p + 1)$-vector variable with $\mathbf{x}_i' = (1, x_{1i}, x_{2i}, \ldots, x_{pi})$, $\beta' = (\beta_0, \beta_1, \ldots, \beta_p)$ is a $(p + 1)$ unknown regression vector, $\varepsilon_i, i = 1, \ldots, n$ is a random error which is distributed as $N(0, \sigma^2)$, with σ^2 unknown, and ε_is are uncorrelated from observation to observation. That is, $y_i \ i = 1, \ldots, n$, is a random variable distributed as $N(\mathbf{x}_i'\beta, \sigma^2)$. Situations arise in which we would like to check if there is a change at location k in the regression model. Then, we test the null hypothesis:

$$H_0 : \mu_{y_i} = \mathbf{x}_i'\beta \qquad \text{for } i = 1, \ldots, n,$$

versus the alternative hypothesis:

$$H_1 : \mu_{y_i} = \mathbf{x}_i'\beta_1 \quad \text{for } i = 1, \ldots, k,$$

$$\text{and} \quad \mu_{y_i} = \mathbf{x}_i'\beta_2 \quad \text{for } i = k + 1, \ldots, n,$$

where $k, k = p + 1, \ldots, n - p$ is the location of the change point, and β, β_1, and β_2 are unknown regression parameters. In the following sections we give a method to locate the change point k.

4.3.1 Informational Approach

An alternate approach for this multiple regression change point problem is to use the Schwarz information criterion, SIC. It is a simple approach that reduces computations in comparison with the likelihood-ratio procedure approach.

Let

$$\mathbf{y} = \begin{pmatrix} y_1 \\ y_2 \\ \vdots \\ y_n \end{pmatrix}, \qquad X = \begin{pmatrix} 1 & x_{11} & \cdots & x_{p1} \\ 1 & x_{12} & \cdots & x_{p2} \\ \vdots & \vdots & \vdots & \vdots \\ 1 & x_{1n} & \cdots & x_{pn} \end{pmatrix} \equiv \begin{pmatrix} \mathbf{x}_1' \\ \mathbf{x}_2' \\ \vdots \\ \mathbf{x}_n' \end{pmatrix}, \quad \text{and} \quad \beta = \begin{pmatrix} \beta_0 \\ \beta_1 \\ \vdots \\ \beta_p \end{pmatrix};$$

then the null hypothesis H_0 corresponds to the model

$$\mu_{\mathbf{y}} = X\beta,$$

where

$$\mu_{\mathbf{y}} = \begin{pmatrix} \mu_{y_1} \\ \mu_{y_2} \\ \vdots \\ \mu_{y_n} \end{pmatrix}.$$

Obviously, the likelihood function under H_0 in matrix notation is

$$L_0(\beta, \sigma^2) = f(y_1, y_2, \ldots, y_n; \beta, \sigma^2)$$
$$= (2\pi)^{-n/2}(\sigma^2)^{-n/2} \exp\{-(\mathbf{y} - X\beta)'(\mathbf{y} - X\beta)/2\sigma^2\},$$

and the MLEs of β, and σ^2 are, respectively,

$$\mathbf{b} \triangleq \widehat{\beta} = (X'X)^{-1}X'\mathbf{y},$$
$$\widehat{\sigma}^2 = \frac{1}{n}(\mathbf{y} - X\mathbf{b})'(\mathbf{y} - X\mathbf{b}).$$

Then, the maximum likelihood under H_0 is

$$L_0(\widehat{\beta}, \widehat{\sigma}^2) \equiv L_0(\mathbf{b}, \widehat{\sigma}^2)$$
$$= (2\pi)^{-n/2} \left[\frac{1}{n}(\mathbf{y} - X\mathbf{b})'(\mathbf{y} - X\mathbf{b}) \right]^{-n/2} e^{-n/2}.$$

Therefore, under H_0 the Schwarz information criterion, denoted by $\mathrm{SIC}(n)$, is obtained as

$$\mathrm{SIC}(n) = -2\log L_0(\mathbf{b}, \widehat{\sigma}^2) + (p+2)\log n$$
$$= n\log[(\mathbf{y} - X\mathbf{b})'(\mathbf{y} - X\mathbf{b})] + n(\log 2\pi + 1) + (p + 2 - n)\log n.$$

Let

$$\mathbf{y}_1 = \begin{pmatrix} y_1 \\ y_2 \\ \vdots \\ y_k \end{pmatrix}, \qquad \mathbf{y}_2 = \begin{pmatrix} y_{k+1} \\ y_{k+2} \\ \vdots \\ y_n \end{pmatrix},$$

$$X_1 = \begin{pmatrix} 1 & x_{11} & \cdots & x_{p1} \\ 1 & x_{12} & \cdots & x_{p2} \\ \vdots & \vdots & \vdots & \vdots \\ 1 & x_{1k} & \cdots & x_{pk} \end{pmatrix} \equiv \begin{pmatrix} \mathbf{x}_1' \\ \mathbf{x}_2' \\ \vdots \\ \mathbf{x}_k' \end{pmatrix},$$

$$
X_2 = \begin{pmatrix} 1 & x_{11} & \cdots & x_{p(k+1)} \\ 1 & x_{12} & \cdots & x_{p(k+2)} \\ \vdots & \vdots & \vdots & \vdots \\ 1 & x_{1k} & \cdots & x_{pn} \end{pmatrix} \equiv \begin{pmatrix} \mathbf{x}'_{k+1} \\ \mathbf{x}'_{k+2} \\ \vdots \\ \mathbf{x}'_n \end{pmatrix}
$$

$$
\beta_1 = \begin{pmatrix} \beta_0 \\ \beta_1 \\ \vdots \\ \beta_p \end{pmatrix}, \qquad \beta_2 = \begin{pmatrix} \beta_0^* \\ \beta_1^* \\ \vdots \\ \beta_p^* \end{pmatrix}, \quad \text{and} \quad \beta = \begin{pmatrix} \beta_0 \\ \beta_1 \\ \vdots \\ \beta_p \end{pmatrix},
$$

for $k = p + 1, \ldots, n - p$; then the alternative hypothesis H_1 corresponds to the following models.

$$
\mu_{\mathbf{y_1}} = X_1 \beta_1 \quad \text{and} \quad \mu_{\mathbf{y_2}} = X_2 \beta_2, \quad \text{for } k = p + 1, \ldots, n - p,
$$

where

$$
\mu_{\mathbf{y_1}} = \begin{pmatrix} \mu_{y_1} \\ \mu_{y_2} \\ \vdots \\ \mu_{y_k} \end{pmatrix} \quad \text{and} \quad \mu_{\mathbf{y_2}} = \begin{pmatrix} \mu_{y_{k+1}} \\ \mu_{y_{k+2}} \\ \vdots \\ \mu_{y_n} \end{pmatrix}.
$$

In this case, the likelihood function is found to be

$$
\begin{aligned}
L_1(\beta_1, \beta_2, \sigma^2) &= f(y_1, y_2, \ldots, y_n; \beta_1, \beta_2, \sigma^2) \\
&= (2\pi)^{-n/2} (\sigma^2)^{-n/2} \exp\{-(\mathbf{y}_1 - X_1\beta_1)'(\mathbf{y}_1 - X_1\beta_1)/2\sigma^2\} \\
&\quad \cdot \exp\{-(\mathbf{y}_2 - X_2\beta_2)'(\mathbf{y}_2 - X_2\beta_2)/2\sigma^2\},
\end{aligned}
$$

and the MLEs of the parameters are, respectively,

$$
\mathbf{b}_1 \triangleq \widehat{\beta}_1 = (X_1'X_1)^{-1}X_1'\mathbf{y}_1,
$$
$$
\mathbf{b}_2 \triangleq \widehat{\beta}_2 = (X_2'X_2)^{-1}X_2'\mathbf{y}_2,
$$
$$
\widehat{\sigma}^2 = \frac{1}{n}[(\mathbf{y}_1 - X_1\mathbf{b}_1)'(\mathbf{y}_1 - X_1\mathbf{b}_1) + (\mathbf{y}_2 - X_2\mathbf{b}_2)'(\mathbf{y}_2 - X_2\mathbf{b}_2)].
$$

Then the maximum likelihood is

$$
L_1(\widehat{\beta}_1, \widehat{\beta}_2, \widehat{\sigma}^2) \equiv L_1(\mathbf{b}_1, \mathbf{b}_2, \widehat{\sigma}^2)
$$

$$
= (2\pi)^{-n/2} \left\{ \frac{1}{n}[(\mathbf{y}_1 - X_1\mathbf{b}_1)'(\mathbf{y}_1 - X_1\mathbf{b}_1) \right.
$$

$$
\left. + (\mathbf{y}_2 - X_2\mathbf{b}_2)'(\mathbf{y}_2 - X_2\mathbf{b}_2)] \right\}^{-n/2} e^{-n/2}.
$$

Therefore, under H_1 the Schwarz information criterion, denoted by $\mathrm{SIC}(k)$ for $k = p+1, \ldots, n-p$, is obtained as

$$
\begin{aligned}
\mathrm{SIC}(k) &= -2 \log L_1(\mathbf{b}_1, \mathbf{b}_2, \widehat{\sigma}^2) + (2p+3) \log n \\
&= n \log[(\mathbf{y}_1 - X_1\mathbf{b}_1)'(\mathbf{y}_1 - X_1\mathbf{b}_1) + (\mathbf{y}_2 - X_2\mathbf{b}_2)'(\mathbf{y}_2 - X_2\mathbf{b}_2)] \\
&\quad + n(\log 2\pi + 1) + (2p + 3 - n) \log n.
\end{aligned}
$$

According to the principle of information criterion in model selection, H_0 will be accepted if $\mathrm{SIC}(n) \leq \min_{p+1 \leq k \leq n-p} \mathrm{SIC}(k)$, and H_1 will be accepted if $\mathrm{SIC}(n) > \min_{p+1 \leq k \leq n-p} \mathrm{SIC}(k)$. When H_1 is accepted, the estimated position of the switching linear model will be $\widehat{k}$ such that

$$
\mathrm{SIC}(\widehat{k}) = \min_{p+1 \leq k \leq n-p} \mathrm{SIC}(k).
$$

4.3.2 Bayesian Approach

Holbert (1982) also investigated the switching multiple linear regression models from a Bayesian point of view. In the Bayesian setting, a change in the regression model is assumed to have taken place. The problem is to search where this change point is located. It is more appropriate to present the problem in the following setting.

$$
\mu_{\mathbf{y_i}} = \mathbf{x}_i'\beta_1 \quad \text{for } i = 1, \ldots, k \quad \text{and} \quad \mu_{\mathbf{y_i}} = \mathbf{x}_i'\beta_2 \quad \text{for} \quad i = k+1, \ldots, n,
$$

for $k = p+1, \ldots, n-p$.
Let

$$
\mathbf{y} = \begin{pmatrix} \mathbf{y}_1 \\ \mathbf{y}_2 \end{pmatrix}, \qquad \beta = \begin{pmatrix} \beta_1 \\ \beta_2 \end{pmatrix}, \quad \text{and} \quad R = \frac{1}{\sigma^2},
$$

where $\mathbf{y}_1, \mathbf{y}_2, \beta_1$, and β_2 are defined in the previous section. We first assign a discrete uniform prior to the change point position k:

$$
\pi_0(k) = \begin{cases} \frac{1}{n-2p}, & k = p+1, \ldots, n-p \\ 0, & \text{otherwise} \end{cases}.
$$

We also assume that the $2(p+1)$ parameter vector β and the parameter R are jointly independent of the change point position k. Finally, we assume that the parameter R has a prior distribution which is gamma with parameters a and b, and the conditional prior of β given $R = r$ is a $2(p+1)$-dimensional normal distribution with mean vector β^*, and covariance matrix $(1/r)\tau^{-1}$, where τ is $(p+1) \times (p+1)$ positive definite; that is,

$$\pi_0(R) = \begin{cases} \frac{b^a}{\Gamma(b)} r^{a-1} e^{-br}, & r > 0 \\ 0, & \text{otherwise} \end{cases},$$

and

$$\pi_0(\beta | R = r) = \frac{r^{p+1} |\tau|^{p+1}}{(2\pi)^{p+1}} \exp\left\{ -\frac{r}{2}(\beta - \beta^*)'\tau(\beta - \beta)^* \right\}.$$

Therefore, the joint prior of β and R can be written as

$$\pi_0(\beta, R) = \frac{b^a}{\Gamma(b)} r^{a-1} e^{-br} \cdot \frac{r^{p+1} |\tau|^{p+1}}{(2\pi)^{p+1}} \exp\left\{ -\frac{r}{2}(\beta - \beta^*)'\tau(\beta - \beta^*) \right\}$$

$$\propto r^{a+p} \exp\left\{ (-r)\left[b + \frac{1}{2}(\beta - \beta^*)'\tau(\beta - \beta^*) \right] \right\}.$$

By introducing the $n \times (2p + 2)$ matrix $X(k)$:

$$X(k) = \begin{pmatrix} X_1 & 0_1 \\ 0_2 & X_2 \end{pmatrix},$$

where X_1, X_2 are defined in the previous section, 0_1 is a $k \times (p + 1)$ zero matrix, and 0_2 is a $(n - k) \times (p + 1)$ zero matrix. In this case, the likelihood function is

$$L_1(\beta, R, k) = L_1(\beta_1, \beta_2, R, k)$$

$$= f(y_1, y_2, \ldots, y_n; \beta_1, \beta_2, R, k)$$

$$= (2\pi)^{-n/2} r^{n/2} \exp\{ -r(\mathbf{y}_1 - X_1\beta_1)'(\mathbf{y}_1 - X_1\beta_1)/2 \}$$

$$\cdot \exp\{ -r(\mathbf{y}_2 - X_2\beta_2)'(\mathbf{y}_2 - X_2\beta_2)/2 \}$$

$$= (2\pi)^{-n/2} r^{-n/2} \exp\{ -r(\mathbf{y} - X(k)\beta)'(\mathbf{y} - X(k)\beta)/2 \},$$

hence, the joint posterior density of the parameters is

$$\pi_1(\beta, R, k) = f(\beta, R, k | y_1, y_2, \ldots, y_n)$$

$$\propto L_1(\beta, R, k)\pi_0(\beta, R)\pi_0(k)$$

$$\propto r^{a+p+n/2} \cdot \exp\left\{ (-r)\left[b + \frac{1}{2}(\beta - \beta^*)'\tau(\beta - \beta^*) \right.\right.$$

$$\left.\left. + \frac{1}{2}(\mathbf{y} - X(k)\beta)'(\mathbf{y} - X(k)\beta) \right] \right\}.$$

Integrating $\pi_1(\beta, R, k)$ with respect to β and R, and simplifying, we obtain the posterior density of the change point k as

$$\pi_1(k) = f(k|y_1, y_2, \ldots, y_n)$$
$$\propto D(k)^{-a^*}|X(k)'X(k) + \tau|^{-1/2}, \quad \text{for } k = p+1, \ldots, n-p,$$

where

$$a^* = a + 1 + n/2,$$

$$D(k) = b + \frac{1}{2}\{[\mathbf{y} - \widehat{\mathbf{y}}(k)]'[\mathbf{y} - \widehat{\mathbf{y}}(k)] + [\widehat{\beta}(k) - \beta^*]'w(k)[\widehat{\beta}(k) - \beta^*]\},$$

$$w(k) = X(k)'X(k)[X(k)'X(k) + \tau]^{-1}\tau,$$

$$\widehat{\beta}(k) = [X(k)'X(k)]^{-1}X(k)'\mathbf{y},$$

$$\widehat{\mathbf{y}}(k) = X(k)\widehat{\beta}(k).$$

Interested readers are referred to Chin Choy (1977), and Chin Choy and Broemeling (1980) for the details.

Chapter 5
Gamma Model

5.1 Problem

In the previous chapters, we introduced the multiple change-point problem for both univariate and multivariate Gaussian models. Now, let us turn our attention away from Gaussian models, and study another important model, the gamma distribution.

Suppose that $x_1, x_2, \ldots, x_n$ is a sequence of independent random variables from gamma distributions with parameters $(\theta_1, \xi), (\theta_2, \xi), \ldots$, and (θ_n, ξ), respectively, where ξ is known, and the pdf of $X_i's$ is

$$f(x, \xi, \theta_i) = \frac{1}{\theta_i^\xi \Gamma(\xi)} x^{\xi-1} e^{-(x/\theta_i)}, \qquad \xi, \theta_i > 0, x > 0, i = 1, \ldots, n.$$

We are interested in testing

$$H_0 : \theta_1 = \theta_2 = \cdots = \theta_n = \theta_0 \tag{5.1}$$

against the alternative:

$$H_1 : \theta_1 = \cdots = \theta_k = \theta_0 \neq \theta_{k+1} = \cdots = \theta_n = \theta_0 + \delta > 0, \tag{5.2}$$

where k is the unknown position of a change point, θ_0 unknown, and $|\delta| > 0$.

The above formulation of the change point problem was posted by Kander and Zacks (1966) when they studied the model from an exponential family. Later, Hsu (1979) adopted their assumption and their general result, and studied the change point problem under the same formulation for the gamma model. In this chapter, we base our discussion on the above-mentioned authors' work, and present the following sections of interest.

5.2 A Solution

Assume that the location of the change point k has an equal chance to fall at any of the possible points $j = 1, 2, \ldots, n - 1$. That is, a prior is put on the location of the change point:

$$\pi_n(j) = \begin{cases} \frac{1}{n-1}, & j = 1, 2, \ldots, n-1 \\ 0, & \text{otherwise} \end{cases} \tag{5.3}$$

Under (5.2), the likelihood function is

$$L_1(\theta_0, \delta)$$

$$= f(x_1, \ldots, x_n; \theta_0, \delta)$$

$$= \sum_{j=1}^{n-1} \pi_n(j) f(x_1, \ldots, x_n; \theta_0, \delta | j)$$

$$= \frac{1}{n-1} \sum_{j=1}^{n-1} \left[\left(\prod_{i=1}^{j} \frac{1}{\theta_i^\xi \Gamma(\xi)} x_i^{\xi-1} e^{-(x_i/\theta_i)} \right) \left(\prod_{i=j+1}^{n} \frac{1}{\theta_i^\xi \Gamma(\xi)} x_i^{\xi-1} e^{-(x_i/\theta_i)} \right) \right]$$

$$= \frac{1}{n-1} \sum_{j=1}^{n-1} \left[\left(\prod_{i=1}^{j} \frac{1}{\theta_i^\xi \Gamma(\xi)} x_i^{\xi-1} e^{-(x_i/\theta_i)} \right) \right.$$

$$\left. \cdot \left(\prod_{i=j+1}^{n} \frac{1}{(\theta_0 + \delta)^\xi \Gamma(\xi)} x_i^{\xi-1} e^{-(x_i/(\theta_0+\delta))} \right) \right]$$

$$= \frac{1}{n-1} \frac{\prod_{i=1}^{n} x_i^{\xi-1}}{\Gamma^n(\xi)} \sum_{j=1}^{n-1} \exp\left[\sum_{i=1}^{j} \left(-\frac{x_i}{\theta_0} - \ln \theta_0^\xi \right) \right]$$

$$\cdot \exp\left[\sum_{i=j+1}^{n} \left(-\frac{x_i}{\theta_0 + \delta} - \ln(\theta_0 + \delta)^\xi \right) \right].$$

The Taylor expansion of $-(x_i/(\theta_0 + \delta)) - \ln(\theta_0 + \delta)^\xi$ for $(\delta/\theta_0) \to 0$ is

$$-\frac{x_i}{\theta_0 + \delta} - \ln(\theta_0 + \delta)^\xi = -\frac{x_i}{\theta_0} - \ln \theta_0^\xi + \delta \left(\frac{x_i}{\theta_0^2} - \frac{\xi}{\theta_0} \right) + o\left(\frac{\delta}{\theta_0} \right).$$

Then,

$$
L_1(\theta_0, \delta) = \frac{1}{n-1} \frac{\prod\limits_{i=1}^{n} x_i^{\xi-1}}{\Gamma^n(\xi)} \sum_{j=1}^{n-1} \left\{ \exp\left[\sum_{i=1}^{j} \left(-\frac{x_i}{\theta_0} - \ln\theta_0^\xi \right) \right] \right.
$$

$$
\left. \cdot \exp\left[\sum_{i=j+1}^{n} \left(-\frac{x_i}{\theta_0} - \ln\theta_0^\xi + \frac{x_i\delta}{\theta_0^2} - \frac{\xi\delta}{\theta_0} + o(\delta) \right) \right] \right\}
$$

Under (5.1), the likelihood function is

$$
L_0(\theta_0) = \frac{\prod\limits_{i=1}^{n} x_i^{\xi-1}}{\Gamma^n(\xi)} \exp\left[\sum_{i=1}^{n} \left(-\frac{x_i}{\theta_0} - \ln\theta_0^\xi \right) \right]
$$

Then, the ratio of L_1 to L_0 is

$$
\Lambda = \frac{L_1(\theta_0, \delta)}{L_0(\theta_0)} = \frac{1}{n-1} \sum_{j=1}^{n-1} \exp\left[\sum_{i=j+1}^{n} \frac{\delta x_i}{\theta_0^2} - \frac{\xi\delta}{\theta_0} + o(\delta) \right]
$$

$$
= \sum_{j=1}^{n-1} \frac{1}{n-1} \left\{ 1 + \sum_{i=j+1}^{n} \left[\frac{\delta x_i}{\theta_0^2} - \frac{\xi\delta}{\theta_0} \right] + (n-j)o(\delta) \right\}
$$

$$
= \sum_{j=1}^{n-1} \frac{1}{n-1} \left\{ 1 + \sum_{i=j+1}^{n} \left[\frac{\delta x_i}{\theta_0^2} - \frac{\xi\delta}{\theta_0} \right] + o(\delta) \right\}
$$

$$
= 1 + \frac{1}{n-1} \sum_{j=1}^{n-1} \sum_{i=j+1}^{n} \left[\frac{\delta x_i}{\theta_0^2} - \frac{\xi\delta}{\theta_0} \right] + o(\delta)
$$

$$
= 1 + \frac{\delta}{\theta_0} \left[\frac{1}{(n-1)\theta_0} \sum_{j=1}^{n-1} \sum_{i=j+1}^{n} x_i - \frac{n\xi}{2} \right] + o(\delta).
$$

Clearly, Λ is a monotone function of $(1/\theta_0)\sum_{j=1}^{n-1}\sum_{i=j+1}^{n} x_i$. Then we can choose the likelihood-ratio based test statistic as

$$
\lambda = \frac{1}{\theta_0} \sum_{j=1}^{n-1} \sum_{i=j+1}^{n} x_i = \frac{1}{\theta_0} \sum_{i=2}^{n} (i-1)x_i
$$

$$= \frac{1}{\theta_0} \sum_{i=1}^{n-1} i x_{i+1}$$

$$= \frac{1}{\theta_0} \sum_{i=1}^{n} (i-1) x_i.$$

When θ_0 is unknown, it is estimated under (5.1) by its MLE

$$\widehat{\theta}_0 = \frac{\overline{x}}{\xi} = \frac{\sum_{i=1}^{n} x_i}{n\xi}.$$

Then, the test statistic becomes

$$\lambda = \frac{n\xi}{\sum_{i=1}^{n} x_i} \sum_{i=1}^{n} (i-1) x_i.$$

For simplicity, let $n\xi = 1/(n-1)$; hence the test statistic finally becomes:

$$T = \frac{\sum_{i=1}^{n} (i-1) x_i}{(n-1) \sum_{i=1}^{n} x_i}.$$

Next, we intend to derive the approximate null distribution of T. Under H_0, the joint pdf of $x_1, \ldots, x_n$ is

$$f(x_1, \ldots, x_n) = \frac{1}{\theta_0^{n\xi} \Gamma^n(\xi)} \left(\prod_{i=1}^{n} x_i \right)^{\xi-1} e^{-(\Sigma x_i/\theta_0)}, \qquad x_i > 0, \xi > 0, \theta_0 > 0.$$

Let us carry out the following transformation,

$$y_2 = \frac{x_2}{\sum_{i=1}^{n} x_i},$$

$$y_3 = \frac{x_3}{\sum_{i=1}^{n} x_i},$$

$$\vdots$$

$$y_n = \frac{x_n}{\sum_{i=1}^{n} x_i},$$

$$z = \sum_{i=1}^{n} x_i.$$

Then the inverse transformation is

$$x_1 = z\left(1 - \sum_{j=1}^{n} y_j\right),$$

$$x_2 = zy_2,$$

$$x_3 = zy_3,$$

$$\vdots$$

$$x_n = zy_n.$$

The Jacobian of this transformation is

$$J = \begin{vmatrix} -z & -z & \ldots & -z & 1 - \sum_{j=2}^{n} y_i \\ z & 0 & \ldots & 0 & y_2 \\ 0 & z & \ldots & 0 & y_3 \\ \ldots & \ldots & \ldots & \ldots & \ldots \\ 0 & 0 & \ldots & z & y_n \end{vmatrix} = (-1)^{n-1} z^{n-1}.$$

So, the joint pdf of $y_2, \ldots, y_n, z$ is

$$g(y_2, \ldots, y_n, z)$$

$$= |J| f\left(z\left(1 - \sum_{j=2}^{n} y_i\right), zy_2, \ldots, zy_n\right)$$

$$= z^{n-1} \frac{1}{\theta_0^{n\xi} \Gamma(\xi)} \left[z^{n\xi-1} \left(\prod_{j=2}^{n} y_j\right)^{\xi-1} \left(1 - \sum_{j=2}^{n} y_i\right)^{\xi-1} \right] e^{-(z/\theta_0)}$$

$$= \frac{1}{\theta_0^{n\xi} \Gamma^n(\xi)} z^{n\xi-1} \left(\prod_{j=2}^{n} y_j\right)^{\xi-1} \left(1 - \sum_{j=2}^{n} y_i\right)^{\xi-1} e^{-(z/\theta_0)}.$$

Then, the joint distribution of $y_2, \ldots, y_n$ is

$$g(y_2, \ldots, y_n) = \int_0^{\infty} g(y_2, \ldots, y_n, z) dz$$

$$
= \frac{\Gamma(n\xi) \prod\limits_{j=2}^{n} y_j^{\xi-1} \left(1 - \sum\limits_{j=2}^{n} y_i\right)^{\xi-1}}{\Gamma^n(\xi)} \int_0^z z^{n\xi-1} e^{-(x/\theta_0)} \cdot \frac{1}{\theta_0^{n\xi} \Gamma(n\xi)} dz
$$

$$
= \frac{\Gamma(n\xi)}{\Gamma^n(\xi)} \prod\limits_{j=2}^{n} y_j^{\xi-1} \left(1 - \sum\limits_{j=2}^{n} y_i\right)^{\xi-1}.
$$

Obviously, the joint distribution of $y_2, \ldots, y_n$ is Dirichlet $D(\xi, \ldots, \xi; \xi)$. Therefore, under H_0, the test statistic T is a linear combination of Dirichlet variates, in fact, $T = \sum_{i=2}^{n}((i-1)/(n-1))y_i$. From Johnson and Kotz (1972, p. 233), the mixed moment of $(y_2, \ldots, y_n)$ is given by

$$
E[y_2^{r_2} y_3^{r_3}, \ldots, y_n^{r_n}] = \frac{\Gamma(n\xi)}{\Gamma^{n-1}(\xi)} \cdot \frac{\Gamma(\xi + r_2) \cdots \Gamma(\xi + r_n)}{\Gamma(n\xi + r_2 + \cdots + r_n)},
$$

where $r_2, r_3, \ldots, r_n \geq 0$. We can therefore obtain the following moments of T under H_0.

$$
\mu_1(T) = E(T) = \sum_{i=2}^{n} \frac{i-1}{n-1} E(y_i)
$$

$$
= \sum_{i=2}^{n} \frac{i-1}{n-1} \frac{\Gamma(n\xi)}{\Gamma^{n-1}(\xi)} \cdot \frac{\Gamma^{n-2}(\xi)\Gamma(\xi+1)}{\Gamma(n\xi+1)}
$$

$$
= \sum_{i=2}^{n} \frac{i-1}{n-1} \cdot \frac{1}{n} = \frac{1}{2}.
$$

$$
\mu_2(T) = V_{ar}(T) = E(T - ET)^2
$$

$$
= E\left[\sum_{i=2}^{n} \frac{i-1}{n-1} y_i - \frac{1}{2}\right]^2
$$

$$
= E\left[\left(\sum_{i=2}^{n} \frac{i-1}{n-1} y_i\right)^2 - \sum_{i=2}^{n} \frac{i-1}{n-1} y_i + \frac{1}{4}\right]
$$

$$
= E\left(\sum_{i=2}^{n} \frac{i-1}{n-1} y_i\right)^2 - \frac{1}{4}
$$

$$
= \sum_{i=2}^{n} \left(\frac{i-1}{n-1}\right)^2 E y_i^2 + \sum_{i \neq j} \frac{(i-1)(j-1)}{(n-1)^2} E(y_i y_j) - \frac{1}{4}
$$

$$= \sum_{i=2}^{n} \left(\frac{i-1}{n-1} \right)^2 \cdot \frac{\Gamma(n\xi) \Gamma^{n-2}(\xi) \Gamma(\xi+2)}{\Gamma^{n-1}(\xi) \Gamma(n\xi+2)}$$

$$+ \sum_{i \neq j} \frac{(i-1)(j-1)}{(n-1)^2} \frac{\Gamma(n\xi) \Gamma^{n-2}(\xi) \Gamma(\xi+1)}{\Gamma^{n-1}(\xi) \Gamma(n\xi+2)} - \frac{1}{4}$$

$$= \sum_{i=2}^{n} \left(\frac{i-1}{n-1} \right)^2 + \frac{\xi+1}{n(n\xi+1)} + \sum_{i \neq j} \frac{(i-1)(j-1)}{(n-1)^2} \frac{\xi}{n(n\xi+1)} - \frac{1}{4}$$

$$= \frac{\xi}{n(n\xi+1)} \left[\sum_{i=2}^{n} \left(\frac{i-1}{n-1} \right) \right]^2 + \frac{1}{n(n\xi+1)} \sum_{i=2}^{n} \left(\frac{i-1}{n-1} \right)^2 - \frac{1}{4}$$

$$= \frac{n\xi}{4(n\xi+1)} + \frac{2n-1}{6(n-1)(n\xi+1)} - \frac{1}{4}$$

$$= \frac{(n+1)}{12(n-1)(n\xi+1)}.$$

Hence, we have

$$E(T^2) = \frac{3n\xi(n-1) + 2(2n-1)}{12(n-1)(n\xi+1)}.$$

$$\mu_3(T) = E(T - ET)^3$$

$$= E\left(T - \frac{1}{2} \right)^3$$

$$= E(T^3) - \frac{3}{2} E(T^2) + \frac{3}{4} E(T) - \frac{1}{8}$$

$$= E\left[\sum_{i=2}^{n} \frac{i-1}{n-1} y_i \right]^3 - \frac{3}{2} \frac{3n\xi(n-1) + 2(2n-1)}{12(n-1)(n\xi+1)} + \frac{1}{4}$$

$$= \sum_{i=2}^{n} \left(\frac{i-1}{n-1} \right)^3 E(y_i^3) + 3 \sum_{i \neq k} \frac{(i-1)^2(k-1)}{(n-1)^3} E(y_i^2 y_k)$$

$$+ \sum_{i \neq j \neq k} \frac{(i-1)(j-1)(k-1)}{(n-1)^3} E(y_i y_j y_k)$$

$$- \frac{3n\xi(n-1) + 2(2n-1)}{8(n-1)(n\xi+1)} + \frac{1}{4}$$

$$= \sum_{i=2}^{n} \left(\frac{i-1}{n-1} \right)^3 \frac{\Gamma(n\xi) \Gamma^{n-2}(\xi) \Gamma(\xi+3)}{\Gamma^{n-1}(\xi) \Gamma(n\xi+3)}$$

$$+3\sum_{i\neq k}\frac{(i-1)^2(k-1)}{(n-1)^2}\frac{\Gamma(n\xi)\Gamma^{n-3}(\xi)\Gamma(\xi+2)\Gamma(\xi+1)}{\Gamma^{n-1}(\xi)\Gamma(n\xi+3)}$$

$$+\sum_{i\neq j\neq k}\frac{(i-1)(j-1)(k-1)}{(n-1)^3}\frac{\Gamma(n\xi)\Gamma^{n-4}(\xi)\Gamma^3(\xi+1)}{\Gamma^{n-1}(\xi)\Gamma(n\xi+3)}$$

$$-\frac{3n\xi(n-1)+2(2n-1)}{8(n-1)(n\xi+1)}+\frac{1}{4}$$

$$=\sum_{i=2}^{n}\left(\frac{i-1}{n-1}\right)^3\frac{(\xi+1)(\xi+2)}{n(n\xi+1)(n\xi+2)}$$

$$+3\sum_{i\neq k}\frac{(i-1)^2(k-1)}{(n-1)^3}\frac{\xi(\xi+1)}{n(n\xi+1)(n\xi+2)}$$

$$+\sum_{i\neq j\neq k}\frac{(i-1)(j-1)(k-1)}{n(n\xi+1)(n\xi+2)}\frac{\xi^2}{n(n\xi+1)(n\xi+2)}$$

$$-\frac{3n\xi(n-1)+2(2n-1)}{8(n-1)(n\xi+1)}+\frac{1}{4}.$$

Continuing with more calculations, we have:

$$\mu_3(T)=\frac{\xi^2}{n(n\xi+1)(n\xi+2)}\left[\sum_{i=2}^{n}\frac{i-1}{n-1}\right]^3$$

$$+\frac{3\xi}{n(n\xi+1)(n\xi+2)}\left[\sum_{i=2}^{n}\frac{(i-1)^2}{(n-1)^2}\right]\left[\sum_{k=2}^{n}\frac{k-1}{n-1}\right]$$

$$+\frac{2}{n(n\xi+1)(n\xi+2)}\sum_{i=2}^{n}\frac{(i-1)^3}{(n-1)^3}$$

$$-\frac{3n\xi(n-1)+2(2n-1)}{8(n-1)(n\xi+1)}+\frac{1}{4}$$

$$=\frac{\xi^2}{n(n\xi+1)(n\xi+2)}\left[\frac{n}{2}\right]^3+\frac{3\xi}{n(n\xi+1)(n\xi+2)}$$

$$\cdot\frac{1}{(n-1)^3}\cdot\frac{1}{6}(n-1)n(2n-1)\cdot\frac{n(n-1)}{2}$$

$$+\frac{2}{n(n\xi+1)(n\xi+2)}\cdot\frac{1}{(n-1)^3}\left[\frac{1}{2}(n-1)n\right]^2$$

$$-\frac{3n\xi(n-1)+2(2n-1)}{8(n-1)(n\xi+1)}+\frac{1}{4}$$

$$=0$$

Then, the skewness $\gamma_1(T) = (\mu_3(T))/(\sqrt{\mu_2(T)}) = 0$.

$$\mu_4(T) = E(T - ET)^4 = E\left(T - \frac{1}{2}\right)^4$$

$$= E(T^4) - 2E\left[T - \frac{1}{2}\right]^3 - \frac{3}{2}E(T^2) + E(T) - \frac{3}{16}$$

$$= E\left[\sum_{i=2}^{n}\frac{i-1}{n-1}y_i\right]^4 - \frac{3n\xi(n-1) + 2(2n-1)}{8(n-1)(n\xi+1)} + \frac{5}{16}$$

$$= \sum_{i=2}^{n}\left(\frac{i-1}{n-1}\right)^4 E(y_i^4) + 3\sum_{i\neq j}\frac{(i-1)^2(j-1)^2}{(n-1)^4}E(y_i^2y_i^2)$$

$$+ 4\sum_{i\neq j}\frac{(i-1)^2(j-1)}{(n-1)^4}E(y_i^3y_i)$$

$$+ 6\sum_{i\neq j\neq k}\frac{(i-1)^2(j-1)(k-1)}{(n-1)^4}E(y_i^2y_jy_k)$$

$$+ \sum_{i\neq j\neq k\neq l}\frac{(i-1)(j-1)(k-1)(l-1)}{(n-1)^4}E(y_iy_jy_ky_l)$$

$$- \frac{3n\xi(n-1) + 2(2n-1)}{8(n-1)(n\xi+1)} + \frac{5}{16}$$

$$= \sum_{i=2}^{n}\left(\frac{i-1}{n-1}\right)^4\frac{\Gamma(n\xi)\Gamma^{n-2}(\xi)\Gamma(\xi+4)}{\Gamma^{n-1}(\xi)\Gamma(n\xi+4)}$$

$$+ 4\sum_{i\neq j}\frac{(i-1)^3(j-1)}{(n-1)^4}\frac{\Gamma(n\xi)\Gamma^{n-3}(\xi)\Gamma(\xi+3)\Gamma(\xi+1)}{\Gamma^{n-1}(\xi)\Gamma(n\xi+4)}$$

$$+ 3\sum_{i\neq j}\frac{(i-1)^2(j-1)^2}{(n-1)^4}\frac{\Gamma(n\xi)\Gamma^{n-3}(\xi)\Gamma^2(\xi+2)}{\Gamma^{n-1}(\xi)\Gamma(n\xi+4)}$$

$$+ 6\sum_{i\neq j\neq k}\frac{(i-1)^2(j-1)(k-1)}{(n-1)^4}\frac{\Gamma(n\xi)\Gamma^{n-4}(\xi)\Gamma(\xi+2)\Gamma^2(\xi+1)}{\Gamma^{n-1}(\xi)\Gamma(n\xi+4)}$$

$$+ \sum_{i\neq j\neq k\neq l}\frac{(i-1)(j-1)(k-1)(l-1)}{(n-1)^4}\frac{\Gamma(n\xi)\Gamma^{n-5}(\xi)\Gamma^4(\xi+1)}{\Gamma^{n-1}(\xi)\Gamma(n\xi+4)}$$

$$- \frac{3n\xi(n-1) + 2(2n-1)}{8(n-1)(n\xi+1)} + \frac{5}{16}.$$

More simplification leads to

$$\mu_4(T) = \frac{1}{n(n-1)^4(n\xi+1)(n\xi+2)(n\xi+3)} \left[(\xi+1)(\xi+2)(\xi+3) \sum_{i=2}^{n} (i-1)^4 \right.$$

$$+ 4\xi(\xi+1)(\xi+2)\sum_{i\neq j}(i-1)^3(j-1) + 3\xi(\xi+1)^2\sum_{i\neq j}(i-1)^2(i-1)^2$$

$$\times 6\xi^2(\xi+1) \sum_{i\neq j\neq k}(i-1)^2(j-1)(k-1)$$

$$\left. +\xi^3 \sum_{i\neq j\neq k\neq l}(i-1)(j-1)(k-1)(l-1) \right]$$

$$- \frac{3n\xi(n-1)+2(2n-1)}{8(n-1)(n\xi+1)} + \frac{5}{16}$$

$$= \frac{3(n+1)[5\xi(n-1)n(n+1)+6(3n^2-4)]}{720(n-1)^3(n\xi+1)(n\xi+2)(n\xi+3)}.$$

Hence, the kurtosis

$$\gamma_2(T) = \frac{\mu_4(T)}{[\mu_2(T)]^2} - 3 = \frac{3(n\xi+1)[5\xi(n-1)n(n+1)+6(3n^2-4)]}{5(n-1)(n+1)(n\xi+2)(n\xi+3)} - 3.$$

Note that the skewness of the distribution of T, $\gamma_1(T) = 0$, and kurtosis $\gamma_2(T) \longrightarrow 0$, as $n \longrightarrow \infty$. Then, under H_0, for sufficiently large n, the distribution of $(T-(1/2))/\sqrt{\text{Var}(T)}$ can be approximated by the standard normal distribution.

The test is based on the likelihood ratio, therefore the inherent properties of the test are still valid. For example, it is locally the most powerful test at level α. A practical example of change point analysis for the gamma model is given in Hsu (1979).

5.3 Informational Approach

For the gamma model, we are interested in testing

$$H_0 : \theta_1 = \theta_2 = \cdots = \theta_n = \theta_0$$

against the alternative:

$$H_1 : \theta_1 = \cdots = \theta_k \neq \theta_{k+1} = \cdots = \theta_n,$$

where k is the unknown position of a change point; θ_0, θ_1, and θ_n are unknown. For testing H_0 against H_1, it is equivalent to selecting the best model among n proposed models, as H_0 corresponds to a model of no change point and H_1 corresponds to $n - 1$ models each revealing a change point at the location k, for $k = 1, \ldots, n - 1$.

We derive the informational approach-SIC as follows.

Under H_0, the likelihood function $L_0(\theta_0)$ is

$$L_0(\theta_0) = \prod_{i=1}^{n} \frac{1}{\theta_0^{\xi} \Gamma(\xi)} x_i^{\xi-1} e^{-(x_i/\theta_0)}$$

$$= \frac{\prod_{i=1}^{n} x_i^{\xi-1}}{\Gamma^n(\xi)} \exp\left[\sum_{i=1}^{n} \left(-\frac{x_i}{\theta_0} - \ln \theta_0^{\xi} \right) \right],$$

and the MLE of θ_0 is

$$\widehat{\theta}_0 = \frac{\sum_{i=1}^{n} x_i}{n\xi}.$$

Therefore, denoting the SIC under H_0 by $\text{SIC}(n)$, we have

$$\text{SIC}(n) = -2 \log L_0(\widehat{\theta}_0) + \log n$$

$$= 2n\xi \log \Sigma x_i - 2(\xi - 1) \sum_{i=1}^{n} \log x_i - \log \frac{n e^{2n\xi} \xi^{2n}(\xi)}{(n\xi)^{2n\xi}}.$$

Under H_1 the MLEs of θ_1 and θ_n are obtained as

$$\widehat{\theta}_1 = \frac{\sum_{i=1}^{k} x_i}{k\xi}, \qquad \widehat{\theta}_n = \frac{\sum_{i=k+1}^{n} x_i}{(n-k)\xi}.$$

Therefore, letting SIC under H_1 be denoted by $\text{SIC}(k)$, for $1 \leq k \leq n-1$, we obtain:

$$\text{SIC}(k) = -2 \log L_1(\widehat{\theta}_1, \widehat{\theta}_n) + 2 \log n$$

$$= 2k\xi \log \sum_{i=1}^{k} x_i + 2(n-k)\xi \log \sum_{i=k+1}^{n} x_i - 2(\xi - 1) \sum_{i=1}^{n} \log x_i$$

$$+ \log \frac{n^2 e^{2n\xi} \xi^{2n}(\xi)}{(n\xi)^{2k\xi}[(n-k)\xi]^{2(n-k)\xi}}.$$

According to the minimum information criterion principle, H_0 is not rejected if $\text{SIC}(n) \leq \min_{1 \leq k \leq n-1} \text{SIC}(k)$, and hence it is concluded that there is no change in the scale parameter of the gamma model. H_0 is rejected if $\text{SIC}(n) > \min_{1 \leq k \leq n-1} \text{SIC}(k)$, and therefore it is concluded that there is a change in the scale parameter of the gamma model. The location of the change point is estimated at $\hat{k}$, where $\hat{k}$ is such that

$$\mathrm{SIC}(\hat{k}) = \min_{1 \leq k \leq n-1} \mathrm{SIC}(k).$$

5.4 Bayesian Approach

A similar change point problem with regard to the gamma distribution defined in Section 5.1 was later studied by Diaz (1982). The approach is completely Bayesian, which gives us an alternate way of finding the change point in this situation. The work of Diaz (1982) is presented in detail as follows.

Let $x_1, \ldots, x_n$ be a sequence of independent random variables from gamma distributions. Specifically, let X_i have the pdf $f(x; \theta_2)$ for $j = k + 1, \ldots, n$, where

$$f(x; \theta) = \frac{x^{\xi-1} e^{-(x/\theta)}}{\theta^{\xi} \Gamma(\xi)}, \qquad \xi > 0, \theta > 0, x > 0.$$

ξ is known, θ unknown, and k is the unknown position of the change point. Our interest here is to test the following hypotheses,

$$H_0 : k = n \quad \text{versus} \quad H_1 : 1 \leq k \leq n - 1.$$

The following prior distributions are assumed for k, θ_1, and θ_2.

Let k have prior pdf $g_0(k)$ given by

$$g_0(k) = \begin{cases} p & \text{if } k = n \\ \frac{1-p}{n-1} & \text{if } k \neq n \end{cases},$$

where p is known such that $0 \leq p \leq 1$. When $k = n$, $\theta_1 = \theta_2$, and the prior density of θ_1 is

$$g_1(\theta_1) = \frac{e^{1/\theta_1 \alpha_1}}{\alpha_1^{r_1} \Gamma(r_1) \theta_1^{r_1+1}}, \qquad \theta_1 > 0$$

and if $k \neq n$, θ_1 and θ_2 are assumed independent, the prior density of θ_2 is given by

$$g_2(\theta_2) = \frac{e^{1/\theta_2 \alpha_2}}{\alpha_2^{r_2} \Gamma(r_2) \theta_2^{r_2+1}}, \qquad \theta_2 > 0,$$

and the prior density of θ_1 is the same as in the case of $k = n$. The parameters α_1, r_1, α_2, and r_2 are positive and known. The reason for choosing such $g_1(\cdot)$ and $g_2(\cdot)$ is that they are conjugate priors. Then, the joint prior density of $x_1, \ldots, x_n$ given k, θ_1, and θ_2 is

$$f(x_1,\ldots,x_n|k,\theta_1,\theta_2)$$

$$= \begin{cases} \theta_1^{-n\xi}\Gamma^{-n}(\xi)(\prod_1^n x_i)^{\xi-1}e^{-(1/\theta_1)\Sigma_1^n x_i} & \text{if } k = n \\ \theta_1^{-k\xi}\theta_2^{-(n-k)\xi}\Gamma^{-n}(\xi)(\prod_1^n x_i)^{\xi-1}e^{-(1/\theta_1)\Sigma_1^k x_i-(1/\theta_2)\Sigma_{k+1}^n x_i} & \text{if } 1\leq k\leq n-1 \end{cases}.$$

Therefore, the joint prior density of λ, θ_1, and θ_2 is given by

$$h_0(k,\theta_1,\theta_2|x_1,\ldots,x_n) = \frac{f(x_1,\ldots,x_n|k,\theta_1,\theta_2)g(k,\theta_1,\theta_2)}{\sum_{k=1}^n \int\int f(x_1,\ldots,x_n|k,\theta_1,\theta_2)g(k,\theta_1,\theta_2)d\theta_1 d\theta_2}$$

$$= \begin{cases} h_1, & k = n \\ h_2, & 1 \leq k \leq n-1 \end{cases},$$

where

$$h_1 = \frac{1}{\int_0^\infty \alpha_1^{-r_1}\Gamma^{-1}(r_1)\cdot\theta_1^{-(n\xi+r_1+1)}e^{-(1/\theta_1)\Sigma_1^n x_i-1/(\theta_1\alpha_1)}d\theta_1}$$

$$\cdot p\theta_1^{-n\xi}e^{-(1/\theta_1)\Sigma_1^n x_i}\cdot\alpha_1^{-r_1}\Gamma^{-1}(r_1)\theta_1^{-r_1-1}e^{-1/(\theta_1\alpha_1)},$$

and

$$h_2 = \frac{1}{\sum_{k=1}^{n-1}\frac{1-p}{n-1}\int_0^\infty\int_0^\infty\frac{e^{-(1/\theta_1)\Sigma_1^k x_i-1/(\theta_1\alpha_1)}}{\alpha_1^{r_1}\Gamma(r_1)\theta_1^{k\xi+r_1+1}}\cdot\frac{e^{-(1/\theta_2)\Sigma_{k+1}^n x_i-1/(\theta_2\alpha_2)}}{\alpha_2^{r_2}\Gamma(r_2)\theta_2^{(n-k)\xi+r_2+1}}d\theta_1 d\theta_2}$$

$$\cdot\frac{1-p}{n-1}\frac{e^{-(1/\theta_1)\Sigma_1^k x_i-(1/\theta_2)\Sigma_{k+1}^n x_i}}{\theta_1^{\lambda\xi}\theta_2^{(n-\lambda)\xi}}\cdot\frac{e^{-1/(\theta_1\alpha_1)}}{\alpha_1^{r_1}\Gamma(r_1)\theta_1^{r_1+1}}\frac{e^{-1/(\theta_2\alpha_2)}}{\alpha_2^{r_2}\Gamma(r_2)\theta_2^{r_2+1}}$$

$$+ p\int_0^\infty\frac{1}{\alpha_1^{r_1}\xi(r_1)}\cdot\frac{e^{-(1/\theta_1)\Sigma_1^k x_i-1/(\theta_1\alpha_1)}}{\theta_1^{k\xi+r_1+1}}d\theta_1.$$

After some calculation, $h_0(k,\theta_1,\theta_2|x_1,\ldots,x_n)$ simplifies to

$$h_0(k,\theta_1,\theta_2|x_1,\ldots,x_n)$$

$$= \begin{cases} p\cdot\text{const}\cdot\theta_1^{-(n\xi+r_1+1)}e^{-(1/\theta_1)\Sigma_1^n x_i-1/(\theta_1\alpha_1)}, & k = n \\ \frac{1-p}{n-1}\cdot\text{const}\cdot\theta_1^{-(k\xi+r_1+1)}e^{-(1/\theta_1)\Sigma_1^k x_i-1/(\theta_1\alpha_1)} \\ \cdot\theta_2^{-[(n-\lambda)\xi+r_2+1]}\alpha_2^{-r_2}\Gamma^{-1}(r_2)e^{-(1/\theta_2)\Sigma_{k+1}^n x_i-1/(\theta_2\alpha_2)}, & 1\leq k\leq n-1 \end{cases}.$$

Therefore, the posterior density of the change point k is given by:

$$h(k|\mathbf{x}) = \int_0^\infty \int_0^\infty h_0(k, \theta_1, \theta_2 | x_1, \ldots, x_n) d\theta_1 d\theta_2$$

$$= \begin{cases} h_3, & k = n \\ h_4 & 1 \le k \le n-1 \end{cases},$$

where

$$h_3 = p \cdot \text{const} \cdot \int_0^\infty \theta_1^{-(n\xi+r_1+1)} e^{-(1/\theta_1)(\Sigma_1^n x_i + 1/\alpha_1)} d\theta_1$$

$$= p \cdot \text{const} \cdot \frac{\Gamma(n\xi + r_1)}{\left(\sum_1^n x_i + 1/\alpha_1\right)^{n\xi+r_1}}$$

$$\cdot \int_0^\infty \frac{\left(\sum_1^n x_i + 1/\alpha_1\right)^{n\xi+r_1}}{\Gamma(n\xi + r_1)} \left(\frac{1}{\theta_1}\right)^{n\xi+r_1-1} e^{-(1/\theta_1)(\Sigma_1^n x_i + 1/\alpha_1)} d\left(\frac{1}{\theta_1}\right)$$

$$= p \cdot \text{const} \cdot \Gamma(n\xi + r_1) \left(\sum_1^n x_i + 1/\alpha_1\right)^{-(n\xi+r_1)},$$

and

$$h_4 = \frac{1-p}{n-1} \cdot \text{const} \cdot \int_0^\infty \theta_1^{-(k\xi+r_1+1)} e^{-(1/\theta_1)(\Sigma_1^k x_i + 1/\alpha_1)} d\theta_1$$

$$\cdot \int_0^\infty \theta_2^{-[(n-k)\bar{\xi}+r_2+1]} \alpha_2^{-r_2} \Gamma^{-1}(r_2) e^{-(1/\theta_2)(\Sigma_{k+1}^n x_i + 1/\alpha_2)} d\theta_2$$

$$= \frac{1-p}{n-1} \cdot \text{const} \cdot \Gamma(k\xi + r_1) \left(\sum_1^k x_i + 1/\alpha_1\right)^{-(k\xi+r_1)}$$

$$\cdot \int_0^\infty \frac{\left(\sum_1^k x_i + 1/\alpha_1\right)^{(k\xi+r_1)}}{\Gamma(k\xi + r_1)} \left(\frac{1}{\theta_1}\right)^{k\xi+r_1-1} e^{-(1/\theta_1)(\Sigma_1^k x_i + 1/\alpha_1)} d\left(\frac{1}{\theta_1}\right)$$

$$\cdot \Gamma((n-k)\xi + r_2) \left(\sum_{k+1}^n x_i + 1/\alpha_2\right)^{-[(n-k)\xi+r_2]} \cdot \frac{1}{\alpha_2^{r_2} \Gamma(r_2)}$$

$$\cdot \int_0^\infty \frac{\left(\sum_{k+1}^n x_i + 1/\alpha_2\right)^{(n-\lambda)\xi+r_2}}{\Gamma((n-k)\xi + r_2)} \left(\frac{1}{\theta_2}\right)^{(n-\lambda)\xi+r_2-1}$$

$$\times e^{-(1/\theta_2)(\Sigma_{k+1}^n x_i + 1/\alpha_2)} d\left(\frac{1}{\theta_2}\right)$$

$$= \frac{1-p}{n-1} \cdot \text{const} \cdot \frac{\Gamma(k\xi + r_1)}{\left(\sum_1^k x_i + 1/\alpha_1\right)^{(k\xi + r_1)}}$$

$$\cdot \frac{\Gamma((n-k)\xi + r_2)}{\left(\sum_{k+1}^n x_i + 1/\alpha_2\right)^{(n-\lambda)\xi + r_2}} \cdot \frac{1}{\alpha_2^{r_2} \Gamma(r_2)}.$$

That is,

$$h(k \,|\mathbf{x}) \propto \begin{cases} p \cdot \Gamma(n\xi + r_1)\left(\sum_1^n x_i + 1/\alpha_1\right)^{-(n\xi + r_1)}, & k = n \\[2ex] \dfrac{(1-p)\Gamma(k\xi + r_1)}{(n-1)\left(\sum_1^k x_i + 1/\alpha_1\right)^{(k\xi + r_1)}} \cdot & \\[2ex] \dfrac{\Gamma((n-k)\xi + r_2)}{\left(\sum_{k+1}^n x_i + 1/\alpha_2\right)^{(n-\lambda)\xi + r_2}} \cdot \dfrac{1}{\alpha_2^{r_2}\Gamma(r_2)}, & 1 \le k \le n-1. \end{cases}$$

It is worth noting that, in many practical situations, one may take an improper (noninformative) prior $g(\theta_i) \propto (1/\theta_i)(i = 1,2)$ for θ_1 and θ_2, and assume the independence of θ_1 and θ_2. In this case, the posterior density of the change point k is given by

$$h(\lambda|\mathbf{x}) \propto \begin{cases} p\Gamma(n\xi)\left(\sum_1^n x_i\right)^{-n\xi}, & k = n \\[2ex] \dfrac{(1-p)\Gamma(k\xi)}{(n-1)\left(\sum_1^k x_i\right)^{k\xi}} \dfrac{\Gamma((n-k)\xi)}{\left(\sum_{k+1}^n x_i\right)^{(n-k)\xi}}, & 1 \le k \le n-1 \end{cases}.$$

The examples analyzed by Hsu (1979) were also analyzed by Diaz (1982), and their conclusions matched.

5.5 Application to Stock Market and Air Traffic Data

Hsu (1979) analyzed stock market data and air traffic flow data to illustrate the method given in Section 5.1. Diaz (1982) reanalyzed those two datasets for change point by using the Bayesian approach, and both authors' conclusions matched. Here, those two datasets are analyzed again to illustrate how to implement the SIC procedure to detect and locate the change point.

Example 5.1 The first dataset to be analyzed is given in Appendix A of Hsu (1979). This dataset contains Friday closing values of the Dow-Jones Industrial Average (DJIA) from July 1, 1971 through August 2, 1974. Let P_t be the Friday closing value of DJIA during that period (total 162 values), then according to Hsu (1979), the return series $\{R_t\}$, where $R_t = (P_{t+1} - P_t)/P_t, t = 1,\ldots,161$, is a sequence of independent normal random variables with mean 0 and unknown variance σ^2. Simple derivation shows that R_t^2 is a gamma random variable with shape parameter $\gamma = \frac{1}{2}$, and scale parameter $\lambda = (2\sigma^2)^{-1}$. Then one can find out if there is a change in the scale parameters of the gamma random sequence $\{R_t^2\}$ by testing:

$$H_0 : \lambda_1 = \lambda_2 = \cdots = \lambda_{161} = \lambda_0$$

against the alternative:

$$H_1 : \lambda_1 = \cdots = \lambda_k \neq \lambda_{k+1} = \cdots = \lambda_{161}.$$

The values of $\mathrm{SIC}(n), n = 161$, and $\mathrm{SIC}(k), k = 1, \ldots, 160$, are computed. It is found that $\mathrm{SIC}(161) = -2245.2 > \min_{1 \leq k \leq 160} \mathrm{SIC}(k) = \mathrm{SIC}(89) = -2267.1$. Therefore, there is a change in the scale parameter and the location of the change point in $\{R_t^2\}$ is $\hat{k} = 89$, which corresponds to the week when the Watergate event took place. This conclusion matches those of Hsu (1979) and Diaz (1982).

Example 5.2 The second dataset to be analyzed is given in Appendix B of Hsu (1979). This dataset contains 213 airplane arrival times collected from a low-altitude transitional control sector (near Newark airport) for the period from noon to 8:00 PM on April 30, 1969. Hsu has examined the data and concluded that the interarrival times are independently exponentially distributed. That is, if T_i denotes the arrival time, then $x_i = T_{i+1} - T_i, i = 1, \ldots, 212$, is a gamma random variable with shape parameter $\gamma = 1$ and scale parameter λ. To see if the air traffic densities are constant over time, one can test:

$$H_0 : \lambda_1 = \lambda_2 = \cdots = \lambda_{212} = \lambda_0$$

against the alternative:

$$H_1 : \lambda_1 = \cdots = \lambda_k \neq \lambda_{k+1} = \cdots = \lambda_{212}$$

for the sequence $\{x_i\}$. The values of $\mathrm{SIC}(n)$, for $n = 212$, and $\mathrm{SIC}(k)$, for $k = 1, \ldots, 211$, are computed. It is found that

$$\mathrm{SIC}(212) = 2504.1 < \min_{1 \leq k \leq 211} \mathrm{SIC}(k) = 2505.1.$$

Therefore, there is no change in the scale parameters; that is, the air traffic densities are constant over time. This conclusion again matches those of Hsu (1979) and Diaz (1982).

5.6 Another Type of Change

So far our discussion about the change points has been limited to the sudden changes in a sequence of random variables from different models. Here, we would like to pinpoint briefly a different type of change point problem under the assumption of gamma distribution, and hope this reveals the rich resources of change point problems, and gives the readers a peek at one of the other considerations that may be encountered in some situations.

Let $x_1, \ldots, x_n$ be a sequence of independent random variables from the gamma distributions with parameters $(\theta_1, \xi), (\theta_1, \xi), \ldots,$ and (θ_n, ξ), respectively. As before, ξ is assumed to be known, and the pdf of $X_i's$ is

$$f(x, \xi, \theta_i) = \frac{1}{\theta_i^\xi \Gamma(\xi)} x^{\xi-1} e^{-(x/\theta_i)}, \qquad \xi, \theta_i > 0, x > 0,$$

for $i = 1, \ldots, n$.

We now intend to test H_0 given in (5.1) versus the alternative:

$$H_2 : \theta_i = \theta_0 e^{\{\beta(i-1)\}}, \qquad \beta \neq 0, \qquad i = 1, 2, \ldots, n, \tag{5.4}$$

with $\beta > 0$ meaning a continuous exponential increase in the parameter θ, and $\beta < 0$ meaning a continuous exponential decrease in the parameter θ.

The likelihood function under (5.4) is

$$L_2(\theta_0, \beta) = \frac{1}{\theta_0^{n\xi} e^{(n(n-1)/2)\xi\beta} \Gamma^n(\xi)} \left(\prod_{i=1}^n x_i\right)^{\xi-1} e^{-(1/\theta_0) \sum_{i=1}^n x_i/(e^{\beta(i-1)})};$$

that is,

$$\log L_2(\theta_0, \beta) = -n\xi \log \theta_0 - \frac{n(n-1)}{2}\xi\beta - n \log \Gamma(\xi)$$

$$+ (\xi - 1) \sum_{i=1}^n \log x_i - \frac{1}{\theta_0} \sum_{i=1}^n \frac{x_i}{e^{\beta(i-1)}},$$

and

$$\frac{\partial \log L_2(\theta_0, \beta)}{\partial \beta} = -\frac{n(n-1)}{2}\xi + \frac{1}{\theta_0} \sum_{i=1}^n \frac{(i-1)x_i}{e^{\beta(i-1)}}$$

$$\frac{\partial \log L_2(\theta_0, \beta)}{\partial \theta_0} = -\frac{n\xi}{\theta_0} + \frac{1}{\theta_0^2} \sum_{i=1}^n \frac{x_i}{e^{\beta(i-1)}},$$

which gives

$$\widehat{\theta}_0 = \frac{1}{n\xi} \sum_{i=1}^n \frac{x_i}{e^{\beta(i-1)}}.$$

Note that

$$\frac{\partial^2 \log L_2(\theta_0, \beta)}{\partial \theta_0^2}\bigg|_{\theta_0 = \widehat{\theta}_0} = \frac{n\xi}{\widehat{\theta}_0^2} - \frac{2}{\widehat{\theta}_0^3} \sum_{i=1}^n \frac{x_i}{e^{\beta(i-1)}}$$

$$= \frac{n^3 \xi^3}{\left(\sum_{i=1}^n \frac{x_i}{e^{\beta(i-1)}} \right)^2} - \frac{2n^3 \xi^3}{\left(\sum_{i=1}^n \frac{x_i}{e^{\beta(i-1)}} \right)^2}$$

$$= - \frac{n^3 \xi^3}{\left(\sum_{i=1}^n \frac{x_i}{e^{\beta(i-1)}} \right)^2} < 0,$$

hence $\widehat{\theta}_0 = (1/n\xi) \sum_{i=1}^n (x_i / (e^{\beta(i-1)}))$ is the MLE of θ_0 under (5.4). Now

$$\left. \frac{\partial \log L_2(\theta_0, \beta)}{\partial \beta} \right|_{\theta_0 = \widehat{\theta}_0, \beta = 0} = - \frac{n(n-1)}{2} \xi + n\xi \frac{\sum_{i=1}^n (i-1) x_i}{\sum_{i=1}^n x_i}$$

$$= - \frac{n(n-1)\xi}{2} + n\xi \cdot (n-1) T$$

$$= n(n-1)\xi \left(T - \frac{1}{2} \right),$$

where

$$T = \frac{\sum_{i=1}^n (i-1) x_i}{(n-1) \sum_{i=1}^n x_i}$$

as defined in Section 5.2.

According to Cox and Hinkley (1974),

$$\left. \frac{\partial \log L_2(\theta_0, \beta)}{\partial \beta} \right|_{\theta_0 = \widehat{\theta}_0, \beta = 0},$$

the likelihood-derivative test (LDT), is asymptotically equivalent to the likelihood-ratio test (LRT). Hence, we can make use of the asymptotic distribution of T obtained in Section 5.2 to obtain an asymptotic test for H_0 given by (5.1) versus H_2 given by (5.4).

Chapter 6
Exponential Model

6.1 Problem

Change point problems occur in various situations and scientific disciplines. In earlier chapters of this monograph, the change point problems associated with the univariate normal, multivariate normal, linear regression, and gamma models were discussed. In this chapter, the change point occurring in an exponential model is studied. An exponential model is useful and appropriate in some experimental sciences, therefore, it is desirable to make an inference about a change point for an exponential model.

In the literature, several authors have studied some aspects of a change point occurring in an exponential model. For instance, Worsley (1986) used maximum likelihood methods to test for a change point in a sequence of independent exponential family random variables, with an emphasis on the exponential distribution. Haccou, Meelis, and Geer (1988) investigated the change point problem for a sequence of exponential random variables by using the likelihood-ratio test, and obtained the asymptotic null distribution of the test statistic. Haccou and Meelis (1988) gave a procedure for testing the number of change points in a sequence of exponential random variables based on partitioning of the likelihood according to a hierarchy of subhypothesis. Ramanayake and Gupta (2003) studied the epidemic change using the likelihood-ratio procedure method.

In the following, we introduce several effective methods for detecting the change point in a sequence of exponential random variables based on the work of Worsley (1986), Haccou, Meelis, and Geer (1988), and Haccou and Meelis (1988).

6.2 Likelihood-Ratio Procedure

Let $x_1, x_2, \ldots, x_{n+1}$ be a sequence of independent exponentially distributed random variables. We are interested in testing the null hypothesis:

$$H_0 : x_1, x_2, \ldots, x_{n+1} \sim \text{iid } f(x; \lambda) = \lambda e^{-\lambda x},$$

against the alternative:

$$H_1 : x_1, x_2, \ldots, x_k \sim \text{iid } f(x; \lambda_1) = \lambda_1 e^{-\lambda_1 x}, \quad \text{and}$$

$$x_{k+1}, x_{k+2}, \ldots, x_{n+1} \sim \text{iid } f(x; \lambda_2) = \lambda_2 e^{-\lambda_2 x}, \qquad 1 \le k \le n.$$

Under H_0, the likelihood function is

$$L_0(\lambda) = \prod_{i=1}^{n+1} f(x_i; \lambda) = \lambda^{n+1} e^{-\lambda \sum_{i=1}^{n+1} x_i},$$

and the MLE of λ is easily found to be:

$$\widehat{\lambda} = \frac{n+1}{\sum_{i=1}^{n+1} x_i}.$$

Under H_1, the likelihood function is

$$L_1(\lambda_1, \lambda_2) = \lambda_1^k e^{-\lambda_1 \sum_{i=1}^{k} x_i} \cdot \lambda_2^{n-k} e^{-\lambda_2 \sum_{i=k+1}^{n+1} x_i},$$

and the MLEs of λ_1 and λ_2 are

$$\widehat{\lambda}_1 = \frac{k}{\sum_{i=1}^{k} x_i} \quad \text{and} \quad \widehat{\lambda}_2 = \frac{n-k+1}{\sum_{i=k+1}^{n+1} x_i}.$$

Hence, the maximum likelihood-ratio procedure test statistic is

$$LPT = \frac{L_0(\widehat{\lambda})}{L_1(\widehat{\lambda}_1, \widehat{\lambda}_2)},$$

and

$$2LPT = 2 \log \left[\left(\frac{n+1}{\sum_{i=1}^{n+1} x_i} \right)^{n+1} \left(\frac{\sum_{i=1}^{k} x_i}{k} \right)^{k} \left(\frac{\sum_{i=k+1}^{n+1} x_i}{n-k+1} \right)^{n-k+1} \right]$$

$$= 2 \log \left[\left(\frac{\sum_{i=1}^{k} x_i}{\sum_{i=1}^{n+1} x_i} \right)^{k} \left(\frac{n+1}{k} \right)^{k} \right.$$

$$\cdot \left. \left(\frac{\sum_{i=k+1}^{n+1} x_i}{\sum_{i=1}^{n+1} x_{ii}} \right)^{n-k+1} \left(\frac{n+1}{n-k+1} \right)^{n-k+1} \right]$$

$$= 2 \log \left[\left(\frac{k}{n+1} \right)^{-k} \left(\frac{\sum_{i=1}^{k} x_i}{\sum_{i=1}^{n+1} x_i} \right)^{k} \left(1 - \frac{k}{n+1} \right)^{-(n-k+1)} \right.$$

$$\left. \cdot \left(1 - \frac{\sum_{i=1}^{k} x_i}{\sum_{i=1}^{n+1} x_i} \right)^{n-k+1} \right].$$

Let

$$\beta_n(x; k) = \frac{\sum_{i=1}^{k} x_i}{\sum_{i=1}^{n+1} x_i} \quad \text{and} \quad r_n(k) = \frac{k}{n+1};$$

then,

$2 \log LPT$

$$= 2 \log[r_n^{-k}(k) \beta_n^k(x; k)(1 - \beta_n(x; k))^{n-k+1} (1 - r_n(k))^{-(n-k+1)}]$$

$$= 2 \log[\beta_n(x; k)/r_n(k)]^k + 2 \log[(1 - \beta_n(x; k))/(1 - r_n(k))]^{n-k+1}$$

$$= 2(n+1) \log[\beta_n(x; k)/r_n(k)]^{r_n(k)}$$

$$\quad + 2(n+1) \log[(1 - \beta_n(x; k))/(1 - r_n(k))]^{1 - r_n(k)}$$

$$= 2(n+1)\{r_n(k) \log[\beta_n(x; k)/r_n(k)]$$

$$\quad + (1 - r_n(k)) \log[(1 - \beta_n(x; k))/(1 - r_n(k))]\}.$$

Let

$$f_n(x; k) = -2 \log LPT$$

$$= 2(n+1)\{-r_n(k) \log[\beta_n(x; k)/r_n(k)]$$

$$\quad - (1 - r_n(k)) \log[(1 - \beta_n(x; k))/(1 - r_n(k))]\},$$

and take a second-order Taylor expansion of $f_n(x, k)$ at the point $r_n(k)$ while viewing $f_n(x; k)$ as a function of $\beta_n(x; k)$; we obtain:

$$f_n(x; k) = f_n(x; k)|_{\beta_n(x;k)=r_n(k)}$$

$$+ \left. \frac{\partial f_n}{\partial \beta} \right|_{\beta_n(x;k)=r_n(k)} (\beta_n(x; k) - r_n(k))$$

$$+ \frac{1}{2!} \left. \frac{\partial^2 f_n}{\partial \beta^2} \right|_{\beta_n(x;k)=r_n(k)} (\beta_n(x; k) - r_n(k))^2$$

$$+ \frac{1}{3!} \left. \frac{\partial^3 f_n}{\partial \beta^3} \right|_{\beta_n(x;k)=r_n(k)} (1 - \xi_{2m}(k))^3 (\beta_n(x; k) - r_n(k))^3$$

$$= 2(n+1)\left[-\frac{r_n(k)}{\beta_n(x;k)} + (1 - r_n(k))\frac{1}{1 - \beta_n(x;k)}\right]\Bigg|_{\beta_n(x;k)=r_n(k)}$$

$$\cdot (\beta_n(x;k) - r_n(k))$$

$$+ \frac{1}{2} \cdot 2(n+1)\left[\frac{r_n(k)}{\beta_n^2(x;k)} + \frac{1 - r_n(k)}{(1 - \beta_n(x;k))^2}\right]\Bigg|_{\beta_n(x;k)=r_n(k)}$$

$$\cdot (\beta_n(x;k) - r_n(k))^2$$

$$+ \frac{1}{6} \cdot 2(n+1)\left[-2\frac{2r_n(k)}{\beta_n^3(x;k)} + \frac{2(1 - r_n(k))}{(1 - \beta_n(x;k))^3}\right]\Bigg|_{\beta_n(x;k)=\xi} (1-\theta)^3$$

$$\cdot (\beta_n(x;k) - r_n(k))^3$$

$$= (n+1)(\beta_n(x;k) - r_n(k))^2[r_n(k)(1 - r_n(k)]$$

$$+ \frac{2}{3}(n+1)\left[\frac{1 - r_n(k)}{(1 - \xi)^3} - \frac{r_n(k)}{\xi^3}\right](r_n(k)(1 - r_n(k))^3(1 - \xi_{2,n}(k))^3$$

$$= (n+1)\frac{(\beta_n(x;k) - r_n(k))^2}{[r_n(k)(1 - r_n(k)]} \cdot \left\{1 + \frac{2}{3}(\beta_n(x;k) - r_n(k))\right.$$

$$\left.\cdot[r_n(k)(1 - r_n(k)] \cdot \frac{\xi^3(1 - r_n(k)) - (1 - \xi)^3(r_n(k))}{\xi^3(1 - \xi)^3}(1 - \xi)^3\right\},$$

where $0 < \theta < 1$, and $\xi = r_n(k) + \theta(\beta_n - r_n)$. Hence,

$$f_n(x;k) = (n+1)\frac{(\beta_n(x;k) - r_n(k))^3}{[r_n(k)(1 - r_n(k))]} \cdot \left\{1 + \frac{2}{3}(\beta_n(x;k) - r_n(k))\right.$$

$$\left.\cdot[r_n(k)(1 - r_n(k))]^2\left(\frac{1-\theta}{1-\xi}\right)^3 - r_n^2(k)(1 - r_n(k))\left(\frac{1-\theta}{\xi}\right)^3\right\}.$$

Here,

$$\left(\frac{1-\theta}{1-\xi}\right)^3 \stackrel{\Delta}{=} \frac{1}{(1 - \xi_{2,n}(k))^3}$$

$$\left(\frac{1-\theta}{\xi}\right)^3 \stackrel{\Delta}{=} \frac{1}{\xi_{1,n}^3(k)}$$

$$R_n(k) \stackrel{\Delta}{=} \frac{2}{3}(\beta_n(x;k) - r_n(k))$$

$$\cdot[r_n(k)(1 - r_n(k))]^2\left(\frac{1-\theta}{1-\xi}\right)^3 - r_n^2(k)(1 - r_n(k))\left(\frac{1-\theta}{\xi}\right)^3,$$

where

$$\xi_{2,n}(k) = \frac{\xi - \theta}{1 - \theta} = \frac{r_n(k)\theta(\beta_n(x;k) - r_n(k)) - \theta}{1 - \theta}$$

$$= r_n(k) + \frac{\theta}{1 - \theta}[1 - \beta_n(x;k)]$$

$$\xi_{1,n}(k) = \frac{\xi}{1 - \theta} = \frac{r_n(k) + \theta(\beta - r_n(k))}{1 - \theta}$$

$$= r_n(k) + \frac{\theta}{1 - \theta}\beta_n(x;k);$$

that is, $\xi_{1,n}(k)$ and $\xi_{2,n}(k)$ are between $r_n(k)$ and $\beta_n(x;k)$, and then

$$f_n(x;k) = \frac{(n+1)[\beta_n(x;k) - r_n(k)]^3}{[r_n(k)(1 - r_n(k)](1 + r_n(k))} \cdot (1 + R_n(k)).$$

Let $U_n(k)$ be the kth order statistic of a random sample of size n from unif(0,1), and

$$f(u) = I_{(0,1)}(u) \quad \text{and} \quad F(u) = \begin{cases} 0, & u \le 0 \\ u, & 0 < u < 1 \\ 1, & u \ge 1 \end{cases}$$

be the pdf and cdf of uniform(0,1), respectively. Then, the pdf of $U_n(k)$ is

$$f_{U_n(k)}(x) = \frac{n!}{(k-1)!(n-k)!}[F(x)]^{k-1}[1 - F(x)]^{n-k}f(x)$$

$$= \frac{n!}{(k-1)!(n-k)!}u^{k-1}(1 - u)^{n-k} \cdot 1$$

$$= \frac{\Gamma(n+1)}{\Gamma(k)\Gamma(n+1-k)}u^{k-1}(1 - u)^{n-k}, \qquad 0 < u < 1,$$

which is the pdf of $Beta(k, n+1)$. It is easy to see that

$$\beta_n(x;k) = \frac{\sum_{i=1}^{k} x_i}{\sum_{i=1}^{n+1} x_i} \sim Beta(k, n+1);$$

that is,

$$\beta_n(x;k) \overset{D}{=} U_n(k).$$

Define the following functions.

$$U_n(y) = \begin{cases} U_n(k), & \text{for } \frac{k-1}{n} < y \leq \frac{k}{n} \\ 0, & \text{for } y = 0 \end{cases}$$

$$z_n(y) = \begin{cases} \frac{k}{n+1}, & \text{for } \frac{k-1}{n} < y \leq \frac{k}{n} \\ 0, & \text{for } y \leq 0 \end{cases},$$

$$x_n(y) = (n+1)^{1/2}(U_n(y) - z_n(y)),$$

$$\zeta_n(y) = [z_n(y)(1 - z_n(y))]^{1/2}.$$

Then, consider the process defined by

$$\widetilde{f}_n(y) = [x_n(y)/\zeta_n(y)]^2(1 + R_n(y)), \qquad 0 \leq y \leq 1,$$

with

$$R_n(y) = \frac{2}{3}x_n(y)(n+1)^{-1/2}[\{z_n(y)(1 - z_n(y))^2/(1 - \xi_{2,n}(y))^3\}$$
$$- \{(z_n(y))^2(1 - z_n(y))/(\xi_{1,n}(y))^3\}],$$

and $\xi_{1,n}(y)$ and $\xi_{2,n}(y)$ between $z_n(y)$ and $U_n(y)$. Therefore,

$$\max_{1 \leq k \leq n} f_n(x; k) \overset{D}{=} \max_{0 \leq y \leq 1} \widetilde{f}_n(y).$$

Note that the uniform quantile process is defined as

$$\tilde{U}_n(y) = n^{1/2}(U_n(y) - y), \qquad 0 \leq y \leq 1,$$

and the asymptotic distribution of $f_n(x; k)$ can be derived by using limit theorems concerning the uniform quantile process $\tilde{U}_n(y)$.

The following lemmas are needed for the derivation of the asymptotic distribution of $f_n(x; k)$.

Lemma 6.1

$$\limsup_{n \to \infty} \sup_{\varepsilon_n \leq y \leq 1 - \varepsilon_n} \{(\log \log n)^{-1/2}|x_n(y)/\zeta_n(y)|\} < 5\sqrt{2}, \quad a.s.,$$

where $\varepsilon_n = (\log \log n)^4/n$.

Proof. This is a direct modification of a theorem proved in Csörgö and Révész (1981). □

Lemma 6.2

$$\limsup_{n \to \infty} \sup_{\varepsilon_n \leq y \leq 1 - \varepsilon_n} \{(\log \log n)|R_n(y)|\} = 0, \quad a.s.$$

Proof. Note that

$$R_n(y)$$

$$= \frac{2}{3} z_n(y)(n+1)^{-(1/2)} \left[\left\{ \frac{z_n(y)(1 - z_n(y))^2}{(1 - \xi_{2,n}(y))^3} \right\} - \left\{ \frac{(z_n(y))^2(1 - z_n(y))}{(\xi_{1,n}(y))^3} \right\} \right]$$

$$= \frac{2}{3} \frac{z_n(y)(n+1)^{-(1/2)}}{[z_n(y)(1 - z_n(y))]^{1/2}} \cdot [z_n^{(y)}(1 - z_n(y))]^{1/2} \cdot \left[\left\{ \frac{z_n(y)(1 - z_n(y))^2}{(1 - \xi_{2,n}(y))^3} \right\} \right.$$

$$\left. - \left\{ \frac{(z_n(y))^2(1 - z_n(y))}{(\xi_{1,n}(y))^3} \right\} \right]$$

$$= \frac{2}{3} \frac{z_n(y)}{\xi_n(y)} \left[(z_n(y))^{3/2} \left\{ \frac{(1 - z_n(y))}{(1 - \xi_{2,n}(y))} \right\}^3 \{(1 - z_n(y))(n+1)\}^{-(1/2)} \right.$$

$$\left. - (1 - z_n(y))^{3/2} \left\{ \frac{z_n(y)}{\xi_{1,n}(y)} \right\}^3 \{z_n(y)(n+1)\}^{-(1/2)} \right];$$

that is,

$$R_n(y) = \frac{2}{3} \left[\frac{z_n(y)}{\xi_n(y)} \right] [r_{2,n}(y) - r_{1,n}(y)].$$

Now,

$$0 < r_{1,n}(y) = (1 - z_n(y))^{3/2} (z_n(y)/\xi_{1,n}(y))^3 (z_n(y)(n+1))^{-(1/2)}$$

$$< (\log \log n)^{-2} (z_n(y)/\xi_{1,n}(y))^3 \text{ uniformly in } y_\theta[E_n, 1 - E_n].$$

Because $\xi_{1,n}(y)$ is between $z_n(y)$ and $U_n(y)$,

$$(\log \log n)^{-2} (z_n(y)/\xi_{1,n}(y))^3 = 0\{(\log \log n)^{-2}\}$$

$$\text{for } y \text{ s.t. } z_n(y) < U_n(y);$$

otherwise

$$0 < z_n(y)/\xi_{1,n}(y) \le z_n(y)/(z_n(y) - |U_n(y) - z_n(y)|)$$

$$= 1 + |z_n(y)|/((n+1)^{1/2} z_n(y) - |z_n(y)|),$$

where

$$r_{1,n}(y) = (1 - z_n(y))^{3/2} \left\{ \frac{z_n(y)}{\xi_{1,n}(y)} \right\}^3 \{z_n(y)(n+1)\}^{-(1/2)},$$

$$r_{2,n}(y) = (z_n(y))^{3/2} \left\{ \frac{(1 - z_n(y))}{(1 - \xi_{2,n}(y))} \right\}^3 \{(1 - z_n(y))(n+1)\}^{-(1/2)}. \qquad \square$$

Theorem 6.3 *Let*

$$a(n) = (2 \log \log n)^{1/2}, \quad and \quad b(n) = 2 \log \log n + \frac{1}{2} \log \log \log n - \frac{1}{2} \log \pi;$$

then

$$\lim_{n \to \infty} P \left\{ \max_{1 \le k \le n} a(n)[f_n(x; k) - b(n)] < t \right\} = e^{-2e^{-t}}, \quad -\infty < t < \infty.$$

Proof. The reader is referred to Haccou et al. (1988).   □

It is interesting to note that the asymptotic distribution obtained in this case is the same as we have obtained in Chapters 2 and 3.

6.3 An Alternate Approach

Because of the slow convergence rate of the asymptotic distribution derived in the previous section, to the Gumbel distribution, efforts have been made to search other methods of detecting change points for the exponential model. We are especially interested in detecting multiple change points when the exponential model is assumed. Haccou and Meelis (1988) proposed a procedure that is based on partitioning the likelihood according to a hierarchy of sub-hypotheses. They considered a maximum of a two change points hypothesis against one change point and then one change against none.

The general formulation of a hierarchy of subhypotheses was extensively studied by Hogg (1961). Suppose a parameter vector $\theta \in \Omega$, and Ω^* is a subspace of Ω. To test $H_0 : \theta \in \Omega^*$ versus $H_1 : \theta \in \Omega - \Omega^*$, assume that there are certain intermediate hypotheses we also wish to investigate, or say, the null hypothesis H_0 can be resolved as follows. Let $\{\Omega_i, i = 1, 2, \ldots, k\}$ be a sequence of nested subspaces of Ω, so that

$$\Omega = \Omega_0 \supset \Omega_1 \supset \Omega_2 \supset \cdots \supset \Omega_k = \Omega^*,$$

where each $\Omega_i, i = 1, 2, \ldots, k$, corresponds to an hypothesis:

$$H_0^i : \theta \in \Omega_i \quad \text{versus} \quad H_1^i : \theta \in \Omega_{i-1} - \Omega_i, \qquad i = 1, 2, \ldots, k.$$

Then to test H_0 versus H_1, we first test

$$H_0^1 : \theta \in \Omega_1 \quad \text{versus} \quad H_1^1 : \theta \in \Omega_0 - \Omega_1$$

and if H_0^1 is accepted, we second test

$$H_0^2 : \theta \in \Omega_2 \quad \text{versus} \quad H_1^2 : \theta \in \Omega_1 - \Omega$$

and if H_0^2 is accepted, ..., and if H_0^{i-1} is accepted, we continue to test,

$$H_0^i : \theta \in \Omega_i \quad \text{versus} \quad H_1^i : \theta \in \Omega_{i-1} - \Omega_i$$

$$\vdots$$

$$H_0 = H_0^k : \theta \in \Omega_k = \Omega^* \quad \text{versus} \quad H_1^k : \theta \in \Omega_{k-1} - \Omega_k.$$

So, we reject H_0 if any one of the H_0^is is rejected, and accept H_0 if $H_0^1, \ldots, H_0^k$ are all accepted.

The test statistic used for H_0^i versus H_1^i is the generalized likelihood-ratio (GLR) statistic:

$$\lambda_i = \frac{L(\widehat{\Omega}_i)}{L(\widehat{\Omega}_{i-1})}, \quad \text{for } i = 1, 2, \ldots, k.$$

Then, the GLR for H_0 versus H_1 is given by

$$\lambda = \frac{L(\widehat{\Omega}_k)}{L(\widehat{\Omega}_0)} = \prod_{i=1}^{k} \left[\frac{L(\widehat{\Omega}_i)}{L(\widehat{\Omega}_{i-1})} \right] = \prod_{i=1}^{k} \lambda_i,$$

and

$$-2 \log \lambda = \sum_{i=1}^{k} (-2 \log \lambda_i).$$

Here, $\lambda_1, \ldots, \lambda_k$ are mutually independent test statistics, and if the significance level for each subtest is α_i, then the significance level of H_0 versus H_1 is $1 - \prod_{i=1}^{k}(1 - \alpha_i)$ or $1 - (1 - \alpha)^k$ if $\alpha_1 = \cdots = \alpha_k = \alpha$.

Now, let $x_1, \ldots, x_n$ be an independent sequence, where each x_i has pdf

$$f(x; v_i) = v_i e^{-v_i x}$$

and

$$v_i = v \quad \text{for } \tau_j \leq i \leq \tau_{j+1}, \quad j = 0, \ldots, m.$$

Hence, the location of change points are $\tau_1, \tau_2, \ldots, \tau_m$, so, there are m change points and $m + 1$ unknown parameters $v_1, \ldots, v_{m+1}$.

Let $k_i = \tau_1/n$ and $\rho_i = (v_{i+1})/v_i$, ("$\rho_i = 1$" $\Longleftrightarrow$ no change from i to $i+1$), and $\theta = (k_1, \ldots, k_m, \rho_1, \ldots, \rho_m)$("$\rho_i \neq 1$" $\Longleftrightarrow$ there is a change). Further let Ω_m be the parameter space corresponding to m or fewer change points. Then, after the reparameterization initiated by k_i and ρ_i, we have $\Omega_l \subset \Omega_m$ for $l < m$; that is, now

$$\Omega_m = \left\{ k, e | k_i = \frac{\tau_i}{n} \quad \text{and} \quad \rho_i = \frac{v_{i+1}}{v_i} \right\}.$$

For example,

$$\Omega_0 = \{k, \rho | \rho = 1 \quad \text{and} \quad k = 0 \text{ or } k = 1\}$$

$$\Omega_1 = \{k, \rho | \rho = 1 \quad \text{and} \quad k = 0 \text{ or } k = 1 \text{ or } \rho \neq 1 \quad \text{and} \quad 0 < k < 1\}$$

and $\Omega_0 \subset \Omega_1$; that is, we have $\Omega_{n-1} \supset \Omega_{n-2} \supset \cdots \supset \Omega_2 \supset \Omega_1 \supset \Omega_0$, where each $\Omega_i, i = 1, 2, \ldots, n-1$, corresponds to:

$$H_0^i : \theta \in \Omega_i \quad \text{versus} \quad H_1^i : \theta \in \Omega_{i+1},$$

where H_0^i is the same as H_1^{i-1}. The hypotheses we are interested in here are:

$$H_0 : \theta \in \Omega_0 (\text{no change}) \quad \text{versus} \quad H_1 : \theta \in \Omega_{n-1}(\text{at most } n-1 \text{ changes}).$$

We use the likelihood-ratio procedure test (LPT) statistic as our test statistic.

$$\Lambda = \frac{\max\limits_{k_1,\ldots,k_{n-1}} L(0)}{\max\limits_{k_1,\ldots,k_{n-1}} L(n-1)},$$

where $L(0)$ denotes the maximum likelihood function under H_0, and $L(n-1)$ denotes the maximum likelihood function under H_1. Or, equivalently, we can use $\lambda = -2 \log \Lambda$ to form the test statistic; that is,

$$\lambda = -2 \log \Lambda$$

$$= -2 \left[\max\limits_{k_1,\ldots,k_{n-1}} \log L(0) - \max\limits_{k_1,\ldots,k_{n-1}} \log L(n-1) \right]$$

$$\triangleq 2[\ell(n-1) - \ell(0)].$$

To obtain $\ell(n-1)$, we derive $\ell(m)$ for $0 \leq m \leq n-1$ corresponding to $H_0^m : \theta \in \Omega_m$. Note that

$$L(\theta) = \prod_{j=0}^{m} \prod_{i=k_j+1}^{k_{j+1}} f(x_i, \nu_{j+1})$$

$$= \prod_{j=0}^{m} \prod_{i=k_j+1}^{k_{j+1}} \nu_{j+1} e^{-\nu_{j+1} x_i}$$

$$= \prod_{j=0}^{m} \nu_{j+1}^{k_{j+1}-k_j} e^{-\nu_{j+1} \sum_{i=k_j+1}^{k_{j+1}} x_i},$$

and hence

$$\log L(\theta) = \sum_{j=0}^{m} \left[(k_{j+1} - k_j) \ln \nu_{j+1} - \nu_{j+1} \sum_{i=k_j+1}^{k_{j+1}} x_i \right].$$

$$\frac{\partial \log L(\theta)}{\partial \nu_{j+1}} = \frac{k_{j+1} - k_j}{\nu_{j+1}} - \sum_{i=k_j+1}^{k_{j+1}} x_i$$

$$\stackrel{set}{=} 0.$$

Therefore

$$\widehat{\nu}_{j+1} = \left(\frac{\sum_{i=k_j+1}^{k_{j+1}} x_i}{k_{j+1} - k_j} \right)^{-1}.$$

Because

$$\left. \frac{\partial^2 \log L(\theta)}{\partial \nu_{j+1}^2} \right|_{\widehat{\nu}_{j+1}} = -\frac{k_{j+1} - k_j}{\nu_{j+1}^2} < 0,$$

$\widehat{\nu}_{j+1}$ obtained above is the MLE of ν_{j+1} under Ω_m. Therefore,

$$\ell(m) = \max_{k_1,\dots,k_{n-1}} \log L(m)$$

$$= \max_{k_1,\dots,k_{n-1}} \sum_{j=0}^{m} \sum_{i=k_j+1}^{k_{j+1}} [\log \widehat{\nu}_{j+1} - x_i \widehat{\nu}_{j+1}].$$

Hence, for testing H_0 versus $H_1 : \theta \in \Omega_{n-1}$, we use $\lambda = 2[\ell(m) - \ell(0)]$ as our test statistic.

Suppose now we consider the nested hypotheses with parameter spaces as $\Omega_{n-1} \supset \Omega_{n-2} \supset \cdots \supset \Omega_2 \supset \Omega_1 \supset \Omega_0$. The test statistic λ can be partitioned into

$$\lambda = \sum_{j=1}^{n-1} \lambda_j,$$

where

$$\lambda_1 = 2[\ell(1) - \ell(0)]$$

is the test statistic for testing $H_0 = H_0^0 : \theta \in \Omega_0$ versus $H_1^0 : \theta \in \Omega_1$,

$$\lambda_2 = 2[\ell(2) - \ell(1)]$$

is the test statistic for testing $H_0^1 : \theta \in \Omega_1$ versus $H_1^1 : \theta \in \Omega_2$,

$$\vdots$$

$$\lambda_m = 2[\ell(m) - \ell(m-1)]$$

is the test statistic for testing $H_0^{m-1} : \theta \in \Omega_{m-1}$ versus $H_1^{m-1} : \theta \in \Omega_m$,

$$\vdots$$

and,

$$\lambda_{n-1} = 2[\ell(n-1) - \ell(n-2)]$$

is the test statistic for testing $H_0^{n-2} : \theta \in \Omega_{n-2}$ versus $H_1^{n-2} = H_1 : \theta \in \Omega_{n-1}$.

These tests are performed successively, and we reject H_0 if any one of the H_0^is is rejected and accept H_0 if and only if $H_0^1, \ldots, H_0^{n-1}$ are all accepted. In particular, if we just consider the following hypotheses,

$$H_0^0 : \theta \in \Omega_0 \quad \text{versus} \quad H_1^0 : \theta \in \Omega_1,$$
$$H_0^1 : \theta \in \Omega_1 \quad \text{versus} \quad H_1^1 : \theta \in \Omega_2,$$
$$H_0^2 : \theta \in \Omega_2 \quad \text{versus} \quad H_1^{n-1} : \theta \in \Omega_{n-11},$$

then the test statistic

$$\lambda = \lambda_1 + \lambda_2 + \lambda_3,$$

where

$$\lambda_1 = 2[\ell(1) - \ell(0)],$$
$$\lambda_2 = 2[\ell(2) - \ell(1)],$$

and

$$\lambda_3 = 2[\ell(n-1) - \ell(2)].$$

These four nested hypotheses are indeed testing 0 change against 1 change, 1 change against 2 changes, and 2 changes against more than 2 changes.

To be able to obtain the critical values of $\lambda_i, i = 1, 2, 3$, we take the following approach.

Let d_i be the decision that there are i changes, $i = 0, 1, 2$, and d_3 the decision that there are more than two changes. Define the decision function $\Phi(i|\mathbf{x})$ as

$$\Phi(i|\mathbf{x}) = P[D = d_i|\mathbf{x}]$$
$$= I(\lambda^* \in \Delta_i),$$

where $\mathbf{x} = (x_1, \ldots, x_n)$, D is the decision variable that takes values $d_0, d_1, d_2,$ or d_3, $I(\cdot)$ is the indicator function, $\lambda^* = (\lambda_1, \lambda_2, \lambda_3)$,

$$\Delta_0 = \{\lambda_i \le k_{i,\alpha_i}; i = 1, 2, 3\},$$
$$\Delta_1 = \{\lambda_1 > k_{1,\alpha_1}, \quad \lambda_2 \le k_{2,\alpha_2} \quad \text{and} \quad \lambda_3 \le k_{3,\alpha_3}\},$$
$$\Delta_2 = \{\lambda_2 > k_{2,\alpha_2} \quad \text{and} \quad \lambda_3 \le k_{3,\alpha_3}\},$$

$$\Delta_1 = \{\lambda_3 > k_{3,\alpha_3}\},$$

and k_{i,α_i} is the critical value of λ_i at significance level α_i, $i = 1, 2, 3$.

Based on the above, Haccou and Meelis (1988) obtained simulated critical values of $\lambda_1, \lambda_2, \lambda_3$ for different sample sizes and different $\alpha_1, \alpha_2, \alpha_3$ values. The interested reader can found these values in their paper.

6.4 Informational Approach

As before, the informational approach of change point hypotheses for the exponential model provides an alternative for modern statistical analysis of change point. Let's test the hypothesis

$$H_0 : x_1, \ldots, x_{n+1} \sim \text{iid } f(x; \lambda) = \lambda e^{-\lambda x},$$

versus

$$H_1 : x_1, \ldots, x_k \sim \text{iid } f(x; \lambda_1) = \lambda_1 e^{-\lambda_1 x}$$

$$x_{k+1}, \ldots, x_{n+1} \sim \text{iid } f(x; \lambda_2) = \lambda_2 e^{-\lambda_2 x}, \qquad 1 \le k \le n.$$

From Section 6.2 of this chapter, we have:

$$L_0(\widehat{\lambda}) = \left(\frac{n+1}{\sum_{i=1}^{n+1} x_i} \right)^{n+1} e^{-n-1},$$

$$L_1(\widehat{\lambda}_1, \widehat{\lambda}_2) = \left(\frac{k}{\sum_{i=1}^{k} x_i} \right)^{k} \left(\frac{n-k+1}{\sum_{i=k+1}^{n+1} x_i} \right)^{n-k+1} e^{-n-1};$$

then

$$\text{SIC}(n+1) = -2 \log L_0(\widehat{\lambda}) + \log(n+1)$$

$$= 2(n+1) \log \sum_{i=1}^{n+1} x_i + 2(n+1) - (2n+1) \log(n+1),$$

and

$$\text{SIC}(k) = -2 \log L_1(\widehat{\lambda}_1, \widehat{\lambda}_2) + 2 \log(n+1)$$

$$= 2k \log \sum_{i=1}^{k} x_i + 2(n-k+1) \log \sum_{i=k+1}^{n+1} x_i + 2(n+1)$$

$$- 2k \log k - 2(n-k+1) \log(n-k+1) + 2 \log(n+1);$$

for $k = 1, 2, \ldots, n$. Then, if

$$\min \text{SIC} = \min\{\text{SIC}(n+1), \text{SIC}(k), k = 1, 2, \ldots, n\}$$

is attained at $\text{SIC}(n+1)$, H_0 is accepted; and if min SIC is attained at $\text{SIC}(k)$, for $k = 1, 2, \ldots, n$, H_0 is rejected and there is a change point at $\hat{k}$ when $\hat{k}$ is such that $\text{SIC}(\hat{k}) = \min_{1 \leq k \leq n} \text{SIC}(k)$.

6.5 Application to Air Traffic Data

In this section, examples are given to show the implementation of the SIC method proposed in the previous section.

Example 6.1 The first dataset to be analyzed is given in Appendix B of Hsu (1979). This dataset contains 213 airplanes' arrival times collected from a low-altitude transitional control sector (near Newark airport) for the period from noon to 8:00 PM on April 30, 1969. Hsu has examined the data and concluded that the interarrival times are independently exponentially distributed. That is, if T_i is the arrival time, then $x_i = T_{i+1} - T_i$, $i = 1, \ldots, 212$, is an exponential random variable with parameter β_i. To see if the air traffic densities are constant over time, one can test:

$$H_0 : \beta_1 = \beta_2 = \cdots = \beta_{212} = \beta_0$$

against the alternative:

$$H_1 : \beta_1 = \cdots = \beta_k \neq \beta_{k+1} = \cdots = \beta_{212}$$

for the sequence $\{x_i\}$. After the computation of the SIC values, it is found that $\text{SIC}(212) = 2504.1 < \min_{1 \leq k \leq 211} \text{SIC}(k) = 2505.1$. Therefore, there is no change in the values of the parameters, or the air traffic densities are constant over time. It is noted that this conclusion is the same as given in Hsu (1979) and a gamma model with shape parameter 1 and unknown scale parameter were considered for the data.

Example 6.2 The second dataset to be used is artificial data generated by computer. The following exponential observations are generated by using $\beta = 1$ for observations 1 through 23 and $\beta = 2$ for observations 24 through 40:

0.4065	4.4628	0.2296	0.5782	1.2425	0.8781	1.1375	0.6644	0.3297	0.1719
1.4615	0.7657	0.9852	2.9949	2.6057	1.5973	0.5688	0.1845	0.8614	0.1749
1.2274	1.7082	1.0285	0.0341	0.2643	0.4984	0.3850	0.2327	0.0286	0.4510
1.4931	0.4050	0.2588	0.1450	0.0730	1.0162	1.1289	0.2038	0.0911	0.2731

After the computation of the SIC values, it is found that

$$\mathrm{SIC}(40) = 68.8918 > \min_{1 \le k \le 39} \mathrm{SIC}(k) = \mathrm{SIC}(23) = 63.2195.$$

Therefore, there is a change in the values of the parameters and the location of the change point is 23. This certainly matches the situation.

Chapter 7
Change Point Model for Hazard Function

7.1 Introduction

In the previous chapters, we presented change point analyses for various models. In this chapter, we introduce another change point problem, often encountered in reliability analysis, the problem of estimating the change point in a failure rate or hazard function.

Let $T_1, T_2, \ldots, T_n$ be a random sample of size n from a lifetime distribution with probability distribution function $F(\cdot)$ and density function $f(\cdot)$. The hazard function of F is defined as

$$r(t) = f(t)/(1 - F(t)) \quad \text{for } t \geq 0.$$

It is of our concern whether the hazard function maintains at a rate a for $0 \leq t \leq \tau$, and later keeps at a lower rate b for $t \geq \tau$, where $a > b \geq 0$. Formally, we are interested in investigating if

$$r(t) = \begin{cases} a & \text{for } 0 \leq t \leq \tau \\ b & \text{for } t > \tau \end{cases}, \tag{7.1}$$

where τ is defined as the change point or threshold of the failure rate function $r(t)$.

This change point is different from the discrete change point discussed in the usual literature of the change point models. It is a continuous time in nature, and its "discrete" realization is the change point location we want to estimate.

For testing whether a new leukemia therapy produces a departure from a constant relapse rate after induction of remission, Matthews and Farewell (1982) proposed to study the above model for the relapse rate of leukemia patients and estimated the threshold τ by a numerical method. Nguyen, Rogers, and Walker (1984) estimated the parameters in the above model using the analysis of mixture model. Matthews, Farewell, and Pyke (1985) considered the inference of the change point in hazard models using a score-statistic process. Basu, Ghosh, and Joshi (1988; BGJ) investigated the estimation of such a change point by semiparametric methods. Ghosh and Joshi (1992) further studied the asymptotic distribution of one of the estimators proposed by BGJ. From the simulation studies of the estimates proposed by these authors, one can see that the estimates are not as good as expected for all occasions.

In a review article by Muller and Wang (1994), it was pointed out that the likelihood of the change point model (7.1) of hazard rate functions is unbounded based on the work of Matthews and Farewell (1982), Worsley (1988), and Henderson (1990) unless $a > b$. Such a hazard function is plotted in Figure 7.1.

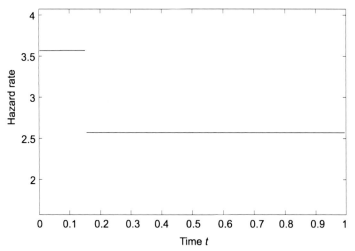

Fig. 7.1 A typical hazard rate function with one change point

Furthermore, it is noted that the likelihood-based estimation of τ does not have a closed form. Therefore, the Bayesian approach for the estimation of the change point τ in the hazard function (7.1) became quite popular and many works were done in the Bayesian framework. For this reason, we present the Bayesian approach of Ghosh, Joshi, and Mukhopadhyay (1993) for the estimation of τ in the following section.

7.2 The Bayesian Approach

For simplicity, we denote the order statistics of the random sample $T_1, T_2, \ldots, T_n$, taken from a lifetime distribution with probability distribution function $F(\cdot)$ and density function $f(\cdot)$, as $t_1, t_2, \ldots, t_n$ with a clear understanding that $t_1 \leq t_2 \leq \cdots \leq t_n$.

Note that there are several classical estimators of the location τ of the change point model (7.1) in the literature.

First, an estimator of τ, denoted $\hat{\tau}$, was obtained in Nguyen et al. (1984) as the value of $\hat{\tau}$ such that the stochastic process $Y_n(\tau)$ approaches 0 at $\hat{\tau}$, where for $0 \leq r \leq n-1$, $Y_n(t_r) = X_n(t_r)$ with

$$X_n(t) = S(t)\left[(n-r)\log\left\{\frac{n}{n-r}\right\} - r\right]n^{-1}$$
$$+ r\frac{M(t)}{n} - Z_n\left[\log\left\{\frac{n}{n-r}\right\}\right]n^{-1},$$

$$M(t) = \sum_{i=r+1}^{n}\frac{t_i}{n-r},$$

$$S^2(t) = \sum_{i=r+1}^{n}\frac{t_i^2}{n-r} - M(t)^2,$$

and

$$Z_n = (t_1 + \cdots + t_n)/n.$$

For $t_r \leq t < t_{r+1}$, $Y_n(t)$ is defined by linear interpolation. For $t \geq t_n$, $Y_n(t)$ is defined by

$$Y_n(t) = X_n(t_n) + \frac{[X_n(t_n) - X_n(t_{n-1})](t - t_n)}{t_n - t_{n-1}}.$$

The estimator $\hat{\tau}$ was shown to be consistent in Nguyen et al. (1984). However, this estimate relies on the asympototic properties of the estimate. As noted in Nguyen et al. (1984), for their model, the likelihood is unbounded as the MLE $\hat{\tau} \to T_n$. For this and other reasons, Ghosh et al. (1993) proposed a Bayesian approach for the estimation of the change position τ.

Let the sample $\{T_1, T_2, \ldots, T_n\}$ be denoted as $\mathbf{D}$, and $T_0 = 0, T_{n+1} = \infty$. It is clear that corresponding to the hazard rate change point model (7.1), the pdf of each of the random variable T_i is given by

$$f_{T_i}(t_i; a, b, \tau) = \begin{cases} ae^{-at_i} & \text{for } 0 \leq t_i \leq \tau \\ be^{-bt_i - (a-b)\tau} & \text{for } t_i > \tau \end{cases}. \tag{7.2}$$

Then, the likelihood function of the parameters a, b, τ over the natural parameter space $(a, b, \tau)|0 < b < a < \infty, 0 < \tau < \infty$ is given by

$$L(a, b, \tau|\mathbf{D}) = \prod_{i=1}^{n} f_{T_i}(t_i; a, b, \tau)$$
$$= a^{R(\tau)}e^{-aQ(\tau)}b^{n-R(\tau)}e^{-b(T-Q(\tau))},$$

where

$$R(\tau) = \sum_{i=1}^{n} I_{[T_i \leq \tau]},$$

$$M(\tau) = \sum_{i=1}^{n} T_i I_{[T_i \leq \tau]},$$

$$Q(\tau) = M(\tau) + (n - R(\tau))\tau,$$

$$T = \sum_{i=1}^{n} T_i.$$

To obtain the Bayesian estimate of the change point location τ, Ghosh et al. (1993) first placed the following prior distribution, $\pi(a, b, \tau)$, on the parameters (a, b, τ) over the natural parameter space $(a, b, \tau)|0 < b < a < \infty$, $0 < \tau < \infty$ as the following.

$$\pi(a, b, \tau) = \frac{1}{ab}, \qquad 0 < b < a < \infty, 0 < \tau < \infty.$$

Then, the joint posterior distribution, $\pi(a, b, \tau|\mathbf{D})$, of (a, b, τ) can be obtained as

$$\pi(a, b, \tau|\mathbf{D}) \propto L(a, b, \tau|\mathbf{D}) \cdot \pi(a, b, \tau)$$
$$= a^{R(\tau)-1} e^{-aQ(\tau)} b^{n-R(\tau)-1} e^{-b(T-Q(\tau))}. \qquad (7.3)$$

Next, we proceed to obtain the joint posterior of (b, τ) by integrating the right-hand side (R.H.S.) of expression (7.3) with respect to a on its natural range (b, ∞). This can be done in two situations as the following.

(I). For $0 < \tau < T_1$.

For this situation, the joint posterior, $\pi(b, \tau|\mathbf{D})$, of (b, τ) was derived in Ghosh et al. (1993) as

$$\pi(b, \tau|\mathbf{D}) = \int_{b}^{\infty} \pi(a, b, \tau|\mathbf{D})da$$

$$\propto \int_{b}^{\infty} \prod_{i=1}^{n} be^{-bt_i - (a-b)\tau} \cdot \frac{1}{ab} da$$

$$= \int_{b}^{\infty} b^n e^{-b \sum_{1}^{n} t_i - an\tau + bn\tau} a^{-1} b^{-1} da$$

$$= \int_{b}^{\infty} a^{-1} e^{-an\tau} b^{n-1} e^{-b(T-n\tau)} da$$

$$= b^{n-1} e^{-b(T-n\tau)} g(b), \qquad (7.4)$$

where

$$g(b) = \int_b^\infty a^{-1} e^{-an\tau} da.$$

(II). For $T_i < \tau < T_{i+1}, i = 1, \ldots, n$.

For this situation, the joint posterior, $\pi(b, \tau|\mathbf{D})$, of (b, τ) was also derived in Ghosh et al. (1993) as

$$\pi(b, \tau|\mathbf{D}) \propto \int_b^\infty a^{R(\tau)-1} e^{-aQ(\tau)} b^{n-R(\tau)-1} e^{-b(T-Q(\tau))} da$$

$$= \frac{(i-1)!}{(Q(\tau))^i} \sum_{j=0}^{i-1} \frac{(Q(\tau))^j}{j!} b^{n-i+j-1} e^{-Tb}. \qquad (7.5)$$

Note that for $T_n \le \tau < \infty$, the R.H.S. of expression (7.5) is clearly,

$$\text{R.H.S. of } (7.5) = \frac{(n-1)!}{T^n} \sum_{j=0}^{n-1} \frac{T^j}{j!} b^{j-1} e^{-Tb}. \qquad (7.6)$$

When integrating (7.6) with respect to b, the first term is

$$\int_0^\infty b^{-1} e^{-Tb} db,$$

which diverges. For this reason, a restriction on the parameter b was imposed by Ghosh et al. (1993). This restriction is that for a positive constant b_0, we require $0 < b_0 \le b$. With this restriction and the above-obtained joint posterior $\pi(b, \tau|\mathbf{D})$ of (b, τ), the posterior density of τ, $\pi(\tau|\mathbf{D})$, was obtained in Ghosh et al. (1993) for the following three cases.

Case 1. When $0 < \tau < T_1$.

In this case, using expression (7.4), the posterior density of τ, $\pi(\tau|\mathbf{D})$ was given as

$$\pi(\tau|\mathbf{D}) \propto \int_{b_0}^\infty b^{n-1} e^{-b(T-n\tau)} g(b) db. \qquad (7.7)$$

Note that the integral given by expression (7.7) does not have a closed form, therefore, the posterior of τ can only be obtained numerically. To assure the existence of $\pi(\tau|\mathbf{D})$ in this case, Ghosh et al. (1993) have shown that $\pi(\tau|\mathbf{D})$ is finite, along with some other properties of $\pi(\tau|\mathbf{D})$ evidenced by Propositions 4.1–4.5 in Ghosh et al. (1993). During the course of establishing these propositions, another restriction, namely, $\tau \ge c > 0$ for a constant c was imposed on τ as if $\tau = 0$; the model is nonidentifiable.

Case 2. When $T_i < \tau < T_{i+1}, i = 1, \ldots, n-1$.

In this case, $\pi(\tau|\mathbf{D})$ was obtained by integrating expression (7.5) with respect to b over the range $[b_0, \infty)$:

$$\pi(\tau|\mathbf{D}) \propto \int_{b_0}^{\infty} \frac{(i-1)!}{(Q(\tau))^i} \sum_{j=0}^{i-1} \frac{(Q(\tau))^j}{j!} b^{n-i+j-1} e^{-Tb} db$$

$$= \frac{(i-1)!}{(Q(\tau))^i} e^{-Tb_0} \sum_{j=0}^{i-1} \left[\left\{ \sum_{k=0}^{n-i+j-1} \frac{(Tb_0)^k}{k!} \right\} \frac{(n-i+j-1)!}{j!} \frac{(Q(\tau))^j}{T^{n-i+j}} \right].$$

(7.8)

Case 3. When $\tau \geq T_n$.

In this case, $\pi(\tau|\mathbf{D})$ was obtained by integrating expression (7.6) with respect to b over the range $[b_0, \infty)$:

$$\pi(\tau|\mathbf{D}) \propto \int_{b_0}^{\infty} \frac{(n-1)!}{T^n} \sum_{j=0}^{n-1} \frac{T^j}{j!} b^{j-1} e^{-Tb} db$$

$$= \frac{(n-1)!}{T^n} \left[h(b_0) + e^{-Tb_0} \sum_{j=1}^{n-1} \left\{ \sum_{k=0}^{j-1} \frac{(Tb_0)^k}{k!} \right\} \frac{1}{j!} \right],$$

(7.9)

where

$$h(b_0) = \int_{b_0}^{\infty} b^{-1} e^{-TB} db.$$

Due to the constancy of $\pi(\tau|\mathbf{D})$ over the interval $[T_n, \infty)$, Ghosh et al. (1993) placed a third restriction on τ: $\tau \leq d < \infty$ for a finite constant d.

Summing all of the considerations above, the prior $\pi(a, b, \tau)$ on the parameters (a, b, τ) should now be completely restated as

$$\pi(a, b, \tau) = \frac{1}{ab}, \qquad 0 < b_0 \leq b < a < \infty, 0 < c \leq \tau \leq d < \infty.$$

Under this prior, the posterior $\pi(\tau|\mathbf{D})$ is given by expressions (7.7), (7.8), and (7.9), respectively, for the three cases of τ. Then, the Bayesian estimate of the change point τ is obtained by either the posterior mean or posterior mode. Ghosh et al. (1993) have given computational details on how to obtain such an estimate using (7.7)–(7.9). Note that the computation itself is quite tedious as is the case for Bayesian computation. So, in the next section, an alternative estimate of τ, given in Chen (2003) is presented in light of the Schwarz information criterion (SIC).

7.3 The Informational Approach

In the literature, the change point of the hazard function is written as an inference problem of testing

$$H_0 : \tau = 0 \quad \text{against } H_1 : \tau > 0.$$

Several test statistics and estimates were proposed for the inference. The most interesting aspect of the inference is to estimate the true change point τ. For this reason, a model selection method using SIC proposed in Chen (2003) is presented herein.

Recall from the previous chapter that the SIC is defined as

$$\text{SIC} = -2 \log L(\hat{\Theta}) + p \log n,$$

where p is the number of free parameters that need to be estimated under the model. Its penalty term $p \log n$ takes the information from the sample size.

Now, we introduce an estimate for τ based on the SIC. If a lifetime random variable T has a hazard function given by (7.1), then it is clear that the probability density function of T is

$$f(t) = \begin{cases} a & \exp\{-at\} & 0 \le t \le \tau \\ b & \exp\{-a\tau - b(t-\tau) & \tau < t < \infty \end{cases}. \tag{7.10}$$

Because a random sample $T_1, T_2, \ldots, T_n$ is available from the lifetime distribution function in (7.2), the order statistics of this sample can be denoted by $T_{(1)} \le T_{(2)} \le \cdots \le T_{(n)}$. Now, let τ_0 be the point such that $T_{(k)} \le \tau_0 < T_{(k+1)}$ for some k with $k = 1, \ldots, n-1$. Then we can truncate the values at τ_0 if such a k can be found and use it to estimate the true τ.

The likelihood function $L(\cdot)$ of a, b, k is given by

$$
\begin{aligned}
L(a, b, k) &= L(a, b, k \mid t_{(1)}, \ldots, t_{(n)}) \\
&= f_{T_{(1)}, \ldots, T_{(n)}}(t_{(1)}, \ldots, t_{(n)}) \\
&= n! \left(\prod_{i=1}^{k} f(t_{(i)}) \right) \left(\prod_{j=k+1}^{n} f(t_{(j)}) \right) \\
&= n! a^k \exp\left\{ -a \sum_{i=1}^{k} t_{(i)} \right\} \cdot b^{n-k} \exp\left\{ -(n-k)a\tau_0 - b \sum_{j=k+1}^{n} (t_{(j)} - \tau_0) \right\},
\end{aligned}
$$

where $t_{(1)}, \ldots, t_{(n)}$ are the sample realizations of the order statistics $T_{(1)} \le T_{(2)} \le \cdots \le T_{(n)}$. The log of the likelihood function is clearly

$$
\begin{aligned}
l(a, b, k) &= \log L(a, b, \tau_0) \\
&= \log n! + k \log a - a \left[\sum_{i=1}^{k} t_{(i)} + (n-k)\tau_0 \right] \\
&\quad + (n-k) \log b - b \left[\sum_{j=k+1}^{n} t_{(j)} - (n-k)\tau_0 \right].
\end{aligned}
$$

For each k, the maximum likelihood estimates (MLEs) of a and b are obtained as

$$\widehat{a} = \frac{k}{(n-k)\tau_0 + \sum_{i=1}^{k} t_{(i)}}$$

and

$$\widehat{b} = \frac{n-k}{\sum_{j=k+1}^{n} t_{(j)} - (n-k)\tau_0}.$$

Define the SIC for the change point hazard function model as

$$\text{SIC}(k; \tau_0) = -2l(\widehat{a}, \widehat{b}, \tau_0) + 3\log n$$

$$= -2(n-k)\log\left\{(n-k) \middle/ \left[\sum_{j=k+1}^{n} t_{(j)} - (n-k)\tau_0\right]\right\}$$

$$- 2k\log\left\{k \middle/ \left[(n-k)\tau_0 + \sum_{i=1}^{k} t_{(i)}\right]\right\} - 2\log n! + 2n + 3\log n,$$

with an empirical choice of τ_0. The next step of the procedure is to find $\widehat{k}$ such that

$$\text{SIC}(\widehat{k}; \tau_0) = \min_{1 \le k \le n-1} \text{SIC}(k; \tau_0). \qquad (7.11)$$

Then the true change point τ of the hazard is estimated by

$$\widehat{\tau} = \frac{\widehat{k}\sum_{j=\widehat{k}+1}^{n} t_{(j)} - (n-\widehat{k})\sum_{i=1}^{\widehat{k}} t_{(i)}}{n(n-\widehat{k})}, \qquad (7.12)$$

which minimizes $\text{SIC}(\widehat{k}; \tau)$ at $\tau = \widehat{\tau}$, and (7.12) is obtained by solving

$$\frac{\partial}{\partial \tau_0}\text{SIC}(\widehat{k}; \tau_0)\bigg|_{\tau_0 = \widehat{\tau}} = 0.$$

This estimate (7.12) is optimal by the model selection principle. Three choices of τ_0 are suggested here:

$$\tau_{01} = t_{(k)}, \qquad \tau_{02} = (t_{(k)} + t_{(k+1)})/2, \quad \text{or} \quad \tau_{03} = 0.618t_{(k)} + 0.382t_{(k+1)},$$

where τ_{03} is designed according to the weighted average with empirically more weight towards the kth order statistic value $t_{(k)}$.

This estimator of the change in hazard function is very appealing to practioners due to its simplicity and ease of computation.

Table 7.1 The Estimates $\hat{\tau}_1$, $\hat{\tau}_2$, $\hat{\tau}_3$ in Chen and $\hat{\tau}$ in GJ

a	b	True τ	$\hat{\tau}_1$	(MSE)	$\hat{\tau}_2$	(MSE)	$\hat{\tau}_3$	(MSE)	$\hat{\tau}$	(MSE)
3	2	.15	.1229	.0033	.1270	.0123	.1276	.0129	.1379	.0078
3	2	.10	.1080	.0382	.1120	.0501	.1044	.0373	.1173	.0092
3	1	.15	.0547	.2017	.0533	.1918	.0533	.1918	.1828	.0119
3	1	.10	.1755	.8167	.1875	.9928	.1967	.0585	.1970	.0409
3	1.5	.15	.0082	.0201	.0152	.0297	.0076	.0203	.1584	.0085
3	1.5	.10	.0301	.0394	.0228	.0299	.0216	.0247	.1475	.0179
2	1.5	.20	.1548	.1932	.1555	.1652	.1461	.1362	.1605	.0197
2	1.5	.15	.1095	.0637	.1075	.0958	.1098	.0957	.1479	.0177
2	1.5	.10	.0751	.0640	.0776	.0796	.0783	.0796	.1538	.0254
2	1	.20	.3074	.0163	.3010	.0167	.3016	.0168	.2229	.0261
2	1	.15	.2144	.0069	.2093	.0066	.2096	.0066	.1923	.0316
2	1	.10	.1693	.0389	.1620	.0392	.1618	.0392	.2135	.0607
2	.5	.20	.5051	.1535	.5007	.1507	.4688	.1335	.3136	.0743
2	.5	.15	.3516	.0715	.3162	.0592	.3104	.0581	.3147	.1325
2	.5	.10	.2143	.0280	.2114	.0261	.2112	.0258	.3225	.2049
1	.5	.30	.6061	.1460	.6083	.2211	.6086	.2216	.4465	.1720
1	.5	.20	.3519	.1286	.3503	.2742	.3224	.1212	.4079	.2444
1	.5	.15	.3243	.7067	.3296	.8190	.3310	.8190	.3825	.2636

7.4 Simulation Results

A simulation study is given to show the feasibility of these new estimates $\hat{\tau}_1$, $\hat{\tau}_2$, and $\hat{\tau}_3$ (with respect to τ_{01}, τ_{02}, and τ_{03}) in comparison with the estimates obtained in Ghosh and Joshi (1992; GJ) and in BGJ. The simulation results are given in Table 7.1, where $\hat{\tau}_1$, $\hat{\tau}_2$, and $\hat{\tau}_3$ are the new estimates proposed in Section 7.2 and $\hat{\tau}$ is the estimate in Table I of GJ. One can observe that the new estimates $\hat{\tau}_1$, $\hat{\tau}_2$, and $\hat{\tau}_3$ are quite compatible with the estimate $\hat{\tau}$ in GJ and $\hat{\tau}_1$, $\hat{\tau}_2$ in BGJ; and they are better than the estimates $\hat{\tau}_3$ and $\hat{\tau}_4$ in BGJ (see the simulation results given in BGJ). However, the new estimates perform better only when $a \geq 3$, $a - b \geq 1$, and $\tau \geq .15$. Thus, it is still desirable to find good estimates for other values of a, b, and τ. A further study for the properties of the new estimates $\hat{\tau}_1$, $\hat{\tau}_2$, and $\hat{\tau}_3$ is also needed.

Chapter 8
Discrete Models

8.1 Introduction

In previous chapters, we have focused on the change point problems for various continuous probability models. In this chapter we study the change point problem for two discrete probability models, namely, binomial and Poisson models.

A survey of the literature of change point analysis reveals that much more work has been done for continuous probability models than for discrete probability models. There are a few notable authors who have contributed to change point analysis for discrete models. Hinkley and Hinkley (1970) studied the inference about the change point for binomial distribution using the maximum likelihood ratio method. Smith (1975) considered a similar problem from a Bayesian's point of view. Pettitt (1980), on the other hand, investigated this problem by means of the cumulative sum statistic. Worsley (1983) discussed the power of the likelihood ratio and cumulative sum tests for the change point problem incurred for a binomial probability model. Fu and Curnow (1990) derived the null and nonnull distributions of the log likelihood ratio statistic for locating the change point in a binomial model.

In this chapter, we discuss the change point problem for a binomial model in Section 8.2, and for a Poisson model in Section 8.3 based on the work of the above-mentioned authors.

8.2 Binomial Model

8.2.1 Likelihood-Ratio Procedure

Suppose that there are c binomial variables, say $x_i \sim \text{bin}(n_i, p_i)$ and $x_i = m_i$ for $i = 1, \ldots, c$, where $x_i = \#$ of successes among n_i trials. Let us test the following null hypothesis,

$$H_0 : p_1 = p_2 = \cdots = p_c = p,$$

against the alternative hypothesis

$$H_1 : p_1 = \cdots = p_k = p \neq p_{k+1} = \cdots = p_c = p'.$$

Denote by $M_k = \sum_{i=1}^{k} m_i$, $N_k = \sum_{i=1}^{k} n_i (k = 1, \ldots, c)$ and $M \equiv M_c$, $N \equiv N_c$, $M_k' = M - M_k$, $N_k' = N - N_k$. Then, under H_0, the likelihood function is

$$L_0(p) = \prod_{i=1}^{c} \binom{n_i}{m_i} p^{m_i}(1-p)^{n_i - m_i}.$$

The following straightforward calculations lead us to obtain the MLE of p:

$$\log L(p) = \sum_{i=1}^{c} \left(\log \binom{n_i}{m_i} + m_i \log p + (n_i - m_i) \log(1-p) \right)$$

$$\frac{\partial \log L(p)}{\partial p} = \sum_{i=1}^{c} \left[m_i \frac{1}{p} + (n_i - m_i) \frac{1}{1-p} \right]$$

$$= \frac{1}{p} \sum_{i=1}^{c} m_i - \frac{1}{1-p} \sum_{i=1}^{c} (n_i - m_i)$$

$$= \frac{1}{p} M - \frac{1}{1-p}(N - M)$$

$$\stackrel{set}{=} 0$$

$$\Rightarrow p(N - M) = (1 - p)M$$

$$\Rightarrow \widehat{p} = \frac{M}{N}.$$

Under H_1, the likelihood function is:

$$L_1(p, p') = \prod_{i=1}^{k} \binom{n_i}{m_i} p^{m_i}(1-p)^{n_i - m_i} \cdot \prod_{j=k+1}^{c} \binom{n_j}{m_j} (p')^{m_j}(1-p')^{n_j - m_j},$$

and the MLEs of p and p' are obtained as follows.

$$\log L_1(p, p') = \sum_{i=1}^{k} \left[\log \binom{n_i}{m_i} + m_i \log p + (n_i - m_i) \log(1-p) \right]$$

$$+ \sum_{j=k+1}^{c} \left[\log \binom{n_j}{m_j} + m_j \log p' + (n_j - m_j) \log(1-p') \right]$$

$$\frac{\partial \log L_1(p, p')}{\partial p} = \sum_{i=1}^{k} \frac{m_i}{p} - \sum_{i=1}^{k} \frac{n_j - m_j}{1 - p}$$

$$\overset{set}{=} 0$$

$$\Rightarrow \widehat{p} = \frac{M_k}{N_k}$$

$$\frac{\partial \log L_1(p, p')}{\partial p'} = \sum_{j=k+1}^{c} \frac{m_j}{p'} - \sum_{j=k+1}^{c} \frac{n_j - m_j}{1 - p}$$

$$= \frac{1}{p'} \left[\sum_{j=1}^{c} m_j - \sum_{j=1}^{k} m_j \right] - \frac{1}{1-p'} \left[\sum_{j=1}^{c} (n_j - m_j) - \sum_{j=1}^{k} (n_j - m_j) \right]$$

$$= \frac{1}{p'} [M - M_k] - \frac{1}{1 - p'} [N - M - (N_k - M_k)]$$

$$\overset{set}{=} 0$$

$$\Rightarrow \widehat{p'} = \frac{M - N_k}{N - N_k} = \frac{M_k'}{N_k'}.$$

Then the log maximum likelihood ratio is obtained as

$$\log \frac{L_0(\widehat{p})}{L_1(\widehat{p}, \widehat{p'})} = \log L_0(\widehat{p}) - \log L_1(\widehat{p}, \widehat{p'})$$

$$= \sum_{i=1}^{c} \left[m_i \log \frac{m}{n} + (n_i - m_i) \log \left(1 - \frac{M}{N} \right) \right]$$

$$= \sum_{i=1}^{k} \left[m_i \log \frac{M_k}{N_k} + (n_i - m_i) \log \left(1 - \frac{M_k}{N_k} \right) \right]$$

$$- \sum_{i=k+1}^{c} \left[m_i \log \frac{M_k'}{N_k'} + (n_i - m_i) \log \left(1 - \frac{M_k'}{N_k'} \right) \right]$$

$$= M \log \frac{M}{N} + (N - M) \log \frac{N - M}{N}$$

$$- M_k \log \frac{M_k}{N_k} - (N_k - M_k) \log \frac{N_k - M_k}{N_k}$$

$$- (M - M_k) \log \frac{M_k'}{N_k'} - [N - M - (N_k - M_k)] \log \frac{N_k' - M_k'}{N_k'}$$

$$= M \log M + (N - M) \log(N - M) - N \log N$$

$$- M_k \log M_k - (N_k - M_k) \log(N_k - M_k) + N_k \log N_k$$

$$- M_k' \ln M_k' - (N_k' - M_k') \ln(N_k' - M_k') + N_k' \ln N_k'.$$

Define $l(n, m) = m \log m + (n - m) \log(n - m) - n \log n$, then

$$L_k \equiv -2 \log \frac{L_0(\widehat{p})}{L_1(\widehat{p}, \widehat{p}')} = 2[l(N_k, M_k) + l(N'_k, M'_k) - l(N, M)],$$

is the $-2 \log$ maximum likelihood ratio, which has chi-square distribution as its asymptotic distribution. Therefore, the change point position $\widehat{k}$ is estimated such that $L = L_{\widehat{k}} = \max_{l \leq k \leq l-1} L_k$, and H_0 is rejected if $L_{\widehat{k}} > C_1$, where C_1 is a constant determined by the null distribution of L and a given significance level α. Later in this chapter, the asymptotic null distribution of L is given.

8.2.2 Cumulative Sum Test (CUSUM)

The cumulative sum test is a frequently used method in change point analysis. In the following, we provide the readers with this alternative approach.

Under the current model, the CUSUM statistic Q_k at time k is the cumulative sum M_k of all successes minus the proportion $r_k M$, where $r_n = N_k/N$ of all successes up to and including time k, divided by the sample standard derivation $\sqrt{N p_0(1 - p_0)}$, where $p_0 = \frac{M}{N}$; that is,

$$Q_k = \frac{M_k - r_k M}{\sqrt{N p_0(1 - p_0)}}$$

for $k = 1, \ldots, c - 1$.

If we let $S_k^2 = r_k(1 - r_k)$, then Q_k^2/S_k^2 is the usual Pearson χ^2 statistic for testing the equality of p and p' conditional on M, and that asymptotically Q_k^2/S_k^2 and L_k are equivalent. Then k is estimated by $\widehat{k}$ such that $Q = Q_{\widehat{k}} = \max_{1 \leq k \leq c-1} |Q_k|$, and we reject H_0 if $Q_{\widehat{k}} > C_2$, where C_2 is some constant to be obtained from the null distribution of Q for a given significance level α.

To be able to obtain the values of C_1 and C_2, we derive the null distributions of L and Q in the following based on the work of Worsley (1983).

8.2.3 Null Distributions of L and Q

First of all, it is easy to verify that M is sufficient for the nuisance parameter p when H_0 is true. Then the conditional distribution of L and Q do not depend on p. Second, M_k and M'_k are sufficient for p and p' if $K = k$, so L and Q depend on M_k and M'_k only. When M is fixed, because $M'_k = M - M_k$, L and Q depend on M_k only. Therefore, we can express the events $\{L_k < x\}$ and $\{Q_k < q\}$ as

$$\{L_k < x\} \overset{\Delta}{=} A_k = \{M_k : a_k \leq M_k \leq b_k\}, \tag{8.1}$$

where $a_k = \inf\{M_k : L_k < x\}$, and $b_k = \sup\{M_k : L_k < x\}$, and

$$\{Q_k < q\} \overset{\Delta}{=} A'_k = \{M_k : a'_k \leq M_k \leq b'_k\}, \tag{8.2}$$

where $a'_k = \inf\{M_k : |Q_k| < q\}$, and $b'_k = \sup\{M_k : |Q_k| < q\}$. Then, $\{L < x\} = \cap_{k=1}^c A_k$, where A'_ks are defined as in (8.2).

Therefore, our goal now is to evaluate $P(\cap_{k=1}^c A_k)$ conditional on $M = m$. Let

$$F_k(\nu) = P\left(\overset{k}{\underset{i=1}{\cap}} A_i \mid M_k = \nu\right), \qquad k = 1, \ldots, c,$$

so that $F_1(\nu) = 1$ if $a_1 \leq \nu \leq b_1$, and $F_c(m) = P(\cap_{k=1}^c A_k)$, where $F_k(\nu)$ is the conditional probability of $\{L < x\}$ or $\{Q < q\}$ given $M_k = \nu$.

A general iterative procedure for evaluating $F_k(\nu)$ is given in Lemma 8.1 below.

Lemma 8.1 *For $k \leq c - 1$, if $p_i = p(i = 1, \ldots, k + 1)$, then*

$$F_{k+1}(\nu) = \sum_{u=a_k}^{b_k} F_k(u) h_k(u, \nu), \qquad a_{k+1} \leq \nu \leq b_{k+1},$$

where, for $0 \leq u \leq N_k$, $0 \leq \nu - u \leq n_{k+1}$,

$$h_k(u, \nu) = \binom{N_k}{u}\binom{n_{k+1}}{\nu - u} \bigg/ \binom{N_{k+1}}{\nu}.$$

Proof. If $a_{k+1} \leq \nu \leq b_{k+1}$, then

$$F_{k+1}(\nu) = P\left(\overset{k+1}{\underset{i=1}{\cap}} A_i \mid M_{k+1} = \nu\right)$$

$$= P\left(\overset{k}{\underset{i=1}{\cap}} A_i \cap A_{k+1} \mid M_{k+1} = \nu\right)$$

$$= P\left(\overset{k}{\underset{i=1}{\cap}} A_i \mid M_{k+1} = \nu\right)$$

$$= \sum_{u=a_k}^{b_k} P\left(\overset{k}{\underset{i=1}{\cap}} A_i \mid M_k = u, M_{k+1} = \nu\right) P\left(M_k = u \mid M_{k+1} = \nu\right).$$

Conditional on M_k and $M, M_1, \ldots, M_k$ are independent of M_{k+1}; then

$$P\left(\overset{k}{\underset{i=1}{\cap}} A_i \mid M_k = u, M_{k+1} = \nu\right) = P\left(\overset{k}{\underset{i=1}{\cap}} A_i \mid M_k = u\right) = F_k(u)$$

and conditional on M_{k+1}, M_k has the hypergeometric distribution with $N_{k+1}, N_k, N_{k+1} - N_k = n_{k+1}$; that is,

$$P(M_k = u \mid M_{k+1} = \nu) = \frac{\binom{N_k}{u}\binom{n_{k+1}}{\nu - u}}{\binom{N_{k+1}}{\nu}}.$$

Therefore,

$$F_{k+1}(\nu) = \sum_{u=a_k}^{b_k} F_k(u) h_k(u, \nu), \qquad a_{k+1} \leq \nu \leq b_{k+1},$$

where

$$h_k(u, \nu) = \frac{\binom{N_k}{u}\binom{n_{k+1}}{\nu - u}}{\binom{N_{k+1}}{\nu}}.$$

$\square$

Lemma 8.1 then can be used iteratively for $k = 1, \ldots, c - 2$ to evaluate $F_k(\nu)$ for $a_k \leq \nu \leq b_k$. Based on these iterations, a final iteration for $k = c-1$ at $\nu = m$ will give the value of $F_c(m)$, where $F_c(m) = P(\cap_{k=1}^c A_k) = P(A) = P(L < x)$ or $P(Q < q)$, depending on the definition of A_k.

It is noted that the above iterative computations can be reduced by using the recurrence properties of the hypergeometric probability function:

$$h_k(0, \nu + 1) = [(n_{k+1} - \nu)/(N_{k+1} - \nu)] h_k(0, \nu),$$

$$h_k(u + 1, \nu) = \frac{(\nu - u)(N_k - u)}{(u + 1)(n_{k+1} - \nu + u + 1)} h_k(0, \nu).$$

8.2.4 Alternative Distribution Functions of L and Q

If there is a change after period k, then Lemma 8.1 can still be used iteratively to find $F_k(\nu)$ for $k = 1, \ldots, k - 1$. Now, consider the sequence from k to $c-1$, conditional on M, and let

$$F_k'(\nu) = P\left(\bigcap_{i=k}^{c-1} A_i \mid M_k = \nu\right), \qquad k = 1, \ldots, c - 1.$$

The following Lemma 8.2 gives an iterative formula to compute $F_k'(\nu)$.

Lemma 8.2 For $k \geq z$, if $p_i = p'(i = k - 1, \ldots, c)$, then

$$F_{k+1}'(\nu) = \sum_{u=a_k}^{b_k} F_k'(u) h_k'(u, \nu), \qquad (a_{k-1} \leq \nu \leq b_{k-1}),$$

where, for $0 \leq m - u \leq N'_k, 0 \leq u - \nu \leq n_k$, and

$$h'_k(u, \nu) = \binom{N'_k}{m - u} \binom{n_k}{u - \nu} \Big/ \binom{N'_{k-1}}{m - \nu}.$$

Proof. If $a_{k-1} \leq \nu \leq b_{k-1}$, then

$$F'_{k-1}(\nu) = P\left(\bigcap_{i=k-1}^{c-1} A_i \mid M_{k-1} = \nu\right)$$

$$= P\left(\bigcap_{i=k}^{c-1} A_i \cap A_{k-1} \mid M_{k-1} = \nu\right)$$

$$= P\left(\bigcap_{i=k}^{c-1} A_i \mid M_{k-1} = \nu\right)$$

$$= \sum_{u=a_k}^{b_k} P\left(\bigcap_{i=k}^{c-1} A_i \mid M_k = u, M_{k-1} = \nu\right) P(M_k = u \mid M_{k-1} = \nu).$$

Conditional on $M, M_{k-1}, M_k, \ldots, M_{c-1}$ are independent of M_{k-1}. Therefore

$$P\left(\bigcap_{i=k}^{c-1} A_i \mid M_k = u, M_{k-1} = \nu\right) = P\left(\bigcap_{i=k}^{c-1} A_i \mid M_k = u\right) = F'_k(u),$$

and conditional on $M_{k-1} = \nu, M_k = u$ has the hypergeometric distribution with parameters $N'_{k-1}, N'_k, n_k = N'_{k-1} - N'_k = N - N_{k-1} - N + N_k = N_k - N_{k-1}$; that is,

$$P(M_k = u \mid M_{k-1} = \nu) = \frac{\binom{N'_k}{m-u} \binom{n_k}{u-\nu}}{\binom{N'_{k-1}}{m-\nu}}.$$

Therefore,

$$F'_{k-1}(\nu) = \sum_{u=a_k}^{b_k} F'_k(u) h'_k(u, \nu),$$

where

$$h'_k(u, \nu) = \frac{\binom{N'_k}{m-u} \binom{n_k}{u-\nu}}{\binom{N'_{k-1}}{m-\nu}}.$$

$\square$

Lemma 8.2 then can be used iteratively for $k = c - 1, \ldots, k + 1$ to find $F'_k(\nu)$. Combining $F_k(\nu)$ and $F'_k(\nu)$, we can calculate $P(\cap_{k=1}^c A_k)$ under the alternative hypothesis H_1.

Theorem 8.3 *Under H_1, conditional on $M = m$,*

$$P\left(\bigcap_{k=1}^{c} A_k\right) = \sum_{\nu=a_k}^{b_k} F_k(\nu)F_k'(\nu)H_k(\nu)\Delta^{-\nu} \Bigg/ \sum_{\nu=0}^{m} H_k(\nu)\Delta^{-\nu},$$

where, for $0 \le \nu \le N_k$, $0 \le m - u \le N_k'$,

$$H_k = \binom{N_k}{\nu}\binom{n_k'}{m-\nu} \Bigg/ \binom{N}{m}$$

and $\Delta = (p'/(1 - p'))/(p/(1 - p))$ is the odds ratio or relation risk.

Proof.

$$P\left(\bigcap_{k=1}^{c} A_k\right) = \sum_{\nu=a_k}^{b_k} P\left(\bigcap_{k=1}^{c} A_k \mid M_k = \nu\right) \cdot P(M_k = \nu).$$

Conditional on $M_k = \nu$ and $M = m$, M_i is independent of M_j for $i < k < j$, and then

$$P\left(\bigcap_{k=1}^{c} A_k \mid M_k = \nu\right) = P\left(\left[\bigcap_{j=k}^{k} A_i\right] \cap \left[\bigcap_{j=k}^{c-1} A_i\right] \mid M_k = \nu\right)$$

$$= P\left(\bigcap_{i=1}^{k} A_i \mid M_k = \nu\right) \cdot P\left(\bigcap_{j=k}^{c-1} A_j \mid M_k = \nu\right)$$

$$= F_k(\nu)F_k'(\nu).$$

Conditional on $M = m$, from Bayes' formula, we have

$$P(M_k = \nu \mid M = m) = \frac{P(M_k = \nu, M = m)}{\sum_{\nu=0}^{m} P(M = m)} = \frac{P(M_k = \nu)}{\sum_{\nu=0}^{m} P(M_k = \nu)}$$

$$= \frac{\dfrac{\binom{N_k}{\nu}p^\nu(1-p)^{N_k-\nu}\binom{n_k'}{m-\nu}p'^{m-\nu}(1-p')^{N_k'-m+\nu}}{\binom{N}{m}p'^m(1-p')^{N_k'-m}(1-p)^{N_k}}}{\displaystyle\sum_{\nu=0}^{m} \dfrac{\binom{N_k}{\nu}p^\nu(1-p)^{N_k-\nu}\binom{n_k'}{m-\nu}p'^{m-\nu}(1-p')^{N_k'-m+\nu}}{\binom{N}{m}p'^m(1-p')^{N_k'-m}(1-p)^{N_k}}}$$

$$= \frac{\dfrac{\binom{N_k}{\nu}\binom{n_k'}{m-\nu}}{\binom{N}{m}}\left(\dfrac{p(1-p')}{p'(1-p)}\right)^\nu}{\displaystyle\sum_{\nu=0}^{m}\dfrac{\binom{N_k}{\nu}\binom{n_k'}{m-\nu}}{\binom{N}{m}}\left(\dfrac{p(1-p')}{p'(1-p)}\right)^\nu}$$

$$= \frac{H_k(\nu)\Delta^{-\nu}}{\sum_{\nu=0}^{m} H_k(\nu)\Delta^{-\nu}}.$$

Therefore, conditional on $M = m$, we get

$$P\left(\bigcap_{k=1}^{c} A_k\right) = \sum_{\nu=a_k}^{b_k} F_k(\nu)F'_k(\nu)H_k(\nu)\Delta^{-\nu} \Big/ \sum_{\nu=0}^{m} H_k(\nu)\Delta^{-\nu}.$$

□

For small M or $M - N$ (say either is less than 200), the alternative distribution of either L or Q can be calculated by using Theorem 8.3.

First, the sequence of bounds a_k, b_k is determined for the value of desired statistic L or Q. Second, the sequences of functions $F_1, \ldots, F_{c-1}$ and $F'_{c-1,\ldots,} F'_1$ are calculated by using Lemma 8.1 and Lemma 8.2, They need only be evaluated between a_k and b_k because the remaining values are zero. Thirdly, given an odds ratio Δ, the alternative distribution of L or Q then can be evaluated at each change point k.

The approximate null distributions of L and Q, as well as the approximate alternative distributions of L and Q were obtained by Worsley (1983), and the readers can refer to that article for some useful details.

8.3 Poisson Model

Let $x_1, \ldots, x_c$ be a sequence of independent random variables from Poisson distribution with mean $\lambda_i, i = 1, 2, \ldots, c$. We now test the following hypotheses H_0 versus H_1 for possible change in mean or variance λ_i. That is, we want to test:

$$H_0 : \lambda_1 = \lambda_2 = \cdots = \lambda_c = \lambda$$

versus the alternative:

$$H_1 : \lambda_1 = \cdots = \lambda_k = \lambda \neq \lambda_{k+1} = \cdots = \lambda_c = \lambda'.$$

8.3.1 Likelihood-Ratio Procedure

Under H_0, the likelihood function is:

$$L_0(\lambda) = \prod_{i=1}^{c} \frac{e^{-\lambda}\lambda^{x_i}}{x_i!}$$

$$= \frac{e^{-c\lambda}\lambda^{\sum_{1}^{c} x_i}}{\prod_{i=1}^{c} x_i!},$$

and the MLE of λ is given by

$$\widehat{\lambda} = \frac{\sum_1^c x_i}{c}.$$

Under H_1, the likelihood function is:

$$L_1(\lambda, \lambda') = \prod_{i=1}^{k} \frac{e^{-\lambda}\lambda^{x_i}}{x_i!} \prod_{i=k+1}^{c} \frac{e^{-\lambda'}\lambda'^{x_i}}{x_i!}$$

$$= \frac{e^{-k\lambda}\lambda^{\sum_1^k x_i}}{\prod_{i=1}^{k} x_i!} \cdot \frac{e^{-(c-k)\lambda'}\lambda'^{\sum_{k+1}^{c} x_i}}{\prod_{k+1}^{c} x_i!},$$

and the MLEs of λ, and λ' are found to be:

$$\widehat{\lambda} = \frac{\sum_1^k x_i}{k} \quad \text{and} \quad \widehat{\lambda}' = \frac{\sum_{k+1}^{c} x_i}{c - k}.$$

Denote by $M_k = \sum_{i=1}^{k} x_i$, $M = M_c$, and $M'_k = M - M_k = \sum_{i=k+1}^{c} x_i$; then under H_0,

$$\widehat{\lambda} = \frac{M}{c},$$

and under H_1,

$$\widehat{\lambda} = \frac{M_k}{c}, \qquad \widehat{\lambda}' = \frac{M'_k}{c - k}.$$

Then,

$$\log \frac{L_0(\widehat{\lambda})}{L_1(\widehat{\lambda}, \widehat{\lambda}')} = \log L_0(\widehat{\lambda}) - \log L_1(\widehat{\lambda}, \widehat{x}) = M(\log M - \log c)$$

$$- [M_k(\log M_k - \log k) + M'_k(\log M'_k - \log(c - k))]$$

$$= -M_k \log \frac{M_k}{k} - M'_k \log \frac{M'_k}{c - k} + M \log \frac{M}{c},$$

and $-2 \log$ maximum likelihood-ratio procedure statistic L_k is:

$$L_k = -2 \log \frac{L_0(\widehat{\lambda})}{L_1(\widehat{\lambda}, \widehat{\lambda}')} = 2 \left\{ M_k \log \frac{M_k}{k} + M'_k \log \frac{M'_k}{c - k} - M \log \frac{M}{c} \right\}.$$

The change point position k is estimated by $\widehat{k}$ s.t.

$$L = L_{\widehat{k}} = \max_{1 \leq k \leq c-1} L_k$$

and H_0 is rejected if $L_{\widehat{k}} < C$, C is some constant that will be determined by the size of the test and the null distribution of L.

8.3.2 Null Distribution of L

Next, we can derive the null distribution of L in a way similar to that in Section 8.2.2 and the work outlined in Worsley (1983), Durbin (1973), and Pettie and Stephens (1977). Again, the parameter we want to estimate is the change point position k, and λ is viewed as an nuisance parameter. Clearly, M is sufficient for λ under H_0; then the conditional distribution of L does not depend on λ. Because M_k and M'_k are sufficient for λ and λ' when $K = k$, then, conditional on M, the likelihood procedure statistic L depends on M_k only. Now we can have the following expression for $\{L < x\}$,

$$\{L < x\} = \cap_k A_k,$$

where $A_k = \{a_k \leq M_k \leq b_n\}$ with $a_k = \inf\{M_k : L_k < x\}$ and $b_k = \sup\{M_k : L_k < x\}$. Then, we have the following two lemmas.

Conditional on $M = m$, let $F_k(\nu) = P(\cap_1^k A_i \mid M_k = \nu)(k = 1, \ldots, c)$, so that $F_1(\nu) = 1$ if $a_1 \leq a_1 \leq \nu \leq b_1$, and $F_c(m) = P(\cap_{k=1}^c A_k)$.

Lemma 8.4 For $k \leq c - 1$ if $\lambda_i = \lambda(i = 1, \ldots, k + 1)$,

$$F_{k+1}(\nu) = \sum_{u=a_k}^{b_k} F_k(u)h_k^*(u, \nu), \qquad a_{k+1} \leq \nu \leq k_{n+1},$$

where, for $0 \leq u \leq \nu \leq m$,

$$h_k^*(u, \nu) = \binom{\nu}{u} \frac{k^u}{(k+1)^\nu}.$$

Proof. Proceed exactly as in the proof of Lemma 8.1 except that, now conditional on $M_{k=1}$ and M, M_k has a binomial distribution with ν independent trials and success probability $k/(k+1)$; that is,

$$P(M_k = u \mid M_{k+1} = \nu) = \binom{\nu}{u} \left(\frac{k}{k+1}\right)^u \cdot \left(1 - \frac{k}{k+1}\right)^{\nu-u}$$

$$= \binom{\nu}{u} \frac{k^u}{(k+1)^\nu} = h^*(u, \nu).$$

$\square$

Now, the conditional null distribution of L is obtained iteratively by using Lemma 8.4.

Under H_1, there is a change after period k; consider the reversed sequence of time periods, and let

$$F'_k(\nu) = P\left(\cap_{i=k}^{c-1} A_i \mid M_k = \nu\right) \qquad k = 1, \ldots, c - 1.$$

We have the following result.

Lemma 8.5 *For $k \geq 2$, if $\lambda_i = \lambda'(i = k - 1, \ldots, c)$,*

$$F'_{k-1}(\nu) = \sum_{u=a_k}^{b_k} F'_k(u) h_k^{**}(u, \nu), \qquad a_{n-1} \leq \nu \leq b_{k-1},$$

where $0 \leq \nu \leq u \leq m$, $0 \leq u \leq m - \nu$,

$$h_k^{**}(u, \nu) = \binom{m - \nu}{u} \frac{(k - 1)^u}{k^{m-\nu}}.$$

Proof. Proceed exactly as in the proof of Lemma 8.2 except that, now conditional on M and M_{k-1}, M_k has a binomial distribution with $m - \nu$ independent trial and success probability $(k - 1)/k$; that is,

$$\begin{aligned}
P(M_k = u \mid M_{k-1} = \nu) &= \binom{m - \nu}{u} \left(\frac{k - 1}{k}\right)^u \left(1 - \frac{k - 1}{k}\right)^{m-\nu-u} \\
&= \binom{m - \nu}{u} \frac{(k - 1)^u}{k^{m-\nu}} = h_k^{**}(u, \nu).
\end{aligned}$$

$\square$

Now, using $F'_k(\nu)$ in Lemma 8.5 combining with $F_k(\nu)$ obtained in Lemma 8.4, the alternative distribution of L can be obtained similarly as in Theorem 8.3, and we give the result in the following Theorem 8.6.

Theorem 8.6 *Under H_1, conditional on $M = m$,*

$$P\left(\bigcap_{k=1}^{c} A_k\right) = \frac{\sum_{u=a_k}^{b_k} F_k(\nu) F'_k(\nu) H_k^*(\nu) \Delta^{*-\nu}}{\sum_{u=a_k}^{b_k} H_k^*(\nu) \Delta^{*-\nu}},$$

where for $0 \leq \nu \leq m$, $0 \leq m - u \leq \nu$, $0 \leq u \leq \nu \leq m$,

$$H_k^*(\nu) = \binom{m}{v} k^\nu (c - k)^{m-\nu} / c^m \quad and \quad \Delta^* = \frac{\lambda'}{\lambda}.$$

8.4 Informational Approach

As the readers may note, throughout this monograph we prefer the use of the information criterion approach for change point analysis because the testing of change point hypotheses can be treated as a model selection problem. The information criterion is an excellent tool for model selection and is computationally effective. In this section we derive the SIC for the change point problem associated with both the binomial model and the Poisson model.

(i) Binomial Model

Here the same assumptions and notations are used as in Section 8.2. Suppose that there are c binomial random variables, say $x_i \sim \text{binomial}(n_i, p_i)$ and $x_i = m_i$ for $i = 1, \ldots, c$, where $x_i = \#$ of successes. Let's test the following null hypothesis,

$$H_0 : p_1 = p_2 = \cdots = p_c = p,$$

against the alternative hypothesis

$$H_1 : p_1 = \cdots = p_k = p \neq p_{k+1} = \cdots = p_c = p'.$$

Under H_0, the maximum likelihood function has been obtained as

$$L_0(\widehat{p}) = \prod_{i=1}^{c} \binom{n_i}{m_i} \widehat{p}^{m_i}(1 - \widehat{p})^{n_i - m_i},$$

where $\widehat{p} = M/N$. Therefore, the Schwarz information criterion under H_0, denoted by $\text{SIC}(c)$, is obtained as

$$
\begin{aligned}
\text{SIC}(c) &= -2\log L_0(\widehat{p}) + \log c \\
&= -2\sum_{i=1}^{c} \log \binom{n_i}{m_i} - 2M \log \frac{M}{N} \\
&\quad - 2(N - M) \log \frac{N - M}{N} + \log c.
\end{aligned}
$$

Under H_1, the maximum likelihood function has been obtained as

$$L_1(\widehat{p}, \widehat{p}') = \prod_{i=1}^{k} \binom{n_i}{m_i} \widehat{p}^{m_i}(1 - \widehat{p})^{n_i - m_i} \cdot \prod_{j=k+1}^{c} \binom{n_j}{m_j} \widehat{p}'^{m_j}(1 - \widehat{p}')^{n_j - m_j},$$

where $\widehat{p} = M_k/N_k$ and $\widehat{p}' = M_k'/N_k'$. Therefore, the Schwarz information criterion under H_1, denoted by $\text{SIC}(k)$, for $k = 1, \ldots, c - 1$, is obtained as

$$
\begin{aligned}
\text{SIC}(k) &= -2\log L_1(\widehat{p}, \widehat{p}') + 2\log c \\
&= -2\sum_{i=1}^{c} \log \binom{n_i}{m_i} - 2M_k \log \frac{M_k}{N_k} - 2(N_k - M_k) \log \frac{N_k - M_k}{N_k} \\
&\quad - 2M_k' \log \frac{M_k'}{N_k'} - 2(N_k' - M_k') \log \frac{N_k' - M_k'}{N_k'} + 2\log c.
\end{aligned}
$$

According to the minimum information criterion principle, we reject H_0 if

$$\text{SIC}(c) > \min_{1 \leq k \leq c - 1} \text{SIC}(k),$$

or if

$$\min_{1 \leq k \leq c-1} \Delta(k) < 0,$$

where

$$
\begin{aligned}
\Delta(k) = {} & M \log M + (N - M) \log(N - M) - N \log N \\
& - M_k \log M_k - (N_k - M_k) \log(N_k - M_k) + N_k \log N_k \\
& - M'_k \log M'_k - (N'_k - M'_k) \log(N'_k - M'_k) + N'_k \log N'_k \\
& - \frac{1}{2} \log \frac{1}{c}.
\end{aligned}
$$

When H_0 is rejected, the estimated change point position is $\widehat{k}$ such that

$$\Delta(\widehat{k}) = \min_{1 \leq k \leq c-1} \Delta(k).$$

(ii) Poisson Model

As stated in Section 8.3, let $x_1, \ldots, x_c$ be a sequence of independent random variables coming from a Poisson distribution with mean λ_i, $i = 1, 2, \ldots, c$; we test the following hypothesis H_0 versus H_1 for possible change in mean or variance λ_i. That is, we test:

$$H_0 : \lambda_1 = \lambda_2 = \cdots = \lambda_c = \lambda$$

versus the alternative:

$$H_1 : \lambda_1 = \cdots = \lambda_k = \lambda \neq \lambda_{k+1} = \cdots = \lambda_c = \lambda'.$$

Under H_0, the maximum likelihood function is given by

$$L_0(\widehat{\lambda}) = \frac{e^{-c\widehat{\lambda}} \widehat{\lambda}^{\sum_1^c x_i}}{\prod_{i=1}^c x_i!},$$

where $\widehat{\lambda} = M/c$ with $M = \sum_1^c x_i$. Therefore, the Schwarz information criterion under H_0, denoted by $\mathrm{SIC}(c)$, is obtained as

$$
\begin{aligned}
\mathrm{SIC}(c) &= -2 \log L_0(\widehat{\lambda}) + \log c \\
&= 2M - 2M \log \left(\frac{M}{c} \right) + 2 \log \prod_{i=1}^c x_i! + \log c.
\end{aligned}
$$

Under H_1, the maximum likelihood function is

$$L_1(\widehat{\lambda}, \widehat{\lambda}') = \frac{e^{-k\widehat{\lambda}} \widehat{\lambda}^{\sum_1^k x_i}}{\prod_1^k x_i!} \cdot \frac{e^{-(c-k)\widehat{\lambda}'} \widehat{\lambda}'^{\sum_{k+1}^c x_i}}{\prod_{k+1}^c x_i!},$$

Table 8.1 Dataset and $\Delta(k)$ Values

Year	i	m_i	n_i	$\Delta(k)$
1960	1	8	2409	1.0399
1961	2	3	2453	-2.0785
1962	3	9	2290	-1.5289
1963	4	12	2171	-0.0234
1964	5	6	2084	-0.9095
1965	6	4	1993	-2.7511*
1966	7	14	2157	-0.6740
1967	8	12	2091	0.1898
1968	9	7	2152	-0.5198
1969	10	5	2007	-2.0086
1970	11	13	2027	-0.6181
1971	12	11	1963	-0.0351
1972	13	12	1982	0.6741
1973	14	9	1974	0.5833
1974	15	6	1932	-0.7964
1975	16	13	1807	0.7130
1976	17	12	1919	-

where $\widehat{\lambda} = M_k/k$, and $\widehat{\lambda}' = M'_k/(c-k)$ with $M_k = \sum_1^c x_i$ and $M'_k = \sum_{k+1}^c x_i$. Therefore, the Schwarz information criterion under H_1, denoted by SIC(k), for $k = 1, \ldots, c-1$, is obtained as

$$\text{SIC}(k) = -2\log L_1(\widehat{\lambda}, \widehat{\lambda}') + 2\log c$$

$$= 2M_k - 2M_k \log \frac{M_k}{k} + 2M'_k - M'_k \log \frac{M'_k}{c-k}$$

$$\times 2\log \prod_{i=1}^c x_i! + 2\log c.$$

According to the minimum information criterion principle, we will reject H_0 if

$$\text{SIC}(c) > \min_{1 \le k \le c-1} \text{SIC}(k),$$

or

$$\min_{1 \le k \le n-1} \left[M\log \frac{M}{c} - M_k \log \frac{M_k}{k} - M'_k \log \frac{M'_k}{c-k} \right] < \frac{1}{2}\log \frac{1}{c},$$

and estimate the change point position by $\widehat{k}$ such that

$$\text{SIC}(\widehat{k}) = \min_{1 \le k \le c-1} \text{SIC}(k).$$

8.5 Application to Medical Data

In Hanify et al. (1981), a set of data, which gives the number of cases of the birth deformity club foot in the first month of pregnancy and the total number of all births for the years from 1960 to 1976 in the northern region of New Zealand was studied. There was a high correlation between the club feet incidence and the amount of 2,4,5-T used in that region. Worsley (1983) analyzed these data by assuming the binomial model, and used both the likelihood-ratio test statistic L and the CUSUM test statistic Q to locate a change point, which was found to be the sixth observation (corresponding to the year 1965 when the herbicide 2,4,5-T was first used in the region).

Here, we use the same dataset, and apply the SIC model selection criterion given in the current section. The calculations show that $\Delta(\widehat{k}) = \min_{1 \le k \le c-1} \Delta(k)$ holds for $k = 6$. Therefore, the change point is successfully located as in Worsley (1983). The calculated $\Delta(k)$ values are listed in Table 7.1 along with the original dataset given in Worsley (1983). The starred $\Delta(k)$ value is the minimum negative $\Delta(k)$, which corresponds to the time point $k = 6$.

Chapter 9
Other Change Point Models

9.1 The Smooth-and-Abrupt Change Point Model

9.1.1 Introduction

Many investigations of change point models consider an abrupt change point or multiple abrupt change points in the parameters of various distributions such as the ones discussed in previous chapters of this monograph. One of the reasons for such consideration is that an abrupt change or multiple abrupt change points are commonly occurring changes in many models across from different disciplines.

In addition to the abrupt change point(s) problem, a smooth change model, mixed with an abrupt change, is also very practical. In a recent paper of the authors (Chen and Gupta, 2007), we studied a smooth change point model mixed with an abrupt change, or the Smooth-and-Abrupt Change Point (SACP) model, for a sequence of normally distributed random variables. The smooth change point problems have been discussed in the literature by some authors. Vilasuso (1996) studied a smooth mean change model that is clearly different from our proposed model. In this section, we present the result of Chen and Gupta (2007) regarding the analysis of the SACP model.

Assume that $X_1, X_2, \ldots, X_n$ is a sequence of normal random variables with parameters (μ_1, σ_1^2), $(\mu_2, \sigma_2^2), \ldots, (\mu_n, \sigma_n^2)$, respectively. Assuming common variances (i.e., $\sigma_1^2 = \sigma_2^2 = \cdots = \sigma_n^2 = \sigma^2$ (unknown)), we are interested in testing the null hypothesis of no change in the means:

$$H_0 : \mu_1 = \mu_2 = \cdots = \mu_n = \mu \tag{9.1}$$

versus the hypothesis of a linear trend change and an abrupt change in the means:

$$H_1 : \mu_i = \begin{cases} \mu, & 1 \leq i \leq k_1 \\ \mu + \beta(i - k_1), & k_1 + 1 \leq i \leq k_2 , \\ \mu, & k_2 + 1 \leq i \leq n \end{cases} \qquad (9.2)$$

where β is the slope of the linear trend change starting at an unknown position k_1 and ending at an unknown position k_2 (the position of the abrupt change). It is a model with a common mean μ before position k_1 and linear trend mean with slope β between positions k_1 and k_2; after position k_2 the model resumes mean μ as before. At position k_2, the change is an abrupt change. Figure 9.1 gives a hypothetical SACP model, which is a simulated random sample of size 60 from a normal distribution with repair structure with the parameters $\mu = 4$, $\sigma = 2$, $\beta = 10$, $k_1 = 20$, and $k_2 = 39$.

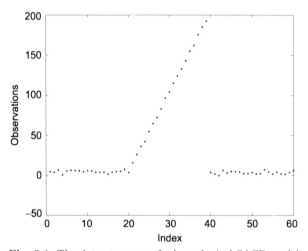

Fig. 9.1 The data structure of a hypothetical SACP model

Note that when $k_1 = 1$, and $k_2 = n$, the SACP model becomes an ordinary linear regression model. When $k_2 = k_1 + 1$, the SACP model becomes a normal model with an outlier at position $k_2 = k_1 + 1$. For these reasons, we assume $1 \leq k_1 \leq n - 2$, $k_1 < k_2 \leq n - 1$, and $n \geq 3$ in the SACP model.

This SACP model is very useful in many disciplines. For instance, in a continuous production process, the measure of a certain product follows a common mean. For some reason, the measure starts to change according to a linear trend. After the measure reaches a certain level, an action has to be taken; then the measure drops back to the original common mean. In genetic studies, gene expression profiles of a cell line may be obtained and biologists may be interested in different gene expression patterns of some genes. Some genes may maintain a constant level of expression before a time point a and then upregulate starting at time point a, and continue to upregulate linearly until another time point b when the gene expression drops back to the original constant level of expression. This biological process can be described by our

SACP model as well. The main goal of a statistical analysis of such a SACP model is to estimate locations of the two change points.

9.1.2 A Bayesian Solution to the SACP Model

Not much investigation has been carried out about the SACP model so far in the literature. In this section, we present a Bayesian approach for analyzing such a SACP model as given in Chen and Gupta (2007).

Again, let $X_1, X_2, \ldots, X_n$, $n \geq 3$, be a sequence of normal random variables with parameters (μ_1, σ_1^2), $(\mu_2, \sigma_2^2), \ldots, (\mu_n, \sigma_n^2)$, respectively. Assume common variances, that is, $\sigma_1^2 = \sigma_2^2 = \cdots = \sigma_n^2 = \sigma^2$, and σ^2 is unknown. We also assume that μ and β are unknown.

When it is known that the change points in such a model exist, we want to find the locations of the change points in the SACP model specified by (9.2). Assume that the two change points (k_1, k_2) are uniformly distributed; that is, the prior distribution of (k_1, k_2) is assumed to be

$$\pi_0(k_1, k_2) = \begin{cases} \frac{2}{(n-1)(n-2)}, & 1 \leq k_1 \leq n-2, k_1+1 \leq k_2 \leq n-1 \\ 0, & \text{otherwise} \end{cases}. \qquad (9.3)$$

We also assign the noninformative prior distributions for σ^2, μ, and β as follows:

$$\pi_0(\sigma^2 | k_1, k_2) \propto \begin{cases} \frac{1}{\sigma^2}, & \sigma^2 > 0 \\ 0, & \text{otherwise} \end{cases}, \qquad (9.4)$$

$$\pi_0(\mu, \beta | \sigma^2, k_1, k_2) \propto \text{constant}. \qquad (9.5)$$

Our main result is presented in the following Theorem 9.1.

Theorem 9.1 For the SACP model specified by (9.2), under the prior distributions (9.3) through (9.5), the posterior density of the two points (k_1, k_2), where $1 \leq k_1 \leq n-2$, $k_1 < k_2 \leq n-1$, and $n \geq 3$, is given by

$$\pi_1(k_1, k_2) = \frac{\pi_1^*(k_1, k_2)}{\sum_{k_1=1}^{n-2} \sum_{k_2=k_1+1}^{n-1} \pi_1^*(k_1, k_2)}, \qquad (9.6)$$

where

$$\pi_1^*(k_1, k_2) = \left\{ \frac{12}{\lambda[2n(2t+1) - 3\lambda]} \right\}^{1/2}$$

$$\cdot \left\{ \sum_{i=1}^{n} X_i^2 - \frac{1}{\varpi} \left(\sum_{j=1}^{t} j X_{j+k_1} \right)^2 - \kappa \right\}^{-((n-2)/2)}.$$

with

$$t = k_2 - k_1, \qquad \lambda = t(t+1), \qquad \varpi = \frac{1}{6}t(t+1)(2t+1),$$

and

$$\kappa = \frac{2\left[(2t+1)n\overline{X} - 3\sum_{j=1}^{t} jX_{j+k_1}\right]^2}{[2n(2t+1) - 3\lambda](2t+1)}.$$

Proof of Theorem 9.1 First of all, the likelihood function for the parameters based on the sample is

$$L_1(\mu, \beta, \sigma^2) = f(X_1, X_2, \ldots, X_n | \mu, \beta, \sigma^2, k_1, k_2)$$

$$= (2\pi)^{-n/2}(\sigma^2)^{-n/2}$$

$$\cdot \exp\left\{-\frac{1}{2\sigma^2}\left[\sum_{i=1}^{n}(X_i - \mu)^2 - 2\beta\sum_{j=1}^{t} jX_{j+k_1} + \lambda\beta\mu + \varpi\beta^2\right]\right\}.$$

From the priors (9.3)–(9.5) and the likelihood function, we obtain the joint posterior density of the parameters (including the unknown change point locations) as

$$\pi_1(\mu, \beta, \sigma^2, k_1, k_2) \propto L_1(\mu, \beta, \sigma^2)\pi_0(\sigma^2 | k_1, k_2).$$

Note that

$$\int_{-\infty}^{\infty} (\sigma^2)^{-n/2-1} \exp\left\{-\frac{1}{2\sigma^2}\left[\sum_{i=1}^{n}(X_i - \mu)^2\right.\right.$$

$$\left.\left. -2\beta\sum_{j=1}^{t} jX_{j+k_1} + \lambda\beta\mu + \varpi\beta^2\right]\right\} d\beta$$

$$\propto (\sigma^2)^{-((n+1)/2)}\varpi^{-1/2} \exp\left\{-\frac{1}{2\sigma^2}\left[\sum_{i=1}^{n}(X_i - \mu)^2\right.\right.$$

$$\left.\left. -\frac{6\left[\sum_{j=1}^{t} jX_{j+k_1} - 3\lambda\mu\right]^2}{36\varpi}\right]\right\}, \tag{9.7}$$

and

$$\int_{-\infty}^{\infty} (\sigma^2)^{-((n+1)/2)}\varpi^{-1/2} \exp\left\{-\frac{1}{2\sigma^2}\left[\sum_{i=1}^{n}(X_i - \mu)^2\right.\right.$$

$$\left.\left. -\frac{6\left[\sum_{j=1}^{t} jX_{j+k_1} - 3\lambda\mu\right]^2}{36\varpi}\right]\right\} d\mu$$

$$\propto (\sigma^2)^{-n/2} \varpi^{-1/2} \left[n - \frac{3\lambda}{2(2t+1)} \right]^{-1/2}$$

$$\cdot \exp\left\{ -\frac{1}{2\sigma^2} \left[\sum_{i=1}^{n} X_i^2 - \frac{1}{\varpi} \left(\sum_{j=1}^{t} jX_{j+k_1} \right)^2 - \kappa \right] \right\}. \qquad (9.8)$$

Integration of $\pi_1(\mu, \beta, \sigma^2, k_1, k_2)$ with respect to μ, β, and σ^2, in light of (9.7) and (9.8), yields the posterior density of the change point locations (k_1, k_2):

$$\pi_1(k_1, k_2) \propto \int_0^\infty \int_{-\infty}^\infty \int_{-\infty}^\infty L_1(\mu, \beta, \sigma^2)\pi_0(\sigma^2|k_1, k_2)d\beta d\mu_1 d\sigma^2$$

$$\propto \left\{ \frac{12}{\lambda[2n(2t+1) - 3\lambda]} \right\}^{1/2}$$

$$\cdot \left\{ \sum_{i=1}^{n} X_i^2 - \frac{1}{\varpi} \left(\sum_{j=1}^{t} jX_{j+k_1} \right)^2 - \kappa \right\}^{-((n-2)/2)}. \qquad (9.9)$$

Let the right-hand side of (9.9) be

$$\pi_1^*(k_1, k_2) = \left\{ \frac{12}{\lambda[2n(2t+1) - 3\lambda]} \right\}^{1/2}$$

$$\cdot \left\{ \sum_{i=1}^{n} X_i^2 - \frac{1}{\varpi} \left(\sum_{j=1}^{t} jX_{j+k_1} \right)^2 - \kappa \right\}^{-((n-2)/2)}.$$

The posterior density is therefore given by (9.6); that is,

$$\pi_1(k_1, k_2) = \frac{\pi_1^*(k_1, k_2)}{\sum_{k_1=1}^{n-2} \sum_{k_2=k_1+1}^{n-1} \pi_1^*(k_1, k_2)}.$$

This completes the proof of Theorem 9.1. $\qquad\square$

As an immediate application of Theorem 9.1, $\pi_1(k_1, k_2)$ can be calculated for all (k_1, k_2) with $1 \le k_1 \le n - 2, k_1 + 1 \le k_2 \le n - 1$. Then the change point locations in the SACP model can be estimated by $(\hat{k}_1, \hat{k}_2)$ such that

$$\pi_1(\hat{k}_1, \hat{k}_2) = \max_{k_1, k_2} \pi_1(k_1, k_2).$$

9.1.3 Empirical Evaluation of the Change Point Location Estimates

To evaluate how close the change point location estimates obtained by the proposed Bayesian approach are to the true change point locations, Chen and Gupta (2007) provided a simulation study. They simulated 10,000 random samples, each of sample size $n = 40$, from a normal distribution with the SACP model structure. Specifically, the two true change point positions are at $k_1 = 14$ and $k_2 = 29$. Other parameters used are $\mu = 4$, $\sigma = 2$, $\beta = 8$. The simulation is done using MATLAB®. The average of $\widehat{k}_1$ and the average of $\widehat{k}_2$ obtained using the posterior density $\pi_1(k_1, k_2)$ given by Equation (9.6) are 13.759 and 28.6843, respectively. Therefore, the posterior density $\pi_1(k_1, k_2)$ given by (9.6) successfully picks up the true change point locations in the SACP model.

9.2 Application of SACP Model to Gene Expression Data

To illustrate the proposed approach for locating the change points in a SACP model, Chen and Gupta (2007) gave an example of using the SACP model to detect the gene expression pattern for a specific gene. The data used are from the yeast *Saccharomyces cerevisiae* microarray experiments of Spellman et al. (1998). There are interesting periodic genes found from the four yeast microarray experiments, namely CDC15, CDC28, alpha, and elution (see Spellman et al., 1998; Wichert Folianos, and Strimmer, 2004; and Chen, 2005). Also, in Chen (2005), genes of patterns other than periodic in these four datasets are identified. We now pick one gene (probe ID: YJR152W), the DAL5 gene, from the CDC15 dataset and analyze it for changes as specified by the SACP model. This gene is observed on 24 equally spaced time points (with a time interval of 20 minutes), and the normalized log expression of this gene can be downloaded from the yeast genome website http://genome-www.stanford.edu/cellcycle. To study if the gene expression of the probe YJR152W is upregulated during a certain period of the time course of the experiment, we test the null hypothesis

$$H_0 : \mu_1 = \mu_2 = \cdots = \mu_{24}$$

versus

$$H_1 : \mu_i = \begin{cases} \mu, & 1 \leq i \leq k_1 \\ \mu + \beta(i - k_1), & k_1 + 1 \leq i \leq k_2 \\ \mu, & k_2 + 1 \leq i \leq 24 \end{cases}.$$

The Bayesian approach is applied to the expression of the DAL5 gene, the maximum posterior value of Equation (9.6) occurs at $\widehat{k}_1 = 11$, and

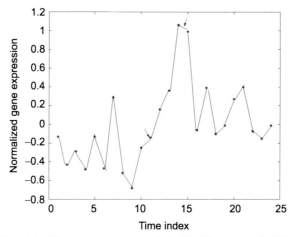

Fig. 9.2 The plot of the gene expression of the DAL5 gene (probe ID: YJR152W)

$\widehat{k}_2 = 14$. These change point location estimates reflect the reality shown in the expression of the DAL5 gene very well. The expression of this DAL5 gene on the 24 observed time points is given in Figure 9.2 and the two changes are pointed to by arrows. Biologically, it is known that the DAL5 gene encodes a necessary component of the allantoate transport system in *Saccharomyces cerevisiae* (Rai et al., 1987), upon a certain time of upregulation (the smooth change segment), and then it drops back (at the second change point) to a steady-state level. These biological explanations of the DAL5 gene exactly verify that the gene expression of the DAL5 gene is reasonably well described by our proposed SACP model.

9.3 The Epidemic Change Point Model for Exponential Distribution

Another frequently studied change point model is the so-called epidemic change point model, first studied by Levin and Kline (1985). Broemeling and Tsurumi (1987) described a number of applications of this model in econometrics. Later, Yao (1993) proposed some test statistics and large derivation approximations to the significance level and powers, for the normal distribution. Aly and Bouzar (1992) have proposed statistics for the exponential family of distributions. The epidemic change point model is popular and practical. For instance, certain flu may break out at a time point, the mortality rate may start to change at that breakout time, endure that change for a certain time, and then drop back to the original constant rate after the flu dies out.

Ramanayake and Gupta (2003) studied the epidemic change point model for a sequence of exponentially distributed random variables and they later

(2004) studied the epidemic change point model for the exponential family of distributions. The epidemic change point model for the exponential distribution can be formally introduced this way (Ramanayake and Gupta, 2003): Let $X_1, \ldots, X_n$ be a sequence of independent exponential random variables with density function,

$$f(x_i, \theta_i) = \frac{1}{\theta_i} e^{-x_i/\theta_i}, \qquad i = 1, \ldots, n,$$

where θ_i and x_i are positive real numbers. An epidemic change model with changes occurring during an unknown period of time (p, q) can be stated as an inference problem of testing the null hypothesis:

$$H_0 : \theta_i = \theta, \qquad i = 1, \ldots, n,$$

versus the alternative hypothesis

$$H_A : \theta_i = \begin{cases} \theta & i \leq p \\ \theta + \delta & p < i \leq q \\ \theta & q < i \leq n \end{cases},$$

where p, q are the unknown change points such that $1 \leq p < q \leq n$ and θ and δ are the unknown parameters such that $\theta, \delta > 0$.

A typical exponentially distributed data pattern with epidemic change is illustrated in Figure 9.3.

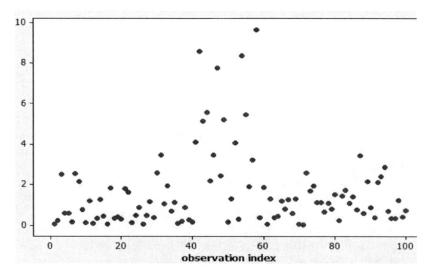

Fig. 9.3 A simulated sequence of observations from an exponential distribution with $\theta = 1, \delta = 4, p = 40, q = 60, n = 100$

Ramanayake and Gupta (2003) studied four test statistics, based on the likelihood ratio, for the detection of an epidemic change in a sequence of independent exponential variables. We present the detailed work of Ramanayake and Gupta (2003) in the following subsections.

9.3.1 A Likelihood-Ratio-Based Statistic T

Suppose that the change points (p, q) are fixed; then the likelihood function under H_0 above is $L_0(\theta; \mathbf{x}) = \theta^{-n} \exp((-S_n)/\theta)$, and the likelihood function H_A above is given by

$$L_A(\theta, \delta, p, q; \mathbf{x}) = \theta^{-n} \exp \frac{-S_n}{\theta} \left(\frac{\theta}{\theta + \delta} \right)^{q-p} \exp \left\{ (S_q - S_p) \left(\frac{1}{\theta} - \frac{1}{\theta + \delta} \right) \right\},$$

where $S_p = \sum_{i=1}^{p} X_i$, $S_q = \sum_{i=1}^{q} X_i$, and $S_n = \sum_{i=1}^{n} X_i$.

As the locations of the changes are usually unknown, we next assume that (p, q) has an equal chance to fall at any possible points $p = 1, \ldots, n - 2$ and $q = p + 1, \ldots, n - 1$. Then the marginal likelihood function under H_A is

$$f_A(\theta, \delta; \mathbf{x}) = \theta^{-n} \exp \frac{-S_n}{\theta} \prod_{p,q} \left(\frac{\theta}{\theta + \delta} \right)^{q-p} \exp \left\{ (S_q - S_p) \left(\frac{1}{\theta} - \frac{1}{\theta + \delta} \right) \right\},$$

and the log likelihood ratio under H_A, to that under H_0, is obtained as

$$\ln R(\theta, \delta, p, q; \mathbf{x}) = (q - p) \ln \left(\frac{\theta}{\theta + \delta} \right) + (S_q - S_p) \frac{1}{\theta} \frac{\delta}{\theta} \left(1 + \frac{\delta}{\theta} \right)^{-1}.$$

Therefore, the log likelihood ratio under H_A, to that under H_0 as $(\delta/\theta) \to 0^+$ can be expressed as

$$\ln L(\theta, \delta; \mathbf{x}) = -N - \sum_{q=2}^{n-1} \sum_{p=1}^{q-1} \left\{ (q - p) \ln \left(\frac{\theta}{\theta + \delta} \right) + (S_q - S_p) \frac{1}{\theta} \frac{\delta}{\theta} \left(1 + \frac{\delta}{\theta} \right)^{-1} \right\}$$

$$= -N + \frac{\delta}{\theta} \left[\sum_{q=2}^{n-1} \sum_{p=1}^{q-1} \left\{ -(q - p) + \frac{(S_q - S_p)}{\theta} \right\} \right] + o_p \left(\frac{\delta}{\theta} \right),$$

where $N = (n - 1)(n - 2)/2$. As the first term, $q - p$, in parentheses is a known constant, an equivalent test statistic was thus proposed as

$$T_0 = \sum_{q=2}^{n-1} \sum_{p=1}^{q-1} \frac{(S_q - S_p)}{\theta} = \frac{1}{\theta} \sum_{q=2}^{n-1} (n - i)(i - 1) X_i.$$

However, in many practical situations it is very unlikely that θ would be known. Thus Ramanayake and Gupta (2003) suggested that θ in the above expression be replaced by the maximum likelihood estimator of θ under H_0, that is, by $\bar{X} = (1/n) \sum_{i=1}^{n} X_i$, and denoted the modified statistic by

$$T = \sum_{i=2}^{n-1} \frac{(n-i)(i-1)X_i}{M \sum_{i=1}^{n} X_i},$$

where $M = (2/n) \sum_{i=1}^{n-1} (n-i)(i-1) = \frac{1}{3}(n-1)(n-2)$. Note that if we let

$$Y_i = \frac{X_i}{\sum_{i=1}^{n} X_i}, \qquad i = 1, \ldots, n-1,$$

then under the null hypothesis we have that $(Y_1, \ldots, Y_{n-1}) \sim$ a Dirichlet distribution, $D_{n-1}(1/2, \ldots, 1/2)$. Hence the moments of $(Y_1, \ldots, Y_{n-1})$ can be written as (according to Johnson and Kotz, 1972, Ch. 40, Sec. 5),

$$\mu_{r_1, \ldots, r_{n-1}} = E(Y_1^{r_1} Y_2^{r_2}, \ldots, Y_{n-1}^{r_{n-1}}) = \frac{\prod_{i=1}^{n-1} 1^{[r_i]}}{(n)^{[\sum_{i=1}^{n-1} r_i]}},$$

where $a^{[r]} = a(a+1) \cdots (a+r-1)$. According to this formula, the moments of Y_is can be obtained as follows.

$$\mu_1 = E(Y_i) = \frac{1}{n}$$

$$\mu_2 = E(Y_i^2) = \frac{3}{n(n+2)}$$

$$\mu_{11} = E(Y_i Y_j) = \frac{1}{n(n+2)}$$

$$\mu_3 = E(Y_i^3) = \frac{1}{n(n+2)(n+4)}$$

$$\mu_{21} = E(Y_i^2 Y_j) = \frac{15}{n(n+2)(n+4)}$$

$$\mu_{111} = E(Y_i Y_j Y_k) = \frac{1}{n(n+2)(n+4)}$$

$$\mu_4 = E(Y_i^4) = \frac{105}{n(n+2)(n+4)(n+6)}$$

$$\mu_{31} = E(Y_i^3 Y_j) = \frac{15}{n(n+2)(n+4)(n+6)}$$

$$\mu_{22} = E(Y_i^2 Y_j^2) = \frac{9}{n(n+2)(n+4)(n+6)}$$

$$\mu_{211} = E(Y_i^2 Y_j Y_k) = \frac{3}{n(n+2)(n+4)(n+6)}$$

$$\mu_{1111} = E(Y_i Y_j Y_k Y_l) = \frac{1}{n(n+2)(n+4)(n+6)}.$$

Therefore, when H_0 is true, the first moments $\mu_1(T), \mu_2(T)$, and the coefficients of skewness and kurtosis $\beta_1(T)$ and $\beta_2(T)$ of the statistic T are obtained as

$$\mu_1(T) = E(T) = \frac{1}{2}$$

$$\mu_2(T) = \text{Var}(T) = \frac{n+1}{10(n-1)(n-2)}$$

$$\beta_1(T) = \frac{\mu_3^2(T)}{\mu_2^3(T)} = \frac{160(n-4)^2}{49(n+1)(n-1)(n-2)}$$

$$= \frac{160}{49}n^{-1} + o(n^{-1})$$

$$\beta_2(T) = \frac{\mu_4(T)}{\mu_2^2(T)} = \frac{3}{7}\frac{7n^3 - 10n^2 - 103n + 250}{(n+1)(n-1)(n-2)}$$

$$= 3 + \frac{12}{7}n^{-1} + o(n^{-1}), \quad \text{as} \quad n \to \infty.$$

From the distribution of T it is noted that T takes values between 0 and 1. And from $\beta_1(T)$ and $\beta_2(T)$, it is noticed that the null distribution of T has positive skewness and kurtosis ($\gamma_2 = \beta_2 - 3$), which tend to zero as $n \to \infty$.

The Null Distribution of Statistic T

For the above-defined test statistic T, the null distribution needs to be derived when T is going to be used in practice as the test statistic for the epidemic change point model. Define the statistics, T_1, as

$$T_1 = \frac{T - 0.5}{\sqrt{\text{Var}(T)}},$$

and obviously, T_1 is the standardized statistic corresponding to statistic T. From the Lyapounov central limit theorem and Slutsky's theorem, one can get that the statistic T_1, under H_0 follows an asymptotic normal distribution as $n \to \infty$. The test statistic is based on the likelihood ratio for small changes in the ratio δ/θ, thus a test that rejects H_0 in favor of H_A that $\delta > 0$, for large values of T_1, will give the locally most powerful one-sided test as $\delta/\theta \to 0^+$.

Also a test that rejects H_0 for large $|T_1|$ is the locally most powerful unbiased test against the two-sided alternatives for small values of $|\delta/\theta|$.

The null cumulative distribution function (CDF), $F_{T_1}(x)$, of the test statistic T_1 was approximated in Ramanayake and Gupta (2003) using a three-term Edgeworth expansion (Johnson and Kotz, 1972) of T_1 as the following.

$$F_{T_1}(x) = \Phi(x) - \left\{ \frac{\sqrt{\beta_1(T_1)}}{6}(x^2 - 1) + \frac{1}{24}(\beta_2(T_1) - 3)(x^3 - 3x) \right.$$

$$\left. + \frac{1}{72}\beta_1(T_1)(x^5 - 10x^3 + 15x) \right\} \phi(x)$$

$$= \Phi(x) - \left\{ \frac{2}{21}\sqrt{\frac{10(n-4)^2}{(n+1)(n-1)(n-2)}}(x^2 - 1) \right.$$

$$+ \frac{1}{14}\frac{(n^2 - 24n + 59)}{(n+1)(n-1)(n-2)}(x^3 - 3x)$$

$$\left. + \frac{20(n-4)^2}{441(n+1)(n-1)(n-2)}(x^5 - 10x^3 + 15x) \right\} \phi(x),$$

where $\Phi(x)$ and $\phi(x)$ are CDF and pdf of the standard normal random distribution, respectively. The critical values c_α, according to the above Edgeworth expansion of the CDF of T_1 for the standardized statistic T_1, can be obtained according to $\int_{c_\alpha}^{\infty} f(x)dx = \alpha$, with the pdf $f(\cdot)$ being obtained using the CDF $F_{T_1}(\cdot)$ above. For selected moderate sample sizes, the critical values were tabulated in Ramanayake and Gupta (2003) and are redisplayed in Table 9.1.

Table 9.1 Approximate Critical Values of the T_1 Test

n	$c_{0.01}$	$c_{0.025}$	$c_{0.05}$	$c_{0.10}$	$c_{0.25}$
25	2.527	2.092	1.729	1.320	0.653
50	2.487	2.065	1.710	1.309	0.656
75	2.463	2.049	1.700	1.304	0.658
100	2.447	2.039	1.693	1.301	0.660
125	2.436	2.031	1.689	1.299	0.661
150	2.427	2.026	1.685	1.298	0.662
175	2.420	2.021	1.682	1.296	0.663
200	2.415	2.018	1.680	1.295	0.664
225	2.410	2.014	1.678	1.295	0.664
250	2.406	2.012	1.676	1.294	0.665
∞	2.326	1.960	1.645	1.282	0.674

Asymptotic Distribution of T Under the Alternative Hypothesis

Under the alternative hypothesis H_A, $X_1, \ldots, X_n$ follow independent exponential distributions with mean θ_i, where

$$
\theta_i = \begin{cases}
\theta & i \leq p \\
\theta + \delta & p + 1 \leq i \leq q, \delta > 0 \\
\theta & q + 1 \leq i \leq n
\end{cases}.
$$

For simplicity in notation, we let $a_i = (n - i)(i - 1)/M$, $i = 1, \ldots, n$, then the statistic T can be rewritten as

$$
T = \frac{\sum_{i=1}^{n} a_i X_i}{\sum_{i=1}^{n} X_i}.
$$

In the following theorem, we give the asymptotic nonnull distribution of the statistics T.

Theorem 9.2 *Suppose that H_A holds and as $n \to \infty$, $p/n \to \lambda_1$ and $q/n \to \lambda_2$ such that $0 < \lambda_1 < \lambda_2 < 1$. Then the statistic $(\sqrt{n}(T - \mu^*))/\sigma^*$ converges to a standard normal distribution as $n \to \infty$, where*

$$
\mu^* = \frac{1}{2} \left\{ \theta - \delta(\lambda_2 - \lambda_1)(2\lambda_2^2 - 3\lambda_2 + 2\lambda_1\lambda_2 - 3\lambda_1 + 2\lambda_1^2) \right\} \left\{ \theta + \delta(\lambda_2 - \lambda_1) \right\}^{-1},
$$

and

$$
\sigma^{*2} = \frac{3}{10} \{ \theta^2 + \delta(2\theta + \delta)(\lambda_2 - \lambda_1)
$$

$$
\cdot (-15\lambda_2^2\lambda_1 + 10\lambda_1\lambda_2 + 10\lambda_2^2 - 15\lambda_2^3 + 6\lambda_2^4 + 6\lambda_1^4 + 10\lambda_1^2 - 15\lambda_1^3
$$

$$
+ 6\lambda_1\lambda_2^3 + 6\lambda_1^2\lambda_2^2 + 6\lambda_2\lambda_1^2 - 15\lambda_1^2\lambda_2) \} \{ \theta + \delta(\lambda_2 - \lambda_1) \}^{-2}.
$$

Proof. From the Lyapounov central limit theorem we have:

$$
\frac{\left\{ \sum_{i=1}^{n} a_i X_i - \sum_{i=1}^{n} a_i \theta_i \right\}^{1/2}}{\left[\sum_{i=1}^{n} a_i^2 \theta_i^2 \right]} \xrightarrow{D} N(0, 1)
$$

and from the weak law of large of numbers we get

$$
\frac{\bar{X}}{E(\bar{X})} \xrightarrow{P} 1,
$$

where

$$
E(\bar{X}) = \frac{1}{n} \sum_{i=1}^{n} \theta_i = \theta + \frac{\delta}{n}(q - p).
$$

Therefore, from Slutsky's theorem, we obtain,

$$\frac{T - \mu_n}{\sigma_n} \xrightarrow{D} N(0, 1),$$

where

$$\mu_n = \frac{\sum_{i=1}^{n} a_i \theta_i}{\sum_{i=1}^{n} \theta_i}$$

$$= \frac{1}{2} \left[\theta - \frac{(q - p)\delta \left(2q^2 - 3nq + 2pq - 2 + 3n - 3np + 2p^2 \right)}{n} \cdot \frac{}{(n - 1)(n - 2)} \right]$$

$$\cdot \left[\theta + \frac{(q - p)\delta}{n} \right]^{-1}$$

and

$$\sigma_n{}^2 = \frac{\sum_{i=1}^{n} a_i{}^2 \theta_i{}^2}{\left[\sum_{i=1}^{n} \theta_i \right]^2}$$

$$= \frac{3}{10} \left[\frac{(n^2 - 2n + 2)}{n(n - 2)(n - 1)} \theta^2 + \frac{(q - p)}{n} (2\delta\theta + \delta^2) \right.$$

$$\times (-10pq + 10npq + 10nq^2 + 15np - 10p^2 + 10np^2 - 15nq^2p - 10q^2$$

$$+ 15nq - 10n + 4 + 10n^2pq + 5n^2 - 15n^2q + 10n^2q^2 - 15nq^3 + 6q^4$$

$$+ 6p^4 - 15n^2p + 10n^2p^2 - 15np^3 + 6pq^3 + 6p^2q^2 + 6qp^3 - 15np^2q)$$

$$\left. \cdot (n - 1)^{-2}(n - 2)^{-2} \right] \left[\theta + \frac{(q - p)\delta}{n} \right]^{-2}.$$

Finally, if we assume that as $n \to \infty$, $q/n = (q(n))/n \to \lambda_2$ and $p/n = (p(n))/n \to \lambda_1$, where $0 < \lambda_1 < \lambda_2 < 1$, we get the desired result.   □

After the null and nonnull distributions of the test statistic T are established, we examine a property of the test and present it in the following Theorem 9.3.

Theorem 9.3 *Test based on T is consistent for testing H_0 versus H_A if*

$$3(\lambda_1 + \lambda_2) - 2(\lambda_1\lambda_2 + \lambda_1{}^2 + \lambda_2{}^2) > 1.$$

Proof. Under the null hypothesis we have that,

$$\frac{\sqrt{n}(T - 0.5)}{\sqrt{0.3}} \xrightarrow{D} N(0, 1).$$

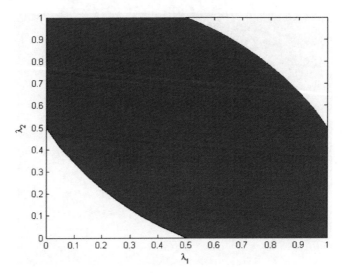

Fig. 9.4 The plot of $3(\lambda_1 + \lambda_2) - 2(\lambda_1\lambda_2 + \lambda_2{}^2 + \lambda_1{}^2) > 1$, indicated by the shaded area

Now let α be the level of the test; then we have that the power of the test at level α can be expressed as

$$\beta(\alpha) = P\left\{ \frac{\sqrt{n}(T - \mu^*)}{\sigma^*} \geq \frac{\sqrt{0.3}}{\sigma^*} z_\alpha - \frac{\sqrt{n}}{\sigma^*}(\mu^* - 0.5) \right\} \to 1,$$

as $n \to \infty$, and if $\mu^* - 0.5 > 0$, we have $\iff 3(\lambda_1 + \lambda_2) - 2(\lambda_1\lambda_2 + \lambda_2{}^2 + \lambda_1{}^2) > 1$, for $\delta > 0$, where z_α is given by the equation, $\alpha = \int_{z_\alpha}^{\infty} \phi(x)dx$. Notice that the condition $3(\lambda_1 + \lambda_2) - 2(\lambda_1\lambda_2 + \lambda_2{}^2 + \lambda_1{}^2) > 1$ holds if λ_1 and λ_2 occur in the middle of the sequence and the shaded area in Figure 9.4. Illustrate the region corresponding to $3(\lambda_1 + \lambda_2) - 2(\lambda_1\lambda_2 + \lambda_2{}^2 + \lambda_1{}^2) > 1$.   □

Note that the parameter space on which the asymptotic nonnull distribution of T converges as indicated in Theorem 9.2 and on which the test statistic T is also consistent as indicated in Theorem 9.3 above is the region R, where $R = \{(\lambda_1, \lambda_2) : 0 < \lambda_1 < \lambda_2 < 1, 3(\lambda_1 + \lambda_2) - 2(\lambda_1\lambda_2 + \lambda_2{}^2 + \lambda_1{}^2) > 1\}$. This region R is illustrated in Figure 9.5.

9.3.2 Likelihood-Ratio Test Statistic

In this subsection, we present the likelihood-ratio test statistics. Note that under H_0 the maximum likelihood estimator (MLE) of θ is clearly

$$\hat{\theta}_0 = \frac{S_n}{n},$$

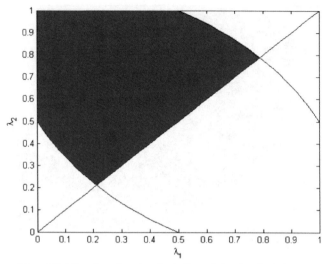

Fig. 9.5 The plot of region R, indicated by the shaded area

and under H_A, the MLE of $(\theta + \delta)$ is easily obtained as

$$\widehat{\theta + \delta} = \frac{S_q - S_p}{q - p}$$

and the MLE of θ is

$$\hat{\theta} = \frac{S_n - (S_q - S_p)}{n - (q - p)}.$$

Hence, the log likelihood ratio of H_A over H_0, for fixed (p, q) is

$$\ln \Lambda_{pq} = n \ln \bar{X}_n - (q - p) \ln \bar{X}_{pq} - (n - q + p) \ln \bar{X}^*_{pq},$$

where,

$$\bar{X}_{pq} = \frac{S_q - S_p}{q - p}, \qquad \bar{X}^*_{pq} = \frac{S_n - (S_q - S_p)}{n - (q - p)} \quad \text{and} \quad \bar{X}_n = \frac{S_n}{n}.$$

Therefore, the likelihood-ratio test (LRT) statistic for unknown (p, q) such that $1 \le p < q \le n$ can be written as $T_2 = \max_{1 \le p < q \le n} \ln \Lambda_{pq}$. In the following, we give the asymptotic null distribution of the LRT T_2.

In obtaining the asymptotic null distribution of T_2 we use the following result.

Theorem 9.4 *Suppose $Y_1, \ldots, Y_n$ are independent random variables from a normal distribution with mean 0 and variance 1. Set $S(k) = \sum_{i=1}^{k} Y_i$ where $i = 1, \ldots, n$. Define*

$$Z_{p,q}^{*\,2} = \frac{\left\{S(q) - S(p) - \frac{q-p}{n}S(n)\right\}^2}{(q-p)\left(1 - \frac{q-p}{n}\right)}$$

for all $1 \le p < q \le n$. Let $n \to \infty$; then for $b = \sqrt{n}c$, with $c \in (0,1)$ fixed, we have

$$P_0\left\{\max_{1 \le p < q \le n} |Z_{pq}^*| \ge b\right\} \sim \frac{b^3}{2\sqrt{2\pi}}(1 - c^3)^{(n/2)-3}\int_0^1 \frac{\nu^2\left[\frac{c}{t(1-t)(1-c^2)}\right]}{t^2(1-t)}\,dt.$$

Here

$$\nu(x) = 2x^{-2}\exp\left\{-2\sum_{n=1}^{\infty}\frac{1}{n}\Phi\left(-x\frac{\sqrt{n}}{2}\right)\right\}, \qquad x > 0.$$

The function $\nu(x)$ can be approximated by $\exp(-0.583x) + o(x^2)$ as $x \to 0$ and Φ denotes the standard normal distribution function.

Proof. See Yao (1993) and Siegmund (1988a). □

Lemma 9.5 *If H_0 holds then $\max_{1 \le p < q \le n} |2\ln\Lambda_{p,q} + Z_{p,q}^{*\,2}| = o_p(1)$.*

Proof. By a second-order Taylor series expansion of $\ln(\bar{X}_n), \ln(\bar{X}_{p,q})$, and $\ln(\bar{X}_{p,q}^*)$ around θ we get,

$$2\ln\Lambda_{p,q} = \frac{1}{\theta^2}\left\{-n(\bar{X}_n - \theta)^2 + (q-p)(\bar{X}_{p,q} - \theta)^2\right.$$

$$\left. + (n-q+p)(\bar{X}_{p,q}^* - \theta)^2\right\} + o_p(1)$$

$$= \frac{(q-p)n}{(n-q+p)}\left\{\left(\frac{\bar{X}_{p,q} - \theta}{\theta}\right) - \left(\frac{\bar{X}_n - \theta}{\theta}\right)\right\}^2 + o_p(1)$$

$$= Z_{p,q}^{*2} + o_p(1)$$

Now by the central limit theorem we get the desired result. □

Theorem 9.6 *Suppose that $n \to \infty$; then for $b = c\sqrt{n}$, with $c \in (0,1)$ fixed, we have*

$$P_0\left\{\max_{1 \le p < q \le n} -2\ln\Lambda_{pq} \ge b^2\right\} \sim \frac{b^3}{2\sqrt{2\pi}}(1 - c^3)^{(n/2)-3}\int_0^1 \frac{\nu^2\left[\frac{c}{t(1-t)(1-c^2)}\right]}{t^2(1-t)}\,dt.$$

Proof. Follows as a consequence of Theorem 3. □

Next we consider the two modified likelihood ratio test statistics proposed by Aly and Bouzar (1992). These two statistics are

$$T_3 = \max_{1 \le p < q \le n}\left[\frac{(q-p)(n-q+p)}{n^2}2\ln\Lambda_{pq}\right]^{1/2},$$

and $T_4 = (2/n^4)\sum_{1 \le p < q \le n}(q-p)(n-q+p)\ln\Lambda_{pq}$.

From Theorem 3.1 of Aly and Bouzar (1992), we have that,

$$T_3 \xrightarrow{D} \sup_{0 \le t < s \le 1} \|B(s) - B(t)\| = E_1$$

and

$$T_4 \xrightarrow{D} \int_{0 \le t < s \le 1} \|B(s) - B(t)\| = E_2,$$

where $B(\cdot)$ is a Brownian bridge process. Hence,

$$P(E_1 > x) = 2 \sum_{r=1}^{\infty} (4r^2 x^2 - 1) \exp\{-2r^2 x^2\}, \qquad x \ge 0,$$

and

$$P(E_2 \le x) = \sqrt{\frac{8}{\pi x}} \sum_{j=0}^{\infty} \exp\left\{\frac{2}{x}(j + 0.5)^2\right\}.$$

Table 9.2 gives the critical values of the distribution of E_1 and E_2 for $\alpha = 0.01, 0.025, 0.05, 0.10, 0.25$.

Table 9.2 Critical Values of the E_1 and E_2 Tests

α	c_{α, E_1}	c_{α, E_2}
0.01	2.0009	1.0737
0.025	1.8624	0.8880
0.05	1.7473	0.7475
0.10	1.6196	0.6070
0.25	1.4205	0.4210

9.3.3 Power Comparisons of the Four Tests

Table 9.3 compares the powers of the statistics T_1, T_2, T_3, and T_4 with the result of 5000 repetition Monte Carlo experiments. In order to keep the table to a reasonable size, only the case of sample size $n = 50$ and $\theta = 1$ with a significance level $\alpha = 0.05$ is considered. Based on the simulation results of Table 9.3, we see that the test based on the likelihood-ratio type statistic T_1, outperforms the others when (p, q) are in the middle of the sequence. This difference is more significant for small δ values. But if both p and q occur either very early or late in the sequence then all the other three tests do better than T_1. This difference is more significant for large values of δ. On the other hand, T_2, T_3, T_4 seem to be quite similar to one another in terms of power. Other simulation results, not reported here, show that these conclusions are unchanged over a range of sample sizes and (δ, θ) values.

Table 9.3 Power of the Tests Based on T_1, T_2, T_3, T_4 with Significance Level of 0.05, Right-Tailed

		$\delta = 1$				$\delta = 3$			
a	b	T_1	T_2	T_3	T_4	T_1	T_2	T_3	T_4
4	20	0.0570	0.3734	0.4340	0.2748	0.0806	0.9516	0.9652	0.9110
4	40	0.4448	0.2970	0.2496	0.1926	0.8594	0.8936	0.7302	0.7792
8	24	0.2464	0.3652	0.4214	0.2586	0.6860	0.9472	0.9636	0.9046
8	40	0.6132	0.3912	0.3674	0.2444	0.9856	0.9614	0.9148	0.8886
12	24	0.3034	0.2894	0.3488	0.2382	0.8006	0.8790	0.9162	0.8474
12	40	0.6780	0.4434	0.4378	0.2732	0.9964	0.9796	0.9686	0.9272
16	24	0.2644	0.1806	0.2332	0.1742	0.7216	0.7004	0.7770	0.7078
16	40	0.6322	0.4492	0.4718	0.2802	0.9906	0.9828	0.9810	0.9328
20	28	0.3054	0.1802	0.2254	0.1604	0.7886	0.7014	0.7782	0.7032
20	40	0.5238	0.4286	0.4774	0.2934	0.9690	0.9766	0.9794	0.9324
24	32	0.2874	0.1838	0.2266	0.1562	0.7656	0.6944	0.7714	0.7062
24	40	0.3810	0.3692	0.4248	0.2612	0.8756	0.9538	0.9674	0.9116
28	36	0.2138	0.1838	0.2296	0.1712	0.6352	0.6960	0.7832	0.7190
28	44	0.1308	0.3674	0.4246	0.2666	0.3348	0.9522	0.9656	0.9056
32	40	0.0476	0.0974	0.1096	0.1008	0.0656	0.8830	0.9270	0.8560

Thus it is recommended that one should use T_1 if the epidemic duration falls around the center of the sequence and use any one of T_2, T_3, T_4 otherwise. However, statistic T_3 is slightly preferred in this situation because it performs better than T_2 and T_4, when the epidemic duration occurs either very early or late in the sequence.

In Ramanayake and Gupta (2003), two data analyses were given to illustrate the use of the proposed four test statistics here. The two datasets used were aircraft arrival times collected from a low-altitude transitional control sector for the period from noon through 8 PM on April 30, 1969 used in Hsu (1979), and the Stanford heart transplant data taken from *The Statistical Analysis of Failure Time Data* by Kalbfleisch and Prentice (1980, Appendix I, pp. 230–232).

9.4 The Epidemic Change Point Model for the Exponential Family

Ramanayake and Gupta (2004) further studied the epidemic change point model for the exponential family. This work provided thorough investigation of the statistical inference problem associated with such an epidemic change point model. We provide the details of the work of Ramanayake and Gupta (2004) here in this section.

Let $X_1, \ldots, X_n$ be a sequence of independent random variables with the density function that belongs to the one-parameter exponential family of the specific form:

$$f(x_i, \theta_i) = \exp\{T(x_i)\theta_i - A(\theta_i) + S(x_i)\}I\{x_i \in C\} \qquad i = 1, \ldots, n, \quad (9.10)$$

where $x_i, \theta_i, T(x_i) \in \Re$ for all $i = 1, \ldots, n$, and $C \subseteq \Re$. Here our interest is to test the null hypothesis of "no change" against the epidemic change in the parameter θ_i. The null and alternative hypotheses can be described in more formal terms as

$$H_0 : \theta_i = \theta_0, \qquad i = 1, \ldots, n$$

$$H_A : \exists \quad p, q \in \mathbf{Z} \quad \text{such that} \quad 1 \le p < q \le n \quad \text{and}$$

$$\theta_i = \begin{cases} \theta_0 & i \le p \\ \theta^* & p < i \le q \\ \theta_0 & q < i \le n \end{cases},$$

where θ_0, θ^* are fixed such that $\theta_0, \theta^* \in \text{int}(\Theta)$.

Using the ideas of Gombay and Hovarth (1996) for the single change in the exponential family, Ramanayake and Gupta (2004) generalized the results to the epidemic change point model for the exponential family. In the following sections, we give the details of the later study.

9.4.1 Derivation of the LRT Statistic

When it comes to the investigation of the hypothesis testing problem for a change point model, it is often convenient to use the likelihood-ratio procedure (LRP) test statistic (see Lehmann, 1986), as mentioned in previous chapters. Although others used this LRT without distinguishing it from the generalized likelihood ratio (GLR) based on the context, we now interchange the terms of LPT and GMLR and in general refer to it as the likelihood-ratio test (LRT) according to the context.

If $(p = p^*, q = q^*)$ are known, the generalized maximum likelihood ratio (GMLR) for testing H_0 against H_A can be written as

$$\Lambda_{p,q} = \frac{\left[\sup_{\theta_0} \prod_{i=1}^{p} f(x_i, \theta_0)\right]\left[\sup_{\theta^*} \prod_{i=p+1}^{q} f(x_i, \theta^*)\right]\left[\sup_{\theta_0} \prod_{i=q+1}^{n} f(x_i, \theta_0)\right]}{\sup_{\theta_0} \prod_{i=1}^{n} f(x_i, \theta_0)}$$

But because p^*, q^* are unknown, we reject the null hypothesis for large values of the statistic,

$$Q_n = \max_{1 \le p < q \le n} \{2 \ln \Lambda_{p,q}\}. \qquad (9.11)$$

Next we define $\hat{\hat{\theta}}_0$ as the maximum likelihood estimator (MLE) of θ_0 under H_0. Let $\hat{\theta}_0$ and $\hat{\theta}^*$ be the MLEs of θ_0 and θ^* under H_A, respectively. Then the logarithm of the GLR for known (p, q) can be written as

$$\ln \Lambda_{p,q} = \sum_{i=1}^{p} \ln f(x_i, \hat{\theta}_0) + \sum_{i=q+1}^{n} \ln f(x_i, \hat{\theta}_0)$$

$$+ \sum_{i=p+1}^{q} \ln f(x_i, \hat{\theta}^*) - \sum_{i=1}^{n} \ln f(x_i, \hat{\hat{\theta}}_0)$$

$$= \hat{\theta}_0 \sum_{i=1}^{p} T(x_i) + \sum_{i=1}^{p} S(x_i) - A(\hat{\theta}_0)p + \hat{\theta}_0 \sum_{i=q+1}^{n} T(x_i) + \sum_{i=q+1}^{n} S(x_i)$$

$$- A(\hat{\theta}_0)(n-q) + \theta_0 \sum_{i=p+1}^{q} T(x_i) + \sum_{i=p+1}^{q} S(x_i) - A(\hat{\theta}^*)(q-p)$$

$$- \hat{\theta}^* \sum_{i=1}^{n} T(x_i) - \sum_{i=1}^{n} S(x_i) + A(\hat{\hat{\theta}}_0)n. \tag{9.12}$$

Next for simplicity in notation, we set

$$B_{p,q} = \frac{1}{(q-p)} \sum_{i=p+1}^{q} T(x_i)$$

$$B_{p,q}^* = \frac{1}{(n-q+p)} \left\{ \sum_{i=1}^{p} T(x_i) + \sum_{i=q+1}^{n} T(x_i) \right\} \quad \text{and} \quad B_n = B_{1,n}.$$

Define $A'(\theta) = (\partial/\partial\theta)A(\theta)$. In order to get a simpler expression for $\Lambda_{p,q}$ we assume the following regularity conditions.

C1. inv $A'(\theta)$, the unique inverse of $A'(\theta)$ exists for each $\theta \in \Theta$.
C2. $\exists\, \epsilon > 0$ such that H''' exists and H'' is positive on

$$T^* = \{\tau : |\tau - (s\tau_1 + (1-s)\tau_2)| \le \epsilon, \quad \text{for some} \quad 0 \le s \le 1\}.$$

C3. $H'(\tau_2)(\tau_1 - \tau_2) + H(\tau_2) - H(\tau_1) < 0$.
C4. $H'(\tau_1)(\tau_2 - \tau_1) + H(\tau_1) - H(\tau_2) < 0$.
C5. For every ϵ such that $0 < \epsilon < 1/2$,

$$\sup_{\epsilon \le s \le 1-\epsilon} \{H(s\tau_1 + (1-s)\tau_2) - sH(\tau_1) - (1-s)H(\tau_2)\} < 0.$$

Here $H(x) = x\, \text{inv}A'(x) - A[\text{inv}A'(x)]$, $\tau_1 = A'(\theta_0)$, $\tau_2 = A'(\theta^*)$, and $\tau = A'(\theta)$. Note that under H_A, $E(B_{p,q}) = \tau_2$ and $E(B_{p,q}^*) = \tau_1$. Thus we have that $\hat{\theta}_0 = \text{inv}A'(B_{p,q}^*)$, $\hat{\theta}^* = \text{inv}A'(B_{p,q})$, and $\hat{\hat{\theta}}_0 = \text{inv}A'(B_n)$. Now in terms of the new notation we can rewrite (9.12) as

$$\ln \Lambda_{p,q} = \hat{\theta}_0(n - q + p)B_{p,q}^* - (n - q + p)A[\operatorname{inv}A'(B_{p,q}^*)] + \hat{\theta}^*(q - p)B_{p,q}$$

$$- (q - p)A[\operatorname{inv}A'(B_{p,q})] - \hat{\theta}_0 nB_n - nA[\operatorname{inv}A'(B_n)] \qquad (9.13)$$

$$= (q - p)H(B_{p,q}) + (n - q + p)H(B_{p,q}^*) - nH(B_n).$$

Next assume that $p^* = p^*(n)$ and $q^* = q^*(n)$ are such that

$$\lim_{n \to \infty} \frac{p^*(n)}{n} = \lambda_1 \quad \text{and} \quad \lim_{n \to \infty} \frac{q^*(n)}{n} = \lambda_2,$$

where $0 < \lambda_1 < \lambda_2 < 1$. Also let,

$$V_{p,q} = \ln \Lambda_{p,q} - \ln \Lambda_{p^*,q^*}, \qquad 1 \le p < q \le n, \qquad (9.14)$$

and

$$\mu^* = (q^* - p^*)H(\tau_2) + (n - q^* + p^*)H(\tau_1). \qquad (9.15)$$

Finally, we denote $\mu_{p,q}$ as the following.
If $1 < p < q \le p^* < q^* < n$,

$$\mu_{p,q} = (q - p)H(\tau_1) + (n - q + p)H\left(\frac{q^* - p^*}{n - q + p}\tau_2 + \frac{n - q + p - q^* + p^*}{n - q + p}\tau_1\right)$$

$$- (q^* - p^*)H(\tau_2) - (n - q^* + p^*)H(\tau_1),$$

if $1 < p \le p^* < q < q^* < n$,

$$\mu_{p,q} = (q - p)H\left(\frac{p^* - p}{q - p}\tau_1 + \frac{q - p^*}{q - p}\tau_2\right) + (n - q + p)H\left(\frac{q^* - q}{n - q + p}\tau_2\right.$$

$$\left. + \frac{n + p - q^*}{n - q + p}\tau_1\right) - (q^* - p^*)H(\tau_2) - (n - q^* + p^*)H(\tau_1),$$

if $1 < p \le p^* < q^* \le q < n$,

$$\mu_{p,q} = (q - p)H\left(\frac{q^* - p^*}{q - p}\tau_2 + \frac{q - p - q^* + p^*}{q - p}\tau_1\right) + (n - q + p)H(\tau_1)$$

$$- (q^* - p^*)H(\tau_2) - (n - q^* + p^*)H(\tau_1),$$

if $1 < p^* \le p < q \le q^* < n$,

$$\mu_{p,q} = (q - p)H(\tau_2) + (n - q + p)H\left(\frac{n - q^* + p^*}{n - q + p}\tau_1 + \frac{q^* - p^* - q + p}{n - q + p}\tau_2\right)$$

$$- (q^* - p^*)H(\tau_2) - (n - q^* + p^*)H(\tau_1),$$

if $1 < p^* \leq p < q^* \leq q < n$,

$$\mu_{p,q} = (q - p)H\left(\frac{q - q^*}{q - p}\tau_1 + \frac{q^* - p}{q - p}\tau_2\right) + (n - q + p)H\left(\frac{p - p^*}{n - q + p}\tau_2\right.$$

$$+ \frac{n - q + p^*}{n - q + p}\tau_1\right) - (q^* - p^*)H(\tau_2) - (n - q^* + p^*)H(\tau_1),$$

and if $1 < p^* < q^* \leq p < q < n$,

$$\mu_{p,q} = (q - p)H(\tau_1) + (n - q + p)H\left(\frac{q^* - p^*}{n - q + p}\tau_2 + \frac{n - q + p - q^* + p^*}{n - q + p}\tau_1\right)$$

$$- (q^* - p^*)H(\tau_2) - (n - q^* + p^*)H(\tau_1). \tag{9.16}$$

With all of the above preparations, we present the investigation on the properties of the LRT statistic Q_n in the following subsections.

9.4.2 Asymptotic Null Distribution of the Statistic Q_n

To obtain the asymptotic null distribution of Q_n we first prove the following lemmas.

Lemma 9.7 *Assume that conditions C1–C2 hold. Then under H_0, we have*

$$\max_{1 \leq p < q \leq n}\left|\frac{2\ln \Lambda_{p,q}}{A''(\theta_0)} - H''(\theta_0)Z_{p,q}^{*}{}^2\right| = o_p(1).$$

Proof. The second-order Taylor series expansion of $\ln \Lambda_{p,q}$ gives

$$\ln \Lambda_{p,q} = (q - p)\left\{H(\theta_0) + H'(\theta_0)(B_{p,q} - \theta_0) + \frac{1}{2}H''(\theta_0)(B_{p,q} - \theta_0)^2\right\}$$

$$+ (n - q + p)\left\{H(\theta_0) + H'(\theta_0)(B_{p,q}^* - \theta_0)\right.$$

$$- n\left\{H(\theta_0) + \frac{1}{2}H''(\theta_0)(B_{p,q}^* - \theta_0)^2\right\}$$

$$+ H'(\theta_0)(B_n - \theta_0) + \frac{1}{2}H''(\theta_0)(B_n - \theta_0)^2\right\} + o_p(1),$$

as $n(B_n - \theta_0)^{2+\gamma} = [\sqrt{n}(B_n - \theta_0)]^2(B_n - \theta_0)^\gamma$, $\gamma \geq 1$ converges in probability to 0, by the central limit theorem (CLT) and the weak law of large numbers, we can write

$$2 \ln \Lambda_{p,q} = H''(\theta_0)\{(q - p)(B_{p,q} - \theta_0)^2 + (n - q + p)(B_{p,q}^* - \theta_0)^2$$
$$- n(B_n - \theta_0)^2\} + o_p(1).$$

Next because $(q - p)B_{p,q} + (n - q + p)B_{p,q}^* = nB_n$, we can write

$$(B_{p,q}^* - \theta_0)^2 = \left\{ \frac{nB_n - (q - p)B_{p,q}}{n - q + p} - \theta_0 \right\}^2$$

$$= \left\{ \frac{n}{n - q + p}(B_n - \theta_0) - \frac{q - p}{n - q + p}(B_{p,q} - \theta_0) \right\}^2$$

$$= \left(\frac{n}{n - q + p} \right)^2 (B_n - \theta_0)^2 + \left(\frac{q - p}{n - q + p} \right)^2 (B_{p,q} - \theta_0)^2$$

$$- \frac{2n(q - p)}{(n - q + p)^2}(B_n - \theta_0)(B_{p,q} - \theta_0).$$

Hence we have that

$$2 \ln \Lambda_{p,q} = H''(\theta_0) \left\{ (q - p)(B_{p,q} - \theta_0)^2 + \frac{n^2}{n - q + p}(B_n - \theta_0)^2 \right.$$

$$+ \frac{(q - p)^2}{n - q + p}(B_{p,q} - \theta_0)^2 - \frac{2(q - p)n}{n - q + p}(B_n - \theta_0)(B_{p,q} - \theta_0)$$

$$\left. - n(B_n - \theta_0)^2 \right\} + o_p(1)$$

$$= H''(\theta_0) \frac{n(q - p)}{n - q + p} \{(B_{p,q} - \theta_0)^2 + (B_n - \theta_0)^2$$

$$- 2(B_n - \theta_0)(B_{p,q} - \theta_0)\} + o_p(1). \tag{9.17}$$

Next note that we can write

$$Z_{p,q}^* = \frac{\left\{ \frac{S(q) - S(p)}{q - p} - \frac{S(n)}{n} \right\}^2}{\frac{n - q + p}{(q - p)n}}$$

$$= \frac{n(q - p)}{n - q + p} \left\{ \left(\frac{S(q) - S(p)}{q - p} \right)^2 + \left(\frac{S(n)}{n} \right)^2 \right.$$

$$\left. - 2 \left(\frac{S(q) - S(p)}{q - p} \right) \left(\frac{S(n)}{n} \right) \right\}.$$

Now by the weak law of large numbers and (9.17), we have the result. $\square$

Theorem 9.8 *Suppose $Y_1, \ldots, Y_n$ are independent random variables from a normal distribution with mean 0 and variance 1. Set $S(k) = \sum_{i=1}^{k} Y_i$ where $k = 1, \ldots, n$. Define*

$$Z_{p,q}^{*}{}^{2} = \frac{\left\{ S(q) - S(p) - \frac{q-p}{n} S(n) \right\}^2}{(q - p)\left(1 - \frac{q-p}{n}\right)}$$

for all $1 \leq p < q \leq n$. Then if $c = b/\sqrt{n}$ converges to a fixed constant between 0 and 1, as $n \to \infty$ and $b \to \infty$ we have

$$P_0 \left\{ \max_{1 \leq p < q \leq n} Z_{p,q}^{*} \geq b \right\} \sim \frac{b^3}{2\sqrt{2\pi}} (1 - c^3)^{(n/2)-3} \int_0^1 \frac{\nu^2 \left[\frac{c}{t(1-t)(1-c^2)} \right]}{t^2 (1 - t)} dt.$$

Here $\nu(x) = 2x^{-2} \exp\{ -2 \sum_{n=1}^{\infty} (1/n) \Phi(-x(\sqrt{n}/2)) \}$, for $x > 0$. The function $\nu(x)$ can be approximated by $\exp(-0.583x) + o(x^2)$ as $x \to 0$ and Φ denotes the standard normal distribution function.

Proof. See Yao (1993) or Siegmund (1988a, b). □

Finally, we can obtain the asymptotic null distribution of the statistic Q_n in the following theorem.

Theorem 9.9 *Assume that conditions C1–C2 hold. Then under H_0, if $c = b/\sqrt{n}$ converges to a fixed constant between 0 and 1, as $n \to \infty$ and $b \to \infty$ we have*

$$P_0 \left\{ \max_{1 \leq p < q \leq n} \frac{2 \ln \Lambda_{p,q}}{A''(\theta_0) H''(\theta_0)} \geq b^2 \right\}$$

$$\sim \frac{b^3}{2\sqrt{2\pi}} (1 - c^3)^{(n/2)-3} \int_0^1 \frac{\nu^2 \left[\frac{c}{t(1-t)(1-c^2)} \right]}{t^2 (1 - t)} dt.$$

Proof. Follows as a consequence of Theorem 9.7 and Lemma 9.8. □

9.4.3 Asymptotic Behavior of the MLEs of the Change Points

If H_0 does not hold, we may want to estimate the locations and the duration of the epidemic change. Let $\hat{p}$ and $\hat{q}$ be the maximum likelihood estimators of p and q, respectively; then essentially it means that $\hat{p}$ and $\hat{q}$ are such that:

$$(\hat{p}, \hat{q}) = \min\{ (p, q) : Q_n = 2 \ln \Lambda_{p,q} \}. \tag{9.18}$$

The following major theorem gives the property of the change point location estimators $(\hat{p}, \hat{q})$.

Theorem 9.10 *Assume that conditions C1–C5 hold. Then under H_A,*

$$|\hat{p} - p^*| = O_p(1) \quad and \quad |\hat{q} - q^*| = O_p(1).$$

In order to prove this major theorem, we first need to establish the following lemmas.

Lemma 9.11 *If $1 < p \le p^* < q \le q^* \le n$ and the conditions of Theorem 9.10 are satisfied, then $\exists c_1, c_2 < 0$ such that $\mu_{p,q} \le c_1(p^* - p) + c_2(q^* - q)$.*

Proof. If $1 < p \le p^* < q \le q^* \le n$, then by (9.16) we have that

$$\mu_{p,q} = (q-p)H(\nu_1) + (n-q+p)H(\nu_2) - (q^* - p^*)H(\tau_2) - (n - q^* + p^*)H(\tau_1),$$

where

$$\nu_1 = \frac{p^* - p}{q - p}\tau_1 + \frac{q - p^*}{q - p}\tau_2 \quad and \quad \nu_2 = \frac{q^* - q}{n - q + p}\tau_2 + \frac{n + p - q^*}{n - q + p}\tau_1.$$

Next define

$$\mu_{1,pq} = (q-p)H(\nu_1) - (p^* - p)H(\tau_1) - (q - p^*)H(\tau_2)$$
$$\mu_{2,pq} = (n - q + p)H(\nu_2) - (q^* - q)H(\tau_2) - (n - q^* + p)H(\tau_1).$$

Then we have $\mu_{p,q} = \mu_{1,pq} + \mu_{2,pq}$. Now by the Taylor series expansion of $H(\nu_1)$ around τ_2 we get

$$\mu_{1,pq} = (q-p)\left\{ H(\tau_2) + H'(\tau_2)(\nu_1 - \tau_2) + \frac{1}{2}H''(\tau_2^*)(\nu_1 - \tau_2)^2 \right\}$$
$$- (p^* - p)H(\tau_1) - (q - p^*)H(\tau_2),$$

where τ_2^* is on the interval connecting τ_2 and ν_1. Next, by the continuity of H'' we can find a constant $c > 0$ such that

$$\left| (q-p)\frac{1}{2}H''(\tau_2^*)(\tau_2 - \nu_1)^2 \right| = \left| \frac{1}{2}H''(\tau_2^*)\frac{(p^* - p)^2}{(q - p)}(\tau_1 - \tau_2)^2 \right|$$
$$\le c\,\frac{(p^* - p)^2}{(q - p)}.$$

Thus we have

$$\mu_{1,pq} \le (p^* - p)\left\{ H(\tau_2) - H(\tau_1) + H'(\tau_2)(\tau_1 - \tau_2) + c\left(\frac{p^* - p}{q - p}\right) \right\}.$$

Now we can choose α small enough such that for $n\alpha \le p \le p^*$,

$$c\left(\frac{p^* - p}{q - p}\right) < \frac{1}{2}|H(\tau_2) - H(\tau_1) + H'(\tau_2)(\tau_1 - \tau_2)|.$$

Thus if $n\alpha \leq p \leq p^*$, it is easy to see that

$$\mu_{1,pq} \leq \frac{1}{2}(p^* - p)\{H(\tau_2) - H(\tau_1) + H'(\tau_2)(\tau_1 - \tau_2)\}.$$

Now by C4 we have that $\mu_{1,pq} \leq (p^* - p)c_1^*$ if $n\alpha \leq p \leq p^*$. Next if $1 \leq p \leq n\alpha$, from C5 we get

$$\mu_{1,pq} = (q - p)\left\{H\left(\frac{p^* - p}{q - p}\tau_1 + \frac{q - p^*}{q - p}\tau_2\right) - \frac{p^* - p}{q - p}H(\tau_1) - \frac{q - p^*}{q - p}H(\tau_2)\right\}$$

$$\leq (q - p)\sup_{\frac{p^* - n\alpha}{q - n\alpha} \leq s \leq \frac{p^*}{q}}\{H(s\tau_1 + (1 - s)\tau_2) - sH(\tau_1) - (1 - s)H(\tau_2)\}$$

$$\leq (q - p)\ c_2^*.$$

Thus we have $\mu_{1,pq} \leq (p^* - p)c_1$ where $c_1 < 0$. Similarly we can prove that $\mu_{p,q} \leq (q^* - q)c_2$, where $c_2 < 0$. Hence we have that,

$$\mu_{p,q} \leq (p^* - p)c_1 + (q^* - q)c_2, \quad \text{where} \quad c_1, c_2 < 0. \qquad \square$$

Lemma 9.12 *If $1 \leq p \leq p^* < q \leq q^* \leq n$ and if conditions of Theorem 9.10 are satisfied then, for every α such that $\frac{1}{2} < \alpha < 1$ we have,*

$$\max_{1 \leq p < p^* < q \leq q^* < n}\frac{V_{p,q} - \mu_{p,q}}{(p^* - p)^\alpha + (q^* - q)^\alpha} = O_p(1).$$

Proof. By (9.13)–(9.16) we can write,

$$V_{p,q} - \mu_{p,q} = (q - p)H(B_{p,q}) + (n - q + p)H(B_{p,q}^*) - (q^* - p^*)H(B_{p^*,q^*})$$
$$- (n - q^* + p^*)H(B_{p^*,q^*}^*) - (q - p)H(\nu_1)$$
$$- (n - q + p)H(\nu_2) + (q^* - p^*)H(\tau_2)$$
$$+ (n - q^* + p^*)H(\tau_1).$$

Taylor series expansions of $H(B_{p,q})$ around $\nu_1, H(B_{p^*,q^*})$ around ν_2, $H(B_{p^*,q^*})$, around τ_2 and $H(B_{p^*,q^*}^*)$, and around τ_1 gives

$$V_{p,q} - \mu_{p,q} = (q - p)\left\{H'(\nu_1)(B_{p,q} - \nu_1) + \frac{1}{2}H''(\nu_1)(B_{p,q} - \nu_1)^2\right\}$$

$$+ (n - q + p)\left\{H'(\nu_2)(B_{p,q}^* - \nu_2) + \frac{1}{2}H''(\nu_2)(B_{p,q}^* - \nu_2)^2\right\}$$

$$- (q^* - p^*) \left\{ H'(\tau_2)(B_{p^*,q^*} - \tau_2) + \frac{1}{2} H''(\tau_1)(B_{p^*,q^*} - \tau_2)^2 \right\}$$

$$- (n - q^* + p^*) \left\{ H'(\tau_1)(B_{p^*,q^*}^* - \tau_1) + \frac{1}{2} H''(\tau_1)(B_{p^*,q^*}^* - \tau_1)^2 \right\} + R_{pq,1}.$$

(9.19)

Note that by the iterated law of logarithm we have,

$$(B_{p,q} - \nu_1)^3 (q - p) = \frac{\left\{ \sum_{i=p+1}^{q} [T(x_i) - \nu_1] \right\}^3}{(q - p)^2}$$

$$= \left\{ \frac{\sum_{i=p+1}^{q} [T(x_i) - \nu_1]}{(q - p)^{2/3}} \right\}^3 = O_p(1).$$

Thus by a similar argument we get,

$$\max_{1 \leq p \leq p^* < q \leq q^* \leq n} |R_{pq,1}| = O_p(1). \tag{9.20}$$

Next let

$$R_{pq,2} = (q - p)\{H'(\nu_1)(B_{p,q} - \nu_1)\} - (q^* - p^*)\{H'(\tau_2)(B_{p^*,q^*} - \tau_2)\}$$

$$= H'(\nu_1) \sum_{i=p+1}^{q} [T(x_i) - \nu_1)] - H'(\tau_2) \sum_{i=p^*+1}^{q^*} [T(x_i) - \tau_2)]$$

$$= H'(\nu_1) \sum_{i=p+1}^{p^*} [T(x_i) - \tau_1)] + [H'(\nu_1) - H'(\tau_2)] \sum_{i=p^*+1}^{q} [T(x_i) - \tau_2)]$$

$$- H'(\tau_2) \sum_{i=q+1}^{q^*} [T(x_i) - \tau_2)].$$

Now, using the iterated law of logarithm again we can write, for $1/2 < \alpha < 1$,

$$\max_{1 \leq p < p^* < q \leq q^* < n} \frac{|R_{pq,2}|}{(p^* - p)^\alpha + (q^* - q)^\alpha} = O_p(1). \tag{9.21}$$

A similar argument gives

$$\max_{1 \leq p < p^* < q \leq q^* < n} \frac{|R_{pq,3}|}{(p^* - p)^\alpha + (q^* - q)^\alpha} = O_p(1), \tag{9.22}$$

where $R_{pq,3} = (q - p)H''(\nu_1)(B_{p,q} - \nu_1)^2 - (q^* - p^*)H''(\tau_2)(B_{p^*,q^*} - \tau_2)^2$.

Next let

$$R_{pq,4} = (n - q + p)\{H'(\nu_2)(B^*_{p,q} - \nu_2)\}$$
$$- (n - q^* + p^*)\{H'(\tau_1)(B^*_{p^*,q^*} - \tau_1)\}$$

$$= H'(\nu_2) \left\{ \sum_{i=1}^{p}[T(x_i) - \tau_1)] + \sum_{i=q+1}^{q^*} [T(x_i) - \tau_2)] \right.$$

$$\left. + \sum_{i=q^*+1}^{n} [T(x_i) - \tau_1)] \right\} + H'(\tau_1) \left\{ \sum_{i=1}^{p}[T(x_i) - \tau_1)] \right.$$

$$\left. - \sum_{i=p+1}^{p^*} [T(x_i) - \tau_1)] + \sum_{i=q^*+1}^{n} [T(x_i) - \tau_1)] \right\}.$$

Hence by the iterated law of logarithm for $\frac{1}{2} < \alpha < 1$ we have

$$\max_{1 \le p < p^* < q \le q^* < n} \frac{|R_{pq,4}|}{(p^* - p)^\alpha + (q^* - q)^\alpha} = O_p(1). \tag{9.23}$$

A similar argument gives

$$\max_{1 \le p < p^* < q \le q^* < n} \frac{|R_{pq,5}|}{(p^* - p)^\alpha + (q^* - q)^\alpha} = O_p(1), \tag{9.24}$$

where

$$R_{pq,5} = (n - q + p)H''(\nu_1)(B^*_{p,q} - \nu_1)^2 - (n - q^* + p^*)H''(\tau_2)(B_{p^*,q^*} - \tau_2)^2.$$

Now by (9.19)–(9.24) we have the result. □

Lemma 9.13 *If* $1 \le p \le p^* < q \le q^* < n$ *and if the conditions of Theorem 9.10 are satisfied, then*

$$\lim_{k_1, k_2 \to \infty} \limsup_{n \to \infty} P\{\hat{p} < p^* - k_1, \hat{q} < q^* - k_2\} = 0.$$

Proof. From Lemma 9.12 we know that, for each $\epsilon > 0, \exists \eta_\epsilon > 0$ such that

$$P\{V_{p,q} - \mu_{p,q} \le [(p^* - p)^\alpha + (q^* - q)^\alpha]\eta_\epsilon, 1 \le p \le p^* < q \le q^* < n\} > 1 - \epsilon, \tag{9.25}$$

for large n. But by Lemma 9.10, we have that $\mu_{p,q} \le c_1(p^* - p) + c_2(q^* - q)$. Thus we get

$$\lim_{n \to \infty} P\{V_{p,q} \le c_{11}(p^* - p) + (c_{22}(q^* - q), \quad 1 \le p \le p^* < q \le q^* < n\} = 1, \tag{9.26}$$

where $c_{11} < 0$ and $c_{22} < 0$. Thus for every $M > 0$,

$$\lim_{k_1,k_2\to\infty} \limsup_{n\to\infty} P\{\max V_{p,q} > -M\} = 0. \qquad (9.27)$$

Also notice that we can write,

$$\{\hat p < p^* - k_1, \hat q < q^* - k_2\} = \left\{ \max_{1\le p<q\le n} V_{p,q} = \max V_{p,q} \right\}$$

$$= \left[\{\max V_{p,q}\} \cap \left\{ \max_{1\le p<q\le n} V_{p,q} > -M \right\} \right]$$

$$\cup \left[\{\max V_{p,q}\} \cap \left\{ \max_{1\le p<q\le n} V_{p,q} \le -M \right\} \right]$$

$$\subseteq \{\max V_{p,q} > -M\} \cup \left\{ \max_{1\le p<q\le n} V_{p,q} \le -M \right\},$$

and by (9.27) the limsup of the first term on the right-hand side tends to 0 as $k_1, k_2 \to \infty$, and the second term is $o_p(1)$ because $V_{p^*,q^*} = 0$. Thus we get that

$$\lim_{k_1,k_2\to\infty} \limsup_{n\to\infty} P\{\hat p < p^* - k_1, \hat q < q^* - k_2\} = 0. \qquad \square$$

The following lemmas are stated without proof, because their proofs are similar to the proof of Lemma 9.15.

Lemma 9.14 *If* $1 < p^* \le p < q^* \le q < n$ *and if the conditions of Theorem* 9.10 *are satisfied, then* $\lim_{k_1,k_2\to\infty} \limsup_{n\to\infty} P\{\hat p < p^* - k_1, \hat q > q^* + k_2\} = 0.$

Lemma 9.15 *If* $1 < p^* \le p < q \le q^* < n$ *and if the conditions of Theorem* 9.10 *are satisfied, then* $\lim_{k_1,k_2\to\infty} \limsup_{n\to\infty} P\{\hat p > p^* + k_1, \hat q > q^* - k_2\} = 0.$

Lemma 9.16 *If* $1 < p \le p^* < q^* \le q < n$ *and if the conditions of Theorem* 9.10 *are satisfied, then* $\lim_{k_1,k_2\to\infty} \limsup_{n\to\infty} P\{\hat p < p^* - k_1, \hat q > q^* + k_2\} = 0.$

Lemma 9.17 *If the conditions of Theorem* 9.10 *are satisfied, then*

$$\limsup_{n\to\infty} P\{1 \le \hat p < \hat q \le p^* < q^* < n\} = 0.$$

Proof. Suppose $1 \le p < q \le p^* < q^* < n$. Then by (9.16) we have that

$$\mu_{p,q} = (n - q + p) \left\{ H\left(\frac{q^* - p^*}{n - q + p} \tau_2 + \frac{n - q + p - q^* + p^*}{n - q + p} \tau_1 \right) \right.$$

$$\left. - \frac{(q^* - p^*)}{n - q + p} H(\tau_2) - \frac{(n - q^* + p^*)}{n - q + p} H(\tau_1) \right\}.$$

Now by C5 we get,

$$\mu_{pq} \leq (n - q + p) \sup_{\frac{q^* - p^*}{n} \leq s \leq \frac{q^* - p^*}{n - p^* + 1}} \{H(s\tau_1 + (1 - s)\tau_2) - sH(\tau_1)$$

$$- (1 - s)H(\tau_2)\}$$

$$\leq (n - q + p)c, \tag{9.28}$$

where $c < 0$. Next by a similar argument as in Lemma 9.12 we can show that

$$\max_{1 \leq p < q \leq p^* < q^* < n} \frac{V_{p,q} - \mu_{p,q}}{(n - q + p)^\alpha} = O_p(1). \tag{9.29}$$

Hence, for every $\epsilon > 0, \exists \eta_\epsilon > 0$ such that

$$P\{V_{p,q} - \mu_{p,q} \leq (n - q + p)^\alpha \eta_\epsilon, 1 \leq p < q \leq p^* < q^* < n\} > 1 - \epsilon$$

for large n. But by (9.28), we get

$$\lim_{n \to \infty} P\{V_{p,q} \leq c^*(n - q + p), 1 \leq p < q \leq p^* < q^* < n\} = 1,$$

where $c^* < 0$. Thus for every $M > 0$,

$$\limsup_{n \to \infty} P\left\{ \max_{1 \leq p < q \leq p^*} V_{p,q} > -M \right\} = 0. \tag{9.30}$$

Also notice that we can write,

$$\{1 \leq \hat{p} < \hat{q} \leq p^* < q^* < n\}$$

$$= \left\{ \max_{1 \leq p < q \leq n} V_{p,q} = \max_{1 \leq p < q \leq p^* < q^* < n} V_{p,q} \right\}$$

$$= \left[\left\{ \max_{1 \leq p < q \leq p^* < q^* < n} V_{p,q} \right\} \cap \left\{ \max_{1 \leq p < q \leq n} V_{p,q} > -M \right\} \right]$$

$$\cup \left[\left\{ \max_{1 \leq p < q \leq p^* < q^* < n} V_{p,q} \right\} \cap \left\{ \max_{1 \leq p < q \leq n} V_{p,q} \leq -M \right\} \right]$$

$$\subseteq \left\{ \max_{1 \leq p < q \leq p^* < q^* < n} V_{p,q} > -M \right\} \cup \left\{ \max_{1 \leq p < q \leq n} V_{p,q} \leq -M \right\},$$

and by (9.30) the limsup of the first term on the right-hand side tends to 0, and the second term is $o_p(1)$ because $V_{p^*,q^*} = 0$. Thus we get the result. $\square$

Lemma 9.18 *If the conditions of Theorem 9.10 are satisfied, then*

$$\limsup_{n \to \infty} P\{1 < p^* < q^* << \hat{p} < \hat{q} < n\} = 0.$$

Proof. By symmetry to the conditions in Lemma 9.17 we get the result. $\square$

After the establishment of the preceding lemmas, we are now in a position to prove the major theorem, Theorem 9.10.

Proof of Theorem 9.10 Note that we can write

$$P\{\hat{p} < p^* - k_1\} = P\{\hat{p} < p^* - k_1, \hat{q} > q^* + k_2\} + P\{\hat{p} < p^* - k_1, \hat{q} \leq q^* + k_2\}$$
$$\leq P\{\hat{p} < p^* - k_1, \hat{q} > q^* + k_2\} + P\{\hat{p} < p^* - k_1, \hat{q} < q^* - k_2\}.$$

Now by Lemmas 9.15 and 9.16 we get

$$\lim_{k_1 \to \infty} \limsup_{n \to \infty} P\{\hat{p} < p^* - k_1\} = 0. \tag{9.31}$$

Similarly, by Lemmas 9.17 and 9.18 we get

$$\lim_{k_1 \to \infty} \limsup_{n \to \infty} P\{\hat{p} > p^* + k_1\} = 0. \tag{9.32}$$

Now by (9.31) and (9.32) we get

$$|\hat{p} - p^*| = O_p(1).$$

By similar arguments we can prove that

$$|\hat{q} - q^*| = O_p(1).$$

This completes the proof of Theorem 9.10. $\square$

9.4.4 Asymptotic Nonnull Distribution of Q_n

In this subsection, we present the derivation of the nonnull distribution of Q_n. We first define the following, in order to prove the theorem that presents the asymptotic nonnull distribution of Q_n. Let $\{Y_i, i > 0\}$ be a sequence of independent and identically distributed (iid) random variables with density function $f(x; \theta_0)$ and let $\{Y_i, i < 0\}$ be a sequence of iid random variables with density function $f(x; \theta^*)$. Also let $\{Z_i, i > 0\}$ be a sequence of iid random variables with density function $f(x; \theta^*)$ and $\{Z_i, i < 0\}$ be iid random variables with density function $f(x; \theta_0)$. Further assume that the four sequences $\{Y_i, i > 0\}, \{Y_i, i < 0\}, \{Z_i, i > 0\}$, and $\{Z_i, i < 0\}$ are independent of one another. We now define $Z_{p,q}$ as the following.

If $k_1 \geq 0, k_2 > 0$,

$$Z_{p,q} = [H'(\tau_2) - H'(\tau_1)] \left\{ \sum_{i=1}^{k_1} (T(Y_i) - \tau_1) - \sum_{i=1}^{k_2} (T(Z_i) - \tau_2) \right\}$$
$$+ k_1 \{ H(\tau_2) - H(\tau_1) + H'(\tau_2)(\tau_1 - \tau_2) \}$$
$$+ k_2 \{ H(\tau_1) - H(\tau_2) + H'(\tau_1)(\tau_2 - \tau_1) \},$$

if $k_1 < 0, k_2 \geq 0$,

$$Z_{p,q} = [H'(\tau_2) - H'(\tau_1)] \left\{ \sum_{i=-k_1}^{-1} (T(Y_i) - \tau_2) - \sum_{i=1}^{k_2} (T(Z_i) - \tau_2) \right\}$$
$$+ (k_2 - k_1) \{ H(\tau_1) - H(\tau_2) + H'(\tau_1)(\tau_2 - \tau_1) \},$$

if $k_1 \leq 0, k_2 < 0$,

$$Z_{p,q} = [H'(\tau_2) - H'(\tau_1)] \left\{ \sum_{i=-k_1}^{-1} (T(Y_i) - \tau_2) - \sum_{i=-k_2}^{-1} (T(Z_i) - \tau_1) \right\}$$
$$- k_1 \{ H(\tau_1) - H(\tau_2) + H'(\tau_2)(\tau_1 - \tau_2) \}$$
$$- k_2 \{ H(\tau_2) - H(\tau_1) + H'(\tau_1)(\tau_2 - \tau_1) \},$$

if $k_1 > 0, k_2 \leq 0$,

$$Z_{p,q} = [H'(\tau_2) - H'(\tau_1)] \left\{ \sum_{i=1}^{k_1} (T(Y_i) - \tau_1) - \sum_{i=-k_1}^{-1} (T(Z_i) - \tau_2) \right\}$$
$$+ (k_1 - k_2) \{ H(\tau_2) - H(\tau_1) + H'(\tau_2)(\tau_1 - \tau_2) \},$$

and if $k_1 = 0, k_2 = 0$,

$$Z_{p,q} = 0.$$

We present the following theorem that will serve as a foundation for the derivation of the nonnull distribution of Q_n.

Theorem 9.19 *Assume that conditions C1–C5 hold. Then under H_A, for every $N_1, N_2 \in \mathbf{N}$,*

$$\left\{ V_{p*-p,q*-q} : \begin{array}{l} p = 0, \pm 1, \ldots, \pm N_1, \\ q = 0, \pm 1, \ldots, \pm N_2, \end{array} 1 \leq p < q \leq n \right\} \overset{D}{\to}$$

$$\left\{ Z_{p,q} : \begin{array}{l} p = 0, \pm 1, \ldots, \pm N_1, \\ q = 0, \pm 1, \ldots, \pm N_2, \end{array} 1 \leq p < q \leq n \right\}.$$

Proof. We consider the following cases.

Case 1. $p^* - p \geq 0, q^* - q \geq 0$.

Let N_1, N_2 be two positive integers. Suppose that $1 \leq p^* - p \leq N_1$ and $0 \leq q^* - q \leq N_2$; then we have that $0 < p \leq p^* < q \leq q^* < n$. Now consider

$$
\begin{aligned}
V_{p,q} - \mu_{p,q} = {} & (q-p)H(B_{p,q}) + (n-q+p)H(B_{p,q}^*) - (q^*-p^*)H(B_{p^*q^*}) \\
& - (n-q^*+p^*)H(B_{p^*q^*}^*) - (q-p)H(\nu_1) + (n-q+p)H(\nu_2) \\
& - (q^*-p^*)H(\tau_2) - (n-q^*+p^*)H(\tau_1).
\end{aligned}
$$

Next we use a Taylor series expansion of $H(B_{p,q})$ around ν_1, $H(B_{p,q}^*)$ around ν_2, $H(B_{p,q})$ around τ_2, and $H(B_{p^*q^*}^*)$ around τ_1 to get, for each $N_1, N_2 \in \mathbf{N}$,

$$
\max_{p*-N_1 \leq p \leq p*, q*-N_2 \leq q \leq q*} |V_{p,q} - \mu_{p,q} - V_{pq,1} - V_{pq,2} - V_{pq,3} - V_{pq,4}| = O_p(n^{-(1/2)}),
\tag{9.33}
$$

where

$$
V_{pq,1} = H'(\nu_1) \sum_{i=p+1}^{q} (T(x_i) - \nu_1) - H'(\tau_2) \sum_{i=p^*+1}^{q^*} (T(x_i) - \tau_2)
$$

$$
V_{pq,2} = \frac{H''(\nu_1)}{2(q-p)} \left[\sum_{i=p+1}^{q} (T(x_i) - \nu_1) \right]^2 - \frac{H''(\tau_2)}{2(q^*-p^*)} \left[\sum_{i=p^*+1}^{q^*} (T(x_i) - \tau_2) \right]^2
$$

$$
V_{pq,3} = H'(\nu_2) \left[\sum_{i=1}^{p}(T(x_i) - \nu_2) + \sum_{i=q+1}^{n} (T(x_i) - \nu_2) \right]^2
$$

$$
- H'(\tau_1) \left[\sum_{i=1}^{p^*}(T(x_i) - \tau_1) + \sum_{i=q^*+1}^{n} (T(x_i) - \tau_1) \right]^2
$$

$$
V_{pq,4} = \frac{H''(\nu_2)}{2(n-q+p)} \left[\sum_{i=1}^{p}(T(x_i) - \nu_2) + \sum_{i=q+1}^{n} (T(x_i) - \nu_2) \right]^2
$$

$$
- \frac{H''(\tau_1)}{2(n-q^*+p^*)} \left[\sum_{i=1}^{p^*}(T(x_i) - \tau_1) + \sum_{i=q^*+1}^{n} (T(x_i) - \tau_1) \right]^2.
$$

But from the central limit theorem (CLT) we get,

$$
\max_{p*-N_1 \leq p \leq p*, q*-N_2 \leq q \leq q*} |V_{pq,2} + V_{pq,4}| = o_p(1).
\tag{9.34}
$$

Next note that

$$
V_{pq,1} = H'(\nu_1) \left\{ \sum_{i=p+1}^{q} T(x_i) - (p^* - p)\tau_1 - (q - p^*)\tau_2 \right\}
$$

$$
- H'(\tau_2) \left\{ \sum_{i=p^*+1}^{q^*} T(x_i) - (q^* - p^*)\tau_2 \right\}
$$

$$
= \{H'(\nu_1) - H'(\tau_2)\} \sum_{i=p^*+1}^{q} (T(x_i) - \tau_2) + H'(\nu_1) \sum_{i=p+1}^{p^*} (T(x_i) - \tau_1)
$$

$$
- H'(\tau_2) \sum_{i=q+1}^{q^*} (T(x_i) - \tau_2),
$$

and

$$
V_{pq,3} = H'(\nu_2) \left\{ \sum_{i=1}^{p} T(x_i) + \sum_{i=q+1}^{n} T(x_i) - (q^* - q)\tau_2 - (n - q^* + p)\tau_1 \right\}
$$

$$
- H'(\tau_1) \left\{ \sum_{i=1}^{p^*} T(x_i) + \sum_{i=q^*+1}^{n} T(x_i) - (n - q^* + p^*)\tau_1 \right\}
$$

$$
= \{H'(\nu_2) - H'(\tau_1)\} \left[\sum_{i=1}^{p} (T(x_i) - \tau_1) + \sum_{i=q+1}^{n} (T(x_i) - \nu_2) \right]
$$

$$
- H'(\tau_1) \sum_{i=p+1}^{p^*} (T(x_i) - \tau_1) + H'(\nu_2) \sum_{i=q+1}^{q^*} (T(x_i) - \tau_2).
$$

Thus by the CLT, for all $p^* - N_1 \le p < p^* < q^* - N_2 \le q \le q^*$ we have

$$
\left| V_{pq,3} + V_{pq,1} - \left\{ H'(\tau_2) \sum_{i=p+1}^{p^*} (T(x_i) - \tau_1) - H'(\tau_2) \sum_{i=q+1}^{q^*} (T(x_i) - \tau_2) \right. \right.
$$

$$
\left. \left. - H'(\tau_1) \sum_{i=p+1}^{p^*} (T(x_i) - \tau_1) + H'(\tau_1) \sum_{i=q+1}^{q^*} (T(x_i) - \tau_2) \right\} \right| = o_p(1),
$$

which gives us

$$\left| V_{pq,3} + V_{pq,1} - \{H'(\tau_2) - H'(\tau_1)\} \left\{ \sum_{i=p+1}^{p^*} (T(x_i) - \tau_1) \right. \right.$$

$$\left. \left. - \sum_{i=q+1}^{q^*} (T(x_i) - \tau_2) \right\} \right| = o_p(1). \tag{9.35}$$

Also note that we can rewrite $\mu_{p,q}$ defined by (9.16) as

$$\mu_{p,q} = (q - p)H(\nu_1) + (n - q + p)H(\nu_2) - (q^* - p^*)H(\tau_2)$$
$$- (n - q^* + p^*)H(\tau_1)$$
$$= \{(q - p) - (q^* - p^*)\}H(\tau_2) - \{(q - p) - (q^* - p^*)\}H(\tau_1)$$
$$+ (p^* - p)H'(\tau_2)(\tau_1 - \tau_2) + \frac{1}{2}\frac{(p^* - p)^2}{q - p}H''(\tau_2^*)(\tau_1 - \tau_2)^2$$
$$+ (q^* - q)H'(\tau_1)(\tau_2 - \tau_1) + \frac{1}{2}\frac{(q^* - q)^2}{n - q + p}H''(\tau_1^*)(\tau_2 - \tau_1)^2.$$

But because H'' is continuous, we can find c' and c such that

$$\left| \frac{1}{2}\frac{(p^* - p)^2}{q - p}H''(\tau_2^*)(\tau_1 - \tau_2)^2 \right| \le c'\frac{(p^* - p)^2}{q - p} = o(1)$$

$$\left| \frac{1}{2}\frac{(q^* - q)^2}{n - q + p}H''(\tau_1^*)(\tau_2 - \tau_1)^2 \right| \le c\frac{(q^* - q)^2}{n - q + p} = o(1)$$

for $p^* - N_1 \le p < p^*$ and $q^* - N_2 < q < q^*$. So we have

$$\max_{p*-N_1 \le p \le p*, q*-N_2 \le q \le q*} |\mu_{p,q} - (q*-q)\{H(\tau_1) - H(\tau_2) + H'(\tau_1)(\tau_2 - \tau_1)\}$$
$$- (p^* - p)\{H(\tau_2) - H(\tau_1) + H'(\tau_2)(\tau_1 - \tau_2)\}| = o(1). \tag{9.36}$$

Finally by putting (9.34), (9.35), and (9.36) together we get

$$\max_{p*-N_1 \le p \le p*, q*-N_2 \le q \le q*} \left| V_{p,q} - (q*-q)\{H(\tau_1) - H(\tau_2) + H'(\tau_1)(\tau_2 - \tau_1)\} \right.$$

$$- (p*-p)\{H(\tau_2) - H(\tau_1) + H'(\tau_2)(\tau_1 - \tau_2)\} - \{H'(\tau_2) - H'(\tau_1)\}$$

$$\times \left\{ \sum_{i=p+1}^{p*} (T(x_i) - \tau_1) - \sum_{i=q+1}^{q*} (T(x_i) - \tau_2) \right\} \Big|$$

$$= o_p(1), \tag{9.37}$$

and therefore we get the desired result.

Case 2. $p^* - p \le 0, q^* - q \ge 0$.

Let N_1, N_2 be two positive integers and suppose that $|p^* - p| \le N_1$ and $|q^* - q| \le N_2$; then we have that $1 \le p^* \le p < q \le q^* \le n$. Then consider

$$V_{p,q} - \mu_{p,q} = (q-p)H(B_{p,q}) + (n-q+p)H(B^*_{p,q}) - (q^* - p^*)H(B_{p^*q^*})$$
$$- (n - q^* - p^*)H(B^*_{p^*q^*}) - (q-p)H(\nu_1) + (n-q+p)H(\zeta_1)$$
$$- (q^* - p^*)H(\tau_2) - (n - q^* - p^*)H(\tau_1),$$

where

$$\zeta_1 = \frac{n - q^* + p^*}{n - q + p}\tau_1 + \frac{q^* - p^* - q + p}{n - q + p}\tau_2.$$

Next if we use the Taylor series expansion, we get

$$\max_{p*\le p\le p*+N_1, q*-N_2\le q<q*} |V_{p,q} - \mu_{p,q} - V_1 - V_2 - V_3 - V_4| = O_p(n^{-(1/2)}), \tag{9.38}$$

where

$$V_1 = H'(\tau_2) \sum_{i=p+1}^{q} (T(x_i) - \tau_2) - H'(\tau_2) \sum_{i=p^*+1}^{q^*} (T(x_i) - \tau_2)$$

$$V_2 = \frac{H''(\tau_2)}{2(q-p)} \left[\sum_{i=p+1}^{q} (T(x_i) - \tau_2) \right]^2 - \frac{H''(\tau_2)}{2(q^* - p^*)} \left[\sum_{i=p^*+1}^{q^*} (T(x_i) - \tau_2) \right]^2$$

$$V_3 = H'(\zeta_1) \left[\sum_{i=1}^{p} (T(x_i) - \zeta_1) + \sum_{i=q+1}^{n} (T(x_i) - \zeta_1) \right]$$
$$- H'(\tau_1) \left[\sum_{i=1}^{p^*} (T(x_i) - \tau_1) + \sum_{i=q^*+1}^{n} (T(x_i) - \tau_1) \right]$$

$$V_4 = \frac{H''(\zeta_1)}{2(n-q+p)} \left[\sum_{i=1}^{p} (T(x_i) - \zeta_1) + \sum_{i=q+1}^{n} (T(x_i) - \zeta_1) \right]^2$$
$$- \frac{H''(\tau_2)}{2(n-q^*+p^*)} \left[\sum_{i=1}^{p^*} (T(x_i) - \tau_1) + \sum_{i=q^*+1}^{n} (T(x_i) - \tau_1) \right]^2.$$

Now by the CLT we get,

$$\max_{p* \leq p \leq p* + N_1, q* - N_2 \leq q < q*} |V_2 + V_4| = o_p(1). \tag{9.39}$$

Next consider

$$V_1 + V_3 = [H'(\zeta_1) - H'(\tau_1)] \left[\sum_{i=1}^{p^*} (T(x_i) - \tau_1) \right.$$

$$\left. + \sum_{i=q^*+1}^{n} (T(x_i) - \tau_1) \right] + [H'(\zeta_1) - H'(\tau_1)]$$

$$\times \left[\sum_{i=p^*+1}^{p} (T(x_i) - \tau_2) + \sum_{i=q+1}^{q^*} (T(x_i) - \tau_2) \right]. \tag{9.40}$$

So we have

$$\max_{p* \leq p \leq p* + N_1, q* - N_2 \leq q < q*} \left| V_1 + V_3 - [H'(\tau_1) - H'(\tau_2)] \right.$$

$$\left. \times \left[\sum_{i=p^*+1}^{p} (T(x_i) - \tau_2) + \sum_{i=q+1}^{q^*} (T(x_i) - \tau_2) \right] \right| = o_p(1). \tag{9.41}$$

Rewrite $\mu_{p,q}$ defined by (9.13) to get

$$\mu_{p,q} = (q - p)H(\tau_2) + (n - q + p)H(\zeta_1)$$

$$+ (q^* - p^*)H(\tau_2) - (n - q^* + p^*)H(\tau_1)$$

$$= \{(q - p) - (q^* - p^*)\}H(\tau_2)$$

$$+ (n - q + p) \left\{ H(\tau_1) + \frac{q^* - q + p^* - p}{n - q + p} H'(\tau_1) \right.$$

$$\times (\tau_2 - \tau_1) + \frac{1}{2} \left(\frac{q^* - q + p^* - p}{n - q + p} \right)^2$$

$$\left. \times H''(\tau_1^*)(\tau_2 - \tau_1) \right\} - (n - q^* + p^*)H(\tau_1)$$

$$= \{(q^* - q) - (p^* - p)\}\{H(\tau_1) - H(\tau_2) + H'(\tau_1)(\tau_2 - \tau_1)\} + o(1).$$

This indicates that

$$\max_{p* \leq p \leq p*+N_1, q*-N_2 \leq q < q*} |\mu_{p,q} - \{(q^* - q) - (p^* - p)\}\{H(\tau_1)$$
$$- H(\tau_2) + H'(\tau_1)(\tau_2 - \tau_1)\}| = o(1). \tag{9.42}$$

Then by combining (9.38)–(9.42) we get the result.

Case 3. $p^* - p \leq 0, q^* - q \leq 0$.
Suppose that $|p^* - p| \leq N_1$ and $|q^* - q| \leq N_2$; then we have that $1 \leq p^* \leq p < q^* \leq q \leq n$. Then consider

$$V_{p,q} - \mu_{p,q} = (q - p)H(B_{p,q}) + (n - q + p)H(B^*_{p,q}) - (q^* - p^*)H(B_{p*q*})$$
$$- (n - q^* - p^*)H(B^*_{p*q*}) - (q - p)H(\eta_1) + (n - q + p)H(\eta_2)$$
$$- (q^* - p^*)H(\tau_2) - (n - q^* - p^*)H(\tau_1),$$

where

$$\eta_1 = \frac{q^* - p}{q - p}\tau_2 + \frac{q - q^*}{q - p}\tau_1 \quad \text{and} \quad \eta_2 = \frac{p - p^*}{n - q + p}\tau_2 + \frac{n - q + p^*}{n - q + p}\tau_1.$$

Similarly as in (9.37) and (9.38) we get

$$\max_{p* \leq p < p*+N_1, q* \leq q < q*+N_2} |V_{p,q} - \mu_{p,q}|$$

$$= H'(\eta_1) \left\{ \sum_{i=p+1}^{q*} (T(x_i) - \tau_2) + \sum_{i=q*+1}^{q} (T(x_i) - \tau_1) \right\}$$

$$+ H'(\eta_2) \left\{ \sum_{i=1}^{p*} (T(x_i) - \tau_1) + \sum_{i=q+1}^{n} (T(x_i) - \tau_1) + \sum_{i=p*+1}^{p} (T(x_i) - \tau_2) \right\}$$

$$- H'(\tau_2) \left\{ \sum_{i=p*+1}^{p} (T(x_i) - \tau_2) + \sum_{i=p+1}^{q*} (T(x_i) - \tau_2) \right\}$$

$$- H'(\tau_1) \left\{ \sum_{i=1}^{p*} (T(x_i) - \tau_1) + \sum_{i=q+1}^{n} (T(x_i) - \tau_2) \right\} + o_p(1)$$

$$= \{H'(\tau_1) - H'(\tau_2)\} \left\{ \sum_{i=p*+1}^{p} (T(x_i) - \tau_2) + \sum_{i=q*+1}^{q} (T(x_i) - \tau_1) \right\} + o_p(1). \tag{9.43}$$

Thus we have

$$\max_{p* \le p < p*+N_1, q* \le q < q*+N_2} \left| V_{p,q} - \mu_{p,q} - \{H'(\tau_2) - H'(\tau_1)\} \right.$$

$$\left. \times \left\{ \sum_{i=p^*+1}^{p} (T(x_i) - \tau_2) + \sum_{i=q^*+1}^{q} (T(x_i) - \tau_1) \right\} \right| = o_p(1). \qquad (9.44)$$

Also notice that we can write

$$\mu_{p,q} = (q - p) \left\{ H(\tau_2) + H'(\tau_2)(\eta_1 - \tau_2) + \frac{1}{2} H''(\tau_2{}^*)(\eta_1 - \tau_2)^2 \right\}$$

$$+ (n - q + p) \left\{ H(\tau_1) + + H'(\tau_1)(\eta_2 - \tau_1) + \frac{1}{2} H''(\tau_1{}^*)(\eta_2 - \tau_1)^2 \right\}$$

$$- (q^* - p^*) H(\tau_2) - (n - q^* + p^*) H(\tau_1) + o_p(1)$$

$$= (q - q^*)\{H(\tau_2) - H(\tau_1) + H'(\tau_2)(\tau_1 - \tau_2)\}$$

$$+ (p - p^*)\{H(\tau_1) - H(\tau_2) + H'(\tau_1)(\tau_2 - \tau_1)\} + o(1). \qquad (9.45)$$

Now by combining (9.42)–(9.45) we get the result.

Case 4. $p^* - p \ge 0, q^* - q \le 0$.
Suppose that $|p^* - p| \le N_1$ and $|q^* - q| \le N_2$; then we have that $1 \le p \le p^* < q^* \le q \le n$. And now by symmetry to Case (2) we get the result. $\qquad \square$

Lemma 9.20 *If conditions of Theorem 9.10 hold, then*

$$n^{-(1/2)}\{\ln \Lambda_{p^*q^*} - \mu^*\} \xrightarrow{D} N(0, \sigma_1{}^2),$$

where

$$\sigma_1^2 = (\lambda_2 - \lambda_1)\{H'(\tau_1) - H'[(\lambda_2 - \lambda_1)\tau_2 + (1 - \lambda_2 + \lambda_1)\tau_1]\}^2 A''(\theta_0)$$

$$+ (1 - \lambda_2 + \lambda_1)\{H'(\tau_1)$$

$$- H'[(\lambda_2 - \lambda_1)\tau_2 + (1 - \lambda_2 + \lambda_1)\tau_1]\}^2 A''(\theta^*) \qquad (9.46)$$

and μ^* is given in (9.15).

Proof. Consider

$$
\begin{aligned}
\ln \Lambda_{p^*q^*} &- \mu^* \\
&= (q^* - p^*)\{H(B_{p^*q^*}) - H(\tau_2)\} \\
&\quad + (n - q^* + p^*)\{H(B^*_{p^*q^*}) - H(\tau_1)\} \\
&\quad - n\left\{H(B_n) - H\left(\frac{q^* - p^*}{n}\tau_2 + \frac{n - q^* + p^*}{n}\tau_1\right)\right\} \\
&= \left\{H'(\tau_2) - H'\left(\frac{q^* - p^*}{n}\tau_2 + \frac{n - q^* + p^*}{n}\tau_1\right)\right\}\left\{\sum_{i=p^*+1}^{q^*}(T(x_i) - \tau_2)\right\} \\
&\quad + \left\{H'(\tau_1) - H'\left(\frac{q^* - p^*}{n}\tau_2 + \frac{n - q^* + p^*}{n}\tau_1\right)\right\} \\
&\quad \times \left\{\sum_{i=1}^{p^*}(T(x_i) - \tau_1) + \sum_{i=q^*+1}^{n}(T(x_i) - \tau_1)\right\} + O_p(1).
\end{aligned}
$$

Now by the CLT we get the result. □

Theorem 9.21 *If the conditions of Theorem 9.10 hold, then*

$$
(\hat{p} - p^*, \hat{q} - q^*) \xrightarrow{D} (\zeta_p, \zeta_q),
$$

where

$$
(\zeta_p, \zeta_q) = \inf\left\{(p, q) : Z_{p,q} = \sup_{-\infty < i_p < i_q < \infty} Z_{i_p, i_q}\right\}.
$$

Proof. By combining Theorems 9.10 and 9.21 with (9.18) we get the result. □

Theorem 9.22 *If the conditions of Theorem 9.10 hold, then*

$$
n^{-(1/2)}\{(Q)_n - 2\mu^*\} \xrightarrow{D} N(0, 4\sigma_1^2).
$$

Proof. By combining Theorems 9.10 and 9.23 and Lemma 9.22 we get the result. □

As the exponential family contains a rich collection of distributions such as the normal and exponential distributions, the results presented here in Section 9.4 are very useful for studying epidemic change point models characterized by a distribution that belongs to the exponential family.

Bibliography

1. Abraham, B. and Wei, W.S. (1984). Inferences about the parameters of a time series model with changing variance. *Metrika*, **31**, 183–194.
2. Akaike, H. (1973). Information theory and an extension of the maximum likelihood principle. *2nd International Symposium of Information Theory*, B.N. Petrov and E. Csaki (Eds.), Akademiai Kiado, Budapest, 267–81.
3. Aly, E.-E.A.A. and Bouzar, N. (1992). On maximum likelihood ratio test for the change point problem. Department of Mathematical Science Center, University of Alberta, Technical Report, **7**.
4. Anderson, T.W. (1984). *An Introduction to Multivariate Statistical Analysis*, 2nd edition. Wiley, New York.
5. Basseville, M. and Benveniste, A. (1986). Detection of abrupt changes in signals and dynamical system, *Lecture Notes in Control and Information Sciences*, **77**, M. Basseville, et al. (Eds.), Springer-Verlag, New York.
6. Basu, A.P., Ghosh, J.K., and Joshi, S.N. (1988). On estimating change-point in a hazard rate. In *Statistical Decision Theory and Related Topics IV*, S.S. Gupta and J.O. Berger (Eds.), Springer-Verlag, New York, **2**, 239–252.
7. Bhattacharya, G.K. and Johnson, R.A. (1968). Non-parametric tests for shift at an unknown time point. *Annals of Mathematical Statistics*, **39**, 1731–1743.
8. Bozdogan, H. (1987). Model selection and Akaike's Information criterion (AIC): The general theory and its analytical extension. *Psychometrika*, **52**, 345–370.
9. Bozdogan, H., Sclove, S.L., and Gupta, A.K. (1994). AIC-Replacements for some multivariate tests of homogeneity with applications in multisample clustering and variable selection. In *Proceedings of the First US/Japan Conference on the Frontiers of Statistical Modeling: An Informational Approach*, V. **2**. Kluwer Academic, Dordrecht, 199–232.
10. Brainerd, B. (1979). Pronouns and genre in Shakespeare's drama. *Computers and the Humanities*, **13**, 3–16.
11. Brodsky, B.E. and Darkhovsky, B.S. (1993). *Nonparametric Methods in Change Point Problems*. Kluwer Academic, Dordrecht.
12. Broemling, L.D. and Tsurumi, H. (1987). *Econometrics and Structural Change, Serie Statistics*, **74**, Marcel Dekker, New York.
13. Brown, R.L., Durbin, J., and Evans, J.M. (1975). Techniques for testing the constancy of regression relationships over time (with discussion). *Journal of the Royal Statistical Society* **B**, 149–192.
14. Carlstein, E., Muller, H.G., and Siegmund, D. (Eds.) (1994). *Change Point Problems*, Proceedings of AMS-IMS-SIAM Summer Research Conference, Mt. Holyoke College, Institute of Mathematical Statistics, Hayward, MA.

15. Charney, M. (1993). *All of Shakespears*. Columbia University Press, New York.
16. Chen, Jiahua and Gupta, A.K. (1998). Information criterion and change point problem in regular models. Department of Mathematics and Statistics, Bowling Green State University, *Technical Report*, No. 98-05.
17. Chen, Jiahua and Gupta, A.K. (2003). Information-theoretic approach for detecting change in the parameters of a normal model. *Mathematical Methods of Statistics*, **12**, 116–130.
18. Chen, Jiahua, Gupta, A.K., and Pan, J. (2006). Information Criterion and change point problem for regular models. *Sankhya*, **68**, 252–182.
19. Chen, J. (1995). *Inference about the Change Points in a Sequence of Gaussian Random Vectors Using Information Criterion*. Department of Mathematics and Statistics, Bowling Green State University, Ph.D Dissertation.
20. Chen, J. (1998). Testing for a change point in linear regression models. *Communications in Statistics-Theory and Methods*, **27**, 2481–2493.
21. Chen, J. (2003). A note on change point analysis in a failure rate. *Journal of Probability and Statistical Science*, **1**, 135–140.
22. Chen, J. (2005). Identification of significant periodic genes in microarray gene expression data, *BMC Bioinformatics*, **6**:286.
23. Chen, J. (2010). Change point methods in genetics. In *Encyclopedia of Statistical Sciences*, S. Kotz, C.B. Read, N. Balakrishinan, and B. Vidakovic (Eds.), Wiley, Hoboken, NJ, 1–7.
24. Chen, J. and Gupta, A.K. (1995). Likelihood procedure for testing change points hypothesis for multivariate Gaussian model, *Random Operators and Stochastic Equations*, **3**, 235–244.
25. Chen, J. and Gupta, A.K. (1997). Testing and locating variance change points with application to stock prices. *Journal of the American Statistical Association*, **92**, 739–747.
26. Chen, J. and Gupta, A.K. (1999). Change point analysis of a Gaussian model. *Statistical Papers*, **40**, 323–333.
27. Chen, J. and Gupta, A.K. (2004), Statistical Inference of Covariance Change points in Gaussian Model, **38**, 17–28.
28. Chen, J. and Gupta, A.K. (2007). A Bayesian approach to the statistical analysis of a smooth-abrupt change point model. *Advances and Applications in Statistics*, **7**, 115–125.
29. Chen, J. and Wang, Y.-P. (2009). A statistical change point model approach for the detection of DNA copy number variations in array CGH data. *IEEE/ACM Transactions on Computational Biology and Bioinformatics*, **6**, 529–541.
30. Chernoff, H. (1973). The use of faces to represent points in k-dimensional space graphically. *Journal of the American Statistical Association*. **68**, 361–368.
31. Chernoff, H. and Zacks, S. (1964). Estimating the current mean of a normal distribution which is subject to changes in time. *Annals of Mathematical Statistics*, **35**, 999–1018.
32. Chin Choy, J.H. (1977). A bayesian analysis of a changing linear model. Ph.D Dissertation. Oklahoma State University. Stillwater, Oklahoma.
33. Chin Choy, J.H. and Broemeling, L.D. (1980). Some Bayesian inferences for a changing linear model, *Technometrics*, **22**, 71–78.
34. Cox, D.R. and Hinkley, D.V. (1979). *Theoretical Statistics*. Chapman and Hall, London.
35. Csörgö, M. and Horváth, L. (1988). Nonparametric methods for the change point problem. In *Handbook of Statistics*, **7**, P.R. Krishnaiah and C.R. Rao (Eds.), John Wiley, New York, 403–425.
36. Csörgö, M. and Révész, P. (1981). Strong Approximation in Probability and Statistics. Academic Press, New York.

37. Darling, D.A. and Erdös, P. (1956). A limit theorem for the maximum of normalized sums of independent random variables. *Duke Mathematics Journal*, **23**, 143–155.

38. Davis, W.W. (1979). Robust Methods for Detection of Shifts of the Innovation Variance of a Time Series. *Technometrics*, **21**, 313–320.

39. Dennis, J.E. and Schnabel, R.B. (1983). *Numerical Methods for Unconstrained Optimization and Nonlinear Equations*. Prentice-Hall, Englewood Cliffs, NJ.

40. Deshayes, J. and Picard, D. (1986). Off-line statistical analysis of change point models using nonparametric and likelihood methods, *Lecture Notes in Control and Information Sciences*, **77** M. Basseville et al. (Eds.), Springer-Verlag, New York, 103–168

41. Diaz, J. (1982). Bayesian detection of a change of scale parameter in sequences of independent gamma random variables. *Journal of Econometrics*, **19**, 23–29.

42. Durbin, J. (1973), Distribution Theory for Tests based on the Sample Distribution Function. Philadelphia: Society for Industrial and Applied Mathematics.

43. Fearnhead, P. and Liu, Z. (2007). On line inference for multiple change point problems. *Journal of Royal Statistical Society, Series B*, **69**, 203–213.

44. Ferreira, P.E. (1975). A Bayesian analysis of a switching regression model: Known number of regimes. *Journal of the American Statistical Association*, **70**, 370–374.

45. Fu, Y. and Curnow, R.N. (1990). Maximum likelihood estimation of multiple change points. *Biometrika*, **77**, 563–573.

46. Gardner, L.A. (1969). On detecting change in the mean of normal variates. *Annals of Mathematical Statistics*, **40**, 116–126.

47. Ghosh, J.K. and Joshi, S.N. (1992). On the asymptotic distribution of an estimate of the change point in a failure rate. *Communications in Statistics – Theory and Methods*, **21**, 3571–3588.

48. Ghosh, J.K., Joshi, S.N., and Mukhopadhyay, C. (1993). A Bayesian approach to the estimation of change-point in a hazard rate. In *Advances in Reliability*, A.P. Basu (Ed.), Elsevier Science, Amsterdam.

49. Giri, N.C. (1977). *Multivariate Statistical Inference*. Academic Press, New York.

50. Goldenshluyer, A., Tsbakov, A., and Zeev, A. (2006). Optimal change-point estimation from indirect observations. *Annals of Statistics*, **34**, 350–372.

51. Gombay, E. and Hovarth, L. (1990). Asymptotic distributions of maximum likelihood tests for change in the mean. *Biometrika*, **77**, 411–414.

52. Gombay, E. and Hovarth, L. (1994). An application of the maximum likelihood test to the change point problem. *Stochastic Processes and Applications*, **50**, 161–171.

53. Gombay, E. and Hovarth, L. (1996). Approximations for the time of change and the power function in change point models. *Journal of Statistical Planning and inference*, **52**, 43–66.

54. Guan, Z. (2004). A semiparametric changepoint model. *Biometrika*, **91**, 161–171.

55. Gupta, A.K., Chattopadhyay, A.K., and Krishnaiah, P.R. (1975). Asymptotic distributions of the determinants of some random matrices. *Communications in Statistics*, **4**, 33–47.

56. Gupta, A.K. and Chen, J. (1996). Detecting changes of mean in multidimensional normal sequences with application to literature and geology. *Computational Statistics*, **11**, 211–221.

57. Gupta, A.K. and Ramanayake, A. (2001). Change point with linear Trend for the exponential distribution distribution. *Journal of Statistical Planning and Inference*, **93**, 181–195.

58. Gupta, A.K. and Tang, J. (1987). On testing homogeneity of variance for Gaussian models. *Journal of Statistical Computation and Simulation*, **27**, 155–173.

59. Gupta, A.K. and Varga, T. (1993). *Elliptically Contoured Models in Statistics*, Kluwer Academic, Dordrecht.

60. Gurevich, G. and Vexler, A. (2005). Change point problems in the model of logistic regression. *Journal of Statistical Planning and Inference*, **131**, 313–331.

61. Haccou, P. and Meelis, E. (1988). Testing for the number of change points in a sequence of exponential random variables. *Journal of Statistical Computation and Simulation*, **30**, 285–298.

62. Haccou, P., Meelis, E., and Geer, S. (1988). The likelihood ratio test for the change point problem for exponentially distributed random variables, *Stochastic Processes and Their Applications*, **27**, 121–139.

63. Hackl, P. (Ed.) (1989). *Statistical Analysis and Forecasting of Economic Structural Change*, Springer-Verlag, New York.

64. Hackl, P. and Westlund, A. (1989). Statistical analysis of structural change: An annotated bibliography, *Empirical Economics*, **14**, 167–172.

65. Hall, C.B., Ying, J., Kuo, L., and Lipton, R.B. (2003). Bayesian and profile log likelihood change point methods for modeling cognitive function over time. *Computational Statistics and Data Analysis*, **42**, 91–109.

66. Hanify, J.A., Metcalf, P., Nobbs, C.L., and Worsley, K.J. (1981). Aerial spraying of 2,4,5-T and human birth malformations: An epidemiological investigation. *Science*, **212**, 349–351.

67. Hannan, E.J. and Quinn, B.G. (1979). The determination of the order of an autoregression. *Journal of Royal Statistical Society*, B **41**, 190–195.

68. Hawkins, D. L. (1989). A U-I approach to retrospective testing for shifting parameters in a linear model. *Communications in Statistics*, **18**, 3117–3134.

69. Hawkins, D.M. (1977). Testing a sequence of observations for a shift in location. *Journal of the American Statistical Association*, **72**, 180–186.

70. Hawkins, D.M. (1992). Detecting shifts in functions of multivariate location and covariance parameters. *Journal of Statistical Planning and Inference*, **33**, 233–244.

71. Henderson, R. (1990). A problem with the likelihood ratio test for a change-point hazard rate model. *Biometrika*, **77**, 835–843.

72. Hinkley, D.V. and Hinkley, E.A. (1970). Inference about the change-point in a sequence of binomial random variables. *Biometika*, **57**, 477–488.

73. Hodgson, G., Hager, J.H., Volik, S., Hariono, S., Wernick, M., Moore, D., Nowak, N., Albertson, D.G., Pinkel, D., Collins, C., Hanahan, D., and Gray, J.W. (2001). Genome scanning with array CGH delineates regional alterations in mouse islet carcinomas. *Nature Genetics*, **29**, 459–464.

74. Hogg, R.V. (1961). On the resolution of statistical hypotheses. *Journal of the American Statistical Association*, **56**, 978–989.

75. Holbert, D. (1982). A Bayesian analysis of a switching linear model. *Journal of Econometrics*, **19**, 77–87.

76. Horuath, L., Huskova, M., Kokoszka, P., and Steinebach, J. (2004). Monitoring changes in linear models. *Journal of Statistical Planning and Inference*, **126**, 225–251.

77. Horváth, L. (1993). The maximum likelihood method for testing changes in the parameters of normal observations. *Annals of Statistics*, **21**, 671–680.

78. Hsu, D.A. (1977). Tests for variance shifts at an unknown time point. *Applied Statistics*, **26**, No.3, 279–284.

79. Hsu, D.A. (1979). Detecting shifts of parameter in gamma sequences with applications to stock price and air traffic flow analysis. *Journal of the American Statistical Association*, **74**, 31–40.

80. Husková, M. and Sen, P.K. (1989). Nonparametric tests for shift and change in regression at an unknown time point. In *Statistical Analysis and Forecasting of Economic Structural Change*, P. Hackl (Ed.), Springer-Verlag, New York, 71–85.

81. Inclán, C. (1993). Detection of multiple changes of variance using posterior odds. *Journal of Business and Economics Statistics*, **11**, 189–300.

82. Inclán, C. and Tiao, G.C. (1994). Use of sums of squares for retrospective detection of changes of variance. *Journal of the American Statistical Association.*, **89**, 913–923.

83. James, B., James, K.L., and Siegmund, D. (1987). Tests for a change-point. *Biometrika*, **74**, 71–83.

84. James, B., James, K.L., and Siegmund, D. (1992). Asymptotic approximations for likelihood ratio tests and confidence regions for a change-point in the mean of a multivariate normal distribution. *Statistica Sinica*, **2**, 69–90.

85. Johnson, N.L. and Kotz, S. (1972). *Distributions in Statistics: Continuous Multivariate Distributions*, John Wiley and Sons, New York.

86. Johnson, R.A. and Wichern, D.W. (1988). *Applied Multivariate Statistical Analysis*, 2nd ed. Prentice Hall, Englewood Cliffs, NJ.

87. Joseph, L. and Wolfson, D.B. (1993). Maximum likelihood estimation in the multipath change-point problem. *Annals of Institute of Statistical Mathematics*, **45**, 511–530.

88. Juruskova, D. (2007). Maximum log-likelihood ratio test for a change in three parameters Weibull distribution. *Journal of Statistical Planning and Inference*, **137**, 1805–1815.

89. Kalbfleisch, J.D. and Prentice, R.L. (1980), *The Statistical Analysis of Failure Time Data*, John Wiley and Sons, New York.

90. Kallioniemi, A., Kallioniemi, O.-P., Sudar, D., Rutovitz, D., Gray, J.W., Waldman, F., and Pinkel, D. (1992). Comparative genomic hybridization for molecular cytogenetic analysis of solid tumors. *Science*, **258**, 818–821.

91. Kander, Z. and Zacks, S. (1966). Test procedures for possible changes in parameters of statistical distributions occurring at unknown time points. *Annals of Mathematical Statistics*, **37**, 1196–1210.

92. Kelly, G.E., Lindsey, J.K., and Thin, A.G. (2004). Models for estimating the change-point in gas exchange data. *Physiological Measurement*, **25**, 1425–1436.

93. Kim, D. (1994). Tests for a change-point in linear regression, *IMS Lecture Notes-Monograph Series*, **23**, 170–176.

94. Kirch, C. and Steinebach, J. (2006). Permutation principles for the change analysis of stochastic processes under strong invariance. *Journal of Computational and Applied Mathematics*, **186**, 64–88.

95. Kitagawa, G. (1979). On the use of AIC for the detection of outliers. *Technometrics*, **21**, 193–199.

96. Krishnaiah, P.R. and Miao, B.Q. (1988). Review about estimation of change points, P.R. Krishnaiah and C.R. Rao (Eds.), *Handbook of Statistics*, **7** (Elsevier, Amsterdam), 375–402.

97. Krishnaiah, P.R., Miao, B.Q., and Zhao, L.C. (1990). Local likelihood method in the problems related to change points. *Chinese Annals of Mathematics*, 11B: 3, 363–375.

98. Lehmann, E.L. (1986). *Testing Statistical Hypotheses*, 2nd ed. Wiley, New York.

99. Leonard, T., Crippen, C., and Aronson, M. (1988). *Day by Day, the Seventies*, V.1, Facts on File, New York.

100. Levin, B. and Kline, J. (1985). The CUSUM test of homogeneity with an application in spontaneous abortion epidemiology. *Statistics in Medicine*, **4**, 469–488.

101. Linn, S.C., West, R.B., Pollack, J.R., Zhu, S., Hernandez-Boussard, T., Nielsen, T.O., Rubin, B.P., Patel, R., Goldblum, J.R., Siegmund, D., Botstein, D., Brown, P.O., Gilks, C.B., and van de Rijn, M. (2003). Gene expression patterns and gene copy number changes in dermatofibrosarcoma protuberans. *American Journal of Pathology*, **163**, 2383–95.

102. Lucito, R., Healy, J., Alexander, J., Reiner, A., Esposito, D., Chi, M., Rodgers, L., Brady, A., Sebat, J., Troge, J., West, J.A., Rostan, S., Nguyen, K.C., Powers, S., Ye, K.Q., Olshen, A., Venkatraman, E., Norton, L., and Wigler, M. (2003). Representational oligonucleotide microarray analysis: A high-resolution method to detect genome copy number variation. *Genome Research*, **13**, 2291–2305.

103. Lucito, R., West, J., Reiner, A., Alexander, D., Esposito, D., Mishra, B., Powers, S., Norton, L., and Wigler, M. (2000). Detecting gene copy number fluctuations in tumor cells by microarray analysis of genomic representations. *Genome Research*, **10**, 1726–1736.

104. Mardia, K.V. (1970). Measures of multivariate skewness and kurtosis with applications. *Biometrika*, **57**, 519–530.

105. Matthews, D.E. and Farewell, V.T. (1982). On testing for a constant hazard against a change-point alternative. *Biometrics*, **38**, 463–468.

106. Matthews, D.E. and Farewell, V.T. (1985). On a singularity in the likelihood for a change-point hazard rate model. *Biometrika*, **72**, 703–704.

107. Matthews, D.E., Farewell, V.T., and Pyke, R. (1985). Asymptotic Score-Statistic Processes and Tests for Constant Hazard Against a Change-Point Alternative. *The Annals of Statistics*, **13**, 583–591

108. Mei, Y. (2006). Sequential change-point detection when unknown parameters are present in the pre-change distribution. *The Annals of Statistics*, **34**, 92–122.

109. Muller, H.G. and Wang, J.-L. (1994). Change-point models for hazard functions. In *Change-Point Problems, IMS Lecture Notes – Monograph Series*, **23**, 224–241.

110. Myers, C.L., Dunham, M.J., Kung, S.Y., and Troyanskaya, O.G. (2004). Accurate detection of aneuploidies in array CGH and gene expression microarray data. *Bioinformatics*, **20**, 3533–3543.

111. Nannya, Y., Sanada, M., Nakazaki, K., Hosoya, N., Wang, L., Hangaishi, A., Kurokawa, M., Chiba, S., Bailey, D.K., Kennedy, G.C., and Ogawa, S. (2005). A robust algorithm for copy number detection using high-density oligonucleotide single nucleotide polymorphism genotyping arrays. *Cancer Research*, **65**, 6071–6079.

112. Nguyen, H.T., Rogers, G.S., and Walker, E.A. (1984). Estimation in change-point hazard rate models. *Biometrika*, **71**, 299–304.

113. Ning, W. and Gupta, A.K. (2009). Change point analysis for generalized lambda distribution. *Commmunication in Statistics-Simulation and Computation*, **38**, 1789–1802.

114. Olshen, A.B., Venkatraman, E.S., Lucito, R., and Wigler, M. (2004). Circular binary segmentation for the analysis of array-based DNA copy number data. *Biostatistics*, **5**, 557–572.

115. Osorio, F. and Galea, M. (2006). Detection of change point in Student-t linear regression models. *Statistical Papers*, **47**, 31–48.

116. Page, E.S. (1954). Continuous inspection schemes. *Biometrika*, **41**, 100–116.

117. Page, E.S. (1955). A test for a change in a parameter occurring at an unknown point. *Biometrika*, **42**, 523–527.

118. Page, E.S. (1957). On problem in which a change in parameter occurs at an unknown points. *Biometrika*, **44**, 248–252.

119. Pan, J. and Chen, Jiahua (2006). Application of modified information criterion to multiple change point problems. *Journal of Multivariate Analysis*, **97**, 2221–2241.

120. Parzen, E. (1992). Comparison change analysis. In *Nonparametric Statistics and Related Topics*, A.K.Md.E. Saleh (Ed.), Elsevier, Amsterdam.

121. Pettitt, A.N. (1980). A simple cumulative sum type statistic for the change-point problem with zero-one observations. *Biometrika*, **67**, 79–84.

122. Pettitt, A.N. and Stephens, M.A. (1977). The Kolmogorov-Smirnov goodness-of-fit statistic with discrete and grouped data. *Technometrics*, **19**, 205–210.

123. Picard, F., Robin, S., Lavielle, M., Vaisse, C., and Daudin, J. (2005). A statistical approach for array CGH data analysis. *BMC Bioinformatics*, **6**:27.

124. Pinkel, D., Seagraves, R., Sudar, D., Clark, S., Poole, I., Kowbel, D., Collins, C., Kuo, W.-L., Chen, C., Zhai, Y., Dairkee, S., Ljjung, B.-M., Gray, J.W., and Albertson, D. (1998). High resolution analysis of DNA copy number variation using comparative genomic hybridization to microarrays. *Nature Genetics*, **20**, 207–211.

125. Poirier, D.J. (1976). *The Econometrics of Structural Change*. North-Holland, New York.

126. Pollack, J.R., Perou, C.M., Alizadeh, A.A., Eisen, M.B., Pergamenschikov, A., Williams, C.F., Jeffrey, S.S., Botstein, D., and Brown, P.O. (1999). Genome-wide analysis of DNA copy-number changes using cDNA microarrays. *Nature Genetics*, **23**, 41–46.

127. Quandt, R.E. (1958). The estimation of the parameters of a linear regression system obeys two separate regimes. *Journal of the American Statistical Association*, **53**, 873–880.

128. Quandt, R.E. (1960). Tests of the hypothesis that a linear regression system obeys two separate regimes. *Journal of the American Statistical Association*, **55**, 324–330.

129. Rai, R., Genbauffe, F., Lea, H.Z., and Cooper, T.G. (1987). Transcriptional regulation of the DAL5 gene in *Saccharomyces cerevisiae*. *Journal of Bacteriology*, **169**, 3521–3524.

130. Ramanayake, A. (1998). *Epidemic Change Point and Trend Analyses for Certain Statistical Models*. Department of Mathematics and Statistics, Bowling Green State University, Ph.D Dissertation.

131. Ramanayake, A. (2004). Tests for a change point in the shape parameter of gamma random variables. *Communication in Statistics-Theory and Methods*, **33**, 821–833.

132. Ramanayake, A. and Gupta, A.K. (2002). Change points with linear treand followed by abrupt change for the exponential dsitributuion. *Journal of Statistical Computation and Simulation*, **74**, 263–278.

133. Ramanayake, A. and Gupta, A.K. (2003). Tests for an epidemic change in a sequence of exponentially distributed random variables. *Biometrical Journal*, **45**, 946–958.

134. Ramanayake, A. and Gupta, A.K. (2004). Epidemic change model for the exponential family. *Communication in Statistics-Theory and Methods*, **33**, 2175–2198.

135. Ramanayake, A. and Gupta, A.K. (2010). Testing for a change point in a sequence of exponential random variables with repeated values. *Journal of Statistical Computation and Simulations*, **80**, 191–199.

136. Rao, C.R. and Wu, Y. (1989). A strongly consistent procedure for model selection in a regression problem. *Biometrika*, **76**, 369–374.

137. Schulze U. (1986). *Mehrphasenregression: Stabilitätsprüfung, Schätrung, Hypothesenprüfung*. Akademie Verlag, Berlin.

138. Schwarz, G. (1978). Estimating the dimension of a model. *Annals of Statistics*, **6**, 461–464.

139. Sen, A.K. and Srivastava, M.S. (1973). On multivariate tests for detecting change in mean. *Sankhyá*, **A35**, 173–186.

140. Sen, A.K. and Srivastava, M.S. (1975a). On tests for detecting change in mean. *Annals of Statistics*, **3**, 98–108.

141. Sen, A.K. and Srivastava, M.S. (1975b). Some one-sided tests on change in level. *Technometrics*, **17**, 61–64.

142. Sen, A.K. and Srivastava, M.S. (1980). On tests for detecting change in the multivariate mean. University of Toronto, Tech. Report No.3.

143. Sen, P.K. and Singer, J.M. (1993). *Large Sample Methods in Statistics: An Introduction with Applications*, Chapman and Hall, New York.

144. Shaban, S.A. (1980). Change point problem and two phase regression: An annotated bibliography, *International Statistical Review*, **48**, 83–93.

145. Shewhart, W.A. (1931). *Economic Control of Quality of Manufactured Products*. D. Van Nostrand, New York.

146. Siegmund, D. (1986). Boundary crossing probabilities and statistical applications. *Annals of Statistics*, **14**, 361–404.

147. Siegmund, D. (1988a). Approximate tail probabilities for the maxima of some random fields. *Annals of Probability*, **16**, 67–80.

148. Siegmund, D. (1988b). Confidence sets in change-point problems. *International Statistical Review*, **56**, 31–48.

149. Sinha, B.K., Rukhin, A., and Ahsanullah, M. (Eds.) (1994). Applied change point problems in statistics, Nova Science, Commack, NY.

150. Smith, A.F.M. (1975). A Bayesian approach to inference about a change-point in a sequence of random variables. *Biometrika*, **62**, 407–416.

151. Snijders, A.M., Nowak, N., Segraves, R., Blackwood, S., Brown, N., Conroy, J., Hamilton, G., Hindle, A.K., Huey, B., Kimura, K., Law, S., Myambo, K., Palmer, J., Ylstra, B., Yue, J.P., Gray, J.W., Jain, A.N., Pinkel, D., and Alberston, D.G. (2001). Assembly of microarrays for genome-wide measurement of DNA copy number. *Nature Genetics*, **29**, 263–264.

152. Spellman, P.T., Sherlock, G., Zhang, M.Q., Iyer, V.R., Anders, K., Eisen, M.B., Brown, P.O., Bostein, D., and Futcher, B. (1998). Comprehensive identification of cell cycle-regulated genes of the yeast *Saccharomyces cerevisiae* by microarray hybridization. *Molecular Biology of the Cell*, **9**, 3273–3297.

153. Spevack, M. (1968). *A Complete and Systematic Concordance to the Works of Shakespeare I, II, III*. Georg Olms, Hildesheim.

154. Srivastava, M.S. and Worsley, K.J. (1986). Likelihood ratio tests for a change in the multivariate normal mean. *Journal of the American Statistical Association*, **81**, 199–204.

155. Sugiura, N. (1978). Further analysis of the data by Akaike's information criterion and the finite corrections. *Communications in Statistics-Theory and Methods*, A**7**, 13–26.

156. Tang, J. and Gupta, A.K. (1984). On the distribution of the product of independent beta random variables. *Statistics and Probability Letters*, **2**, 165–168.

157. Venkatraman, E.S. and Olshen, A.B. (2007). A faster circular binary segmentation algorithm for the analysis of array CGH data. *Bioinformatics*, **23**, 657–663.

158. Vexler, A., Wu, C., Liu, A., Whitcomb, B.W., and Schistervak, E.F. (2009). An extension of change point problem. *Statistics*, **43**, 213–225.

159. Vilasuso, J. (1996). Changes in the duration of economic expansions and contractions in the United States. *Applied Economics Letters*, **3**, 803–806.

160. Vlachonikolis, I.G. and Vasdekis, V.G.S. (1994). On a class of change-point models in covariance structures for growth curves and repeated measurements. *Communication in Statistics-Theory and Methods*, **23**, 1087–1102.

161. Vostrikova, L.J. (1981). Detecting "disorder" in multidimensional random processes. *Soviet Mathematics Doklady*, **24**, 55–59.

162. Wichern, D.W., Miller, R.B., and Hsu, D.A. (1976). Changes of variance in first order autoregressive time series models—with an application. *Applied Statistics*, **25**, 248–356.

163. Wichert, S., Folianos, K., and Strimmer, K. (2004). Identifying periodically expressed transcripts in microarray time series data, *Bioinformatics*, **20**, 5–20.

164. Wolfe, D.A. and Schechtman, E. (1984). Nonparametric statistical procedures for the change point problem. *Journal of Statistical Planning and Inference*, **9**, 389–396.

165. Worsley, K.J. (1979). On the likelihood ratio test for a shift in location of normal populations. *Journal of the American Statistical Association*, **74**, 365–367.

166. Worsley, K.J. (1983). The power of likelihood ratio and cumulative sum tests for a change in a binomial probability. *Biometrika*, **70**, 455–464.

167. Worsley, K.J. (1986). Confidence regions and tests for a change-point in a sequence of exponential family random variables. *Biometrika*, **73**, 91–104.

168. Worsley, K.J. (1988). Exact percentage points of the likelihood for a change-point hazard-rate model. *Biometrics*, **44**, 259–263.

169. Wu, Y. (2007). False alarms and sparse change segment detection by using a CUSUM. *Sequential Analysis*, **26**, 321–334.

170. Wu, Y. (2008). Simultaneous change point analysis and variable solution in a regression problem. *Journal of Multivariate Analysis*, **99**, 2154–2171.

171. Yao, Q. (1993). Tests for change-points with epidemic alternatives. *Biometrika*, **80**, 179–191.

172. Yao, Y.C. and Davis, R.A. (1986). The asymptotic behavior of the likelihood ratio statistics for testing shift in mean in a sequence of independent normal variates. *Sankhyá*, A**48**, 339–353.

173. Zacks, S. (1983). Survey of classical and Bayesian approaches to the change-point problem: Fixed sample and sequential procedures of testing and estimation. *Recent Advances in Statistics*, Academic Press, 245–269.

174. Zacks, S. (1991). Detection and change-point problem, *Handbook of Sequential Analysis*, B.K. Ghosh and P.K. Sen (Eds.), Series Statistics, 118, Marcel Dekker, New York.

175. Zhao, L.C., Krishnaiah, P.R., and Bai, Z.D. (1986a). On detection of the number of signals in presence of white noise. *Journal of Multivariate Analysis*, **20**, 1–25.

176. Zhao, L.C., Krishnaiah, P.R., and Bai, Z.D. (1986b). On detection of the number of signals when the noise covariance matrix is arbitrary. *Journal of Multivariate Analysis*, **20**, 26–49.

Author Index

Subject Index

Printed by Publishers' Graphics LLC USA

2012